Reviews in Plasmonics

Editor
Chris D. Geddes, Ph.D., CSci, CPhys, CChem, MIoP, MRSC

For further volumes:
http://www.springer.com/series/7164

Chris D. Geddes
Editor

Reviews in Plasmonics 2010

Springer

Editor
Chris D. Geddes
University of Maryland
Baltimore County, MD 21202, USA
geddes@umbc.edu

ISSN 1573-8086
ISBN 978-1-4614-0883-3 e-ISBN 978-1-4614-0884-0
DOI 10.1007/978-1-4614-0884-0
Springer New York Dordrecht Heidelberg London

Library of Congress Control Number: 2011941008

Printed on acid-free paper

Springer is part of Springer Science+Business Media (www.springer.com)

Preface

In the last 10 years, we have seen significant growth in plasmonics-related research with many researchers around the world publishing high quality material in now several peer reviewed journals, solely dedicated to the topic. To this end, we launched the "*Plasmonics*" Springer journal in 2005, which after only a few years, now has an ISI impact factor close to 4. This rapid growth of the journal in this area of science reflects the need for plasmonics-based publishing media. To address this ever-growing need, we are now launching a new hard bound review volume, "*Reviews in Plasmonics*," which is solely dedicated to publishing review articles, which are typically considered too lengthy for journal publication.

In this first volume, we have invited notable scientists from around the world to review their findings, including works on nanoparticle synthesis, SPR-based sensors, SERS as well as plasmon-assisted fluorescence, typically referred to as metal-enhanced fluorescence, to name but just a few. We subsequently thank the authors for their most timely and notable contributions and we all hope you find this volume a useful resource.

Finally, we would like to thank Caroleann Aitken, the Institute of Fluorescence manager, for helping compile the volume, as well as Michael Weston of Springer for help in launching the volume.

Baltimore, MD, USA — Dr. Chris D. Geddes, Professor

Contents

Chapter 1
Metal Nanoparticles for Molecular Plasmonics

Andrea Steinbrück, Andrea Csaki, and Wolfgang Fritzsche

1 Introduction

The color induced by metal nanoparticles has attracted people's attention since hundreds of years. Probably, the most famous example for the use of gold nanoparticles for staining is the Lycurgus Cup that was manufactured in the fifth to fourth century BC. This probably accidental application was later extended to a more controlled technique for coloring glass, today still visible in many colorful church windows. A global summary of the historical approaches, especially for gold colloids can be found elsewhere [35]. Michael Faraday (1791–1867) and Gustav Mie (1869–1957) were the most famous scientists who laid the systematic groundwork for synthesis and theoretical calculations of colloidal solutions. Faraday synthesized gold nanoparticles in 1857 and called them "activated gold" [47]. Shortly after that, Graham established the term "colloid" [67]. The theoretical calculations for metal nanoparticles of Mie [142] are still used today by many scientists. Mie worked on the issue of light scattering by small spheres of arbitrary size and material, and calculated the response of such a metal sphere to an external electromagnetic field (light). The Mie theory is based on the solution of Maxwell's equations for (subwavelength-sized) spheres in a nonabsorbing medium. The most important free parameter is the sphere radius. Material properties are considered as frequency-dependent permittivity. Another important fact is that the Mie theory is true only for uncharged spherical particles. It should be mentioned that almost at the same time Debye solved the same

A. Steinbrück
Nano Biophotonics Department, Institute of Photonic Technology,
PO Box 100239, Jena 07702, Germany

Chemistry Division, Los Alamos National Laboratory, Los Alamos, NM 87545, USA

A. Csaki • W. Fritzsche (✉)
Nano Biophotonics Department, Institute of Photonic Technology,
PO Box 100239, Jena 07702, Germany
e-mail: fritzsche@ipht-jena.de

C.D. Geddes (ed.), *Reviews in Plasmonics 2010*, Reviews in Plasmonics,
DOI 10.1007/978-1-4614-0884-0_1, © Springer Science+Business Media, LLC 2012

problem too [36]. Furthermore, since the interest grew for investigations of nanoparticles with variable shape or for bimetallic particles, several models or approximations were calculated by other scientists as well [100] and [116].

As already mentioned, the color of colloidal solutions is an eye-catching property. This color is explained as the result of a collective oscillation of electrons of the conductive electron band near the Fermi level. This effect causes a so-called localized (surface) plasmon (LSP) band in the spectra where absorption reaches a maximum at a certain wavelength of light. The location of the LSP band is characteristic for the material, the size, the shape, and the surrounding medium of a nanoparticle. The size distribution is so mirrored in the spectrum. A narrow size distribution causes a sharp LSP band, whereas a wide distribution reflects in a broad absorption band. In bimetallic particles, the location of the LSP band(s) depends further on the composition and the distribution of the two metals [116].

Only metals with free electrons (Au, Ag, Cu) possess plasmon resonances in the visible spectrum and therefore intense colors. The shape of the particles influences these resonances, as, e.g., elongated nanoparticles (rods) display two distinct plasmon bands related to transverse and longitudinal electron oscillations. The longitudinal oscillation is very sensitive to the aspect ratio of the particles [129].

A growing interest in the optical properties of the nanoparticles is supported by the need for enhanced bioanalytical methods. The potential of surface plasmon resonance in this field was demonstrated by the impressive technical development for bioanalytical applications based on surface plasmon resonance on thin metal layers [107]. This trend is supported by the progress in the controlled synthesis as well as sophisticated bioconjugation of metal nanoparticles in combination with molecular principles; and opened a new field for nanoparticle-based molecular nanotechnology [55]. In addition, the impetus for understanding the mechanisms of optical processes in nanoscale systems represents another strong motivation for work in molecular plasmonics, as this interdisciplinary field between nanoparticle plasmonics and molecular construction is dubbed. This review intends to give an overview about nanoparticle synthesis and related optical properties, as well as about the state of the art of bioconjugation of nanoparticles in order to realize molecular constructs for potential applications.

2 Synthesis of Nanoparticles

Since the days of Faraday, the synthesis of nanoparticles has made substantial progress. Today, people have succeeded to synthesize stable nanoparticles with variable sizes (but a sharp size distribution), shapes, and materials by reliable procedures. Nanoparticles can be produced of many metals such as Au, Ag, Cu, Pt, Pd, Ru, and others. The principles of the synthesis process of nanoparticles have been investigated and are yet much better understood even though there are some questions left to answer. For the characterization of nanoparticles, several methods are used. Transmission electron microscopy (TEM), high-resolution TEM (HRTEM), atomic force microscopy (AFM), scanning electron microscopy (SEM), X-ray diffraction, and other methods image the size, shape, and ultrastructure of the particles; UV–vis spectroscopy is applied to analyze their optical properties.

2.1 *Materials*

2.1.1 Gold Nanoparticles

Schmid and coworkers did pioneering work on well-defined phosphine-stabilized gold clusters [178, 181]. The gold clusters they had synthesized had the size of about 1.4 nm with a narrow size distribution. This work contributed remarkably to the understanding of the properties of small metal particles.

A currently very common method for the synthesis of gold nanoparticles was introduced by Turkevich [207]. He produced 20 nm gold nanoparticles by citrate reduction of $HAuCl_4$ in water. Frens succeeded in the preparation of gold nanoparticles with variable sizes (16–147 nm) by varying the ratio of citrate to gold [54]. In 1994, Brust set another milestone with the synthesis of thiol-stabilized gold nanoparticles by his two-phase-synthesis [21]. One year earlier, the possibility to stabilize gold nanoparticles with thiols was reported by Mulvaney and Giersig [151]. The method is similar to Faraday's synthesis: $AuCl_4^-$ is transferred to toluene by a phase transfer reagent. In the organic phase, $AuCl_4^-$ is reduced by $NaBH_4$ which is immediately seen as the color of the organic phase changes from orange to brown indicating the formation of gold nanoparticles. During the synthesis, the particles are "covered" with a shell of dodecanethiol ligands. The Brust–Schiffrin method allows the synthesis of thermally stable and air-stable gold nanoparticles. The particles show reduced polydispersity and the size can be controlled from 1.5 to 5.2 nm. The work opened the way for functionalization of gold nanoparticles with a variety of thiol ligands [30, 88]. An even narrower size distribution than with alkanethiol should be achievable by the stabilization of Au nanoparticles by dendrons [112].

Another very useful method for the synthesis of nanoparticles is the seeding growth method. It is based on a step-by-step enlargement of before grown particles ("seeds"). Therewith, various sizes with narrow size distribution were synthesized. Gold seed particles of 3 or 12 nm size, respectively, were used to prepare 20–100 nm Au particles by the reduction with citrate, yielding good monodispersity. The use of NH_2OH as reducing agent was tested and resulted in spheres and rods [19]. Jana et al. found that for higher concentrations of seeds, the nanoparticle growth is better controlled. It is also necessary to add the reducing agent slowly to the solution because otherwise the formation of more seeds is promoted instead of growth of the seeds [97]. The method also works on surfaces where the seeds are immobilized [141]. When Au particle monolayers (immobilized on silane-modified glass) are immersed in a growth solution containing Au^+ and NH_2OH, conductive Au films are generated [18]. Currently, such solution-based generated surfaces were tested for extremely sensitive protein bioassay using surface-enhanced Raman spectroscopy (SERS) [199].

Willner and coworkers introduced an enzyme-coupled process to enlarge Au nanoparticles. The reducing agent, H_2O_2, is synthesized by glucose oxidase in the presence of glucose and O_2. This process has potential application as glucose biosensor [228].

2.1.2 Silver Nanoparticles

In principle, the synthesis of silver nanoparticles can be carried out by the same methods mentioned for gold nanoparticles. The citrate method introduced for gold nanoparticles by Turkevich [207] can be adopted when $AuCl_4^-$ is substituted by a silver salt, i.e., silver nitrate [1, 147]. The method is considered to work in principle, but with less control over size and shape distribution [130]. The size of the particles was about 35 nm with a broad size distribution. For more details regarding the growth stages of silver and even gold nanoparticles, see Abid [1]. Utilization of $NaBH_4$ as reducing agent resulted in ca. 6 nm particles with a narrow size distribution. TEM measurements confirmed spherical particles but no aggregates or rods [130]. Moreover, a method where EDTA is used as the reducing agent is applied which yielded 20 nm silver particles [91]. The size distribution was even narrower than with $NaBH_4$, but there was no size control by changing the concentrations of the reactants [108].

In addition, an analogous method to the two-phase reduction method developed by Brust and Schiffrin [21] has also been reported for the preparation of Ag particles [115].

The seeding method as described before can be performed using NH_4OH as reducing agent to deposit silver salt from solution to the silver seeds. With this method, particles with well-defined size and shape can be synthesized.

Beside these methods, the formation of silver particles without addition of reducing agents by irradiation of a silver salt solution with laser light was observed. Here, usually surfactants such as SDS or Tween 20 are added to the solution as stabilizer. When the process is performed without any surfactant, the created particles show a very large size and shape distribution and tend to aggregate quickly. Smaller particles are synthesized when the concentration of sodium dodecyl sulfate is increased and when the laser power is decreased. A mechanism for the synthesis from "embryonic silver particles" that are rapidly formed and grown in solution was proposed. There is a competition between SDS to cover the surface and the particle growth [3, 133]. Henglein intensively studied the processes involved in particle formation, especially for silver nanoparticles [80].

The photo-induced formation of particles is also used for the synthesis of particles made from other metals than silver [135].

Liz-Marzán and coworkers succeeded in the synthesis of stable spherical Ag nanoparticles by using *N*,*N*-dimethylformamide (DMF) as reducing agent for Ag^+ and poly(*N*-vinyl-2-pyrrolidone) (PVP) as stabilizer [164].

Novel cost-effective method for the silver nanoparticle synthesis is the bioreduction using bacteria, whereas this technique provides particles with nonoptimum size distribution [106].

2.1.3 Platinum, Palladium, and Rhodium Nanoparticles

For the synthesis of platinum nanoparticles, the citrate reduction method is used as well, resulting in a particle size of 2–4 nm that can be further grown by hydrogen treatment [206]. By changing the ratio of reducing agent and metal precursor, 10 nm Pt particles can be achieved [6, 81]. Larger platin nanoparticles by seeding method are described by Bigall [14]. This method enables the synthesis of arbitrary particle sizes between 10 and 100 nm.

Alternatively, one can use organic solvents to reduce metal salts; for example, ethanol was used to prepare Pt, Pd, Au, or Rh nanoparticles in the presence of a protecting polymer, such as PVP [85, 118]. Ethylene glycol or larger polyols are used in the so-called polyol method by Figlarz [50]. This method also allows us to prepare Ag nanowires or nanoprisms (see Sect. 2.2) [186].

In 2004, Panigrahi et al. published the synthesis of Au, Ag, Pt, and Pd nanoparticles by the reduction of metal salts by various sugars [159]. The resulting sizes depend strongly on the sugar that is used. Particle sizes of 1, 3, 10, and 20 nm for Au, Pt, Ag, and Pd were prepared with fructose. With glucose or sucrose, larger particles could be synthesized. Another interesting method for the preparation of platinum particles is the use of dendrimers to stabilize the nanoparticles. With this method, particles with a controlled size well below 5 nm could be prepared [221].

A single-step method to synthesize catalytically active, hydrophobic Pt nanoparticles by the spontaneous reduction of aqueous $PtCl_6^{2-}$ by hexadecylaniline at a liquid–liquid interface was published in 2002 by Mandal et al. [137].

Beyond, hydrogen can be used as reducer for the synthesis of Pt [81]. So, a Pt(II) solution can be reduced by hydrogen in the presence or absence of stabilizing sodium citrate and/or sodium hydroxide. Using citrate and hydroxide slowed down the reduction but elongated the lifetime of the colloids [83]. A corresponding method is used to prepare Pd nanoparticles with a narrow size distribution [81].

2.1.4 Copper Nanoparticles

For the synthesis of copper nanoparticles, different methods have been reported. The synthesis of well-dispersed copper nanoparticles was achieved by the reduction of aqueous copper chloride solution with $NaBH_4$ in the nonionic water-in-oil microemulsions [170]. Pure metallic Cu nanoparticles can also be synthesized by the reduction of cupric chloride with hydrazine in the aqueous hexadecyltrimethylammonium bromide (CTAB) solution. Thereby, the adjustment of the pH is crucial for the synthesis. The mean diameter of Cu nanoparticles first decreased and then approached a constant with the increase of hydrazine concentration, whereas the

CTAB concentration had no significant influence on the size of Cu nanoparticles [28, 220]. However, at a low concentration of CTAB, the particles display a higher degree of size diversity (particles between 5 and 20 nm). But with higher concentrations of CTAB, a lower degree of size variation (2–10 nm) can be achieved [10]. Copper and copper(I) oxide nanoparticles protected by self-assembled monolayers (SAM) of thiol, carboxyl, and amine functionalities have been prepared by the controlled reduction of aqueous copper salts using Brust synthesis and yielding particles with 4–8 nm diameters [9]. Comparably to silver nanoparticles, copper nanoparticles offer a high antimicrobial effect [8].

2.1.5 Cobalt Nanoparticles

To produce monodisperse magnetic colloids (ferrofluids) of cobalt nanocrystals, Sun and Murray [195] reduced cobalt chloride in the presence of stabilizing agents. The sizes of the particles are ranging from 2 to 11 nm with only 7% SD.

The size and shape-selective synthesis of Co spheres and rods was investigated by Puntes and Krishnan in 2001 [169]. They found that the principles in CdSe particle synthesis also work for Co. A two-surfactants system was used, whereas the first surfactant differentially adsorbs to the nanocrystal faces providing rod formation and the second promotes the monomer exchange for focusing of the size distribution [169].

2.1.6 Nickel Nanoparticles

For the synthesis of Ni nanoparticles also, the reduction with hydrazine in an aqueous solution of cationic surfactants (CTAB/TC12AB) with an appropriate amount of NaOH, a trace of acetone, and elevated temperature was studied. Particles with a mean diameter of 10–36 nm, increasing with increasing nickel chloride concentration or decreasing hydrazine concentration, were prepared [27].

The use of hydrogen as a reducing agent is practicable for Ni nanoparticle as well. The synthesis is carried out in organic solution in the presence of a nonionic surfactant. Particle size analysis showed a distribution of 1–4 nm, the average diameter was 2.1 nm [62].

2.2 Shape

The synthesis of spherical metal nanoparticles (pentagondodecahedra and/or icosahedra) is well understood in general, but the processes leading to the growth of nonspherically shaped nanoparticles are rather complex. All synthesis resulting in

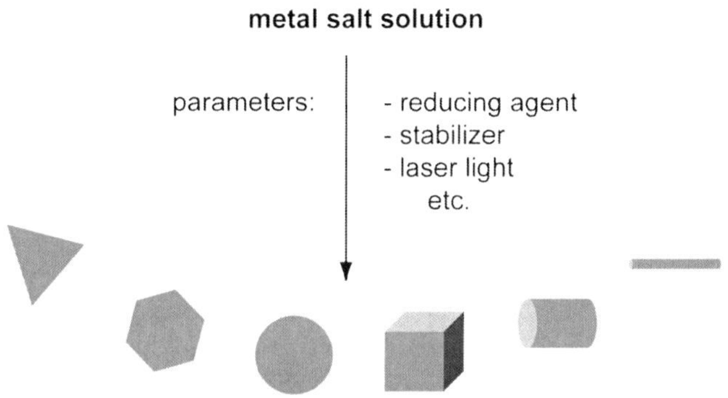

Fig. 1.1 Dependent on various parameters such as type of reducing agent, stabilizers, laser light or others, nanoparticles of different shapes can be realized

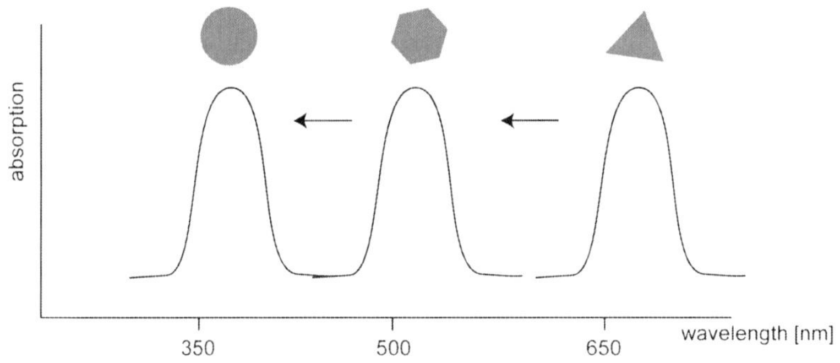

Fig. 1.2 Shift of the surface plasmon resonance during the conversion of triangular-shaped nanoparticles into spheres by heating (Mock 2002)

anisotropic shaped particles are seed-mediated reactions and are dependent on a variety of factors: kinetic parameters, the shape and dimension of the seeds [68], the energy minimum of the crystalline facettes on the seeds, the use of surfactants, and the used reduction agent and silver ions. A systematic overview of the wide range of potential particle shapes for palladium nanoparticles are introduced by the Xia group [222].

Typical surfactant are CTAB [152], poly(sodium styrenesulfonate) (PSSS) [5], Bis(*p*-sulfonatophenyl) phenylphosphine dihydrate dipotassium (BSPP) [101], sodium dodecylsulfate (SDS) [202], bis(2-ethylhexyl) sulfosuccinate (AOT), and PVP [196].

Typical examples for the variety of shapes are: spheres, prisms, rods with different aspect ratios, and cubes with varying edge lengths (Fig. 1.1). In general, beside the size, the shape of a nanoparticle has great influence on the absorption spectrum. This effect is shown in Fig. 1.2 for silver spheres, pentagons, and triangles [145].

2.2.1 Rods

The seeding growth method can be applied for the synthesis of nanorods. For example, 3.5 nm citrate-stabilized Au spheres are used as seeds for the synthesis of Au nanorods with aspect ratios ranging from 4.6 to 18 (with 16 nm short axis each) by varying the ratio of seed to metal salt [23, 99, 152]. Murphy and coworkers improved this method by the substitution of citrate-stabilized Au by CTAB-capped seeds. With this trick, the yield of rod production was increased. The aspect ratios of the rods varied from 1.5 to 10. Ascorbic acid, a milder reducing agent, was used here instead of $NaBH_4$. Absorption measurements showed maxima between 600 and 1,300 nm [98, 154]. Furthermore, experiments were carried out to study the growth processes and the "transformation" of spheres to high aspect ratio rods. It was found that the surfactants applied at the synthesis process play a major role. Two surfactants must be found that differentially adsorb to the nanocrystal faces protecting one face from deposition of further material and another surfactant allowing other faces to grow preferentially [96, 104]. A correlation between both the seed and nanorod diameter has been found. However, on the other hand, the seed diameter could not be proved to correlate to the nanorod length [212]. The purity of the surfactant plays an enormous effect [187]. Changes in the [Br^-] ion amount due to impurity leads to less efficient rod synthesis [58].

In addition, the use of different surfactants and seeding growth has been used also for the synthesis of Co nanorods [169].

Besides, Mokari et al. [146] prepared CdSe nanorods and tetrapods and used Au nanospheres and dithiolhexane to build networks. The Au spheres were attached to the CdSe particles and the interaction of thiols and Au is used to net the CdSe particles with each other [146].

2.2.2 Cubes

Beside nanorods, there were several investigations concerning the formation of nanocubes. Sun and Xia published the preparation of silver nanocubes and hollow gold nanoboxes, respectively [197]. The used method is similar to the so-called polyol method where a metal salt is solved and reduced by ethylene glycol in the presence of PVP. The reaction resulted in slightly truncated nanocubes with edge lengths varying from 70 to 175 nm. By treatment of these nanocubes with gold salt solution, the nanocubes served as templates for the formation of hollow nanoboxes. The silver is oxidized to silver ions and at the same time the gold ions are reduced to gold resulting in highly truncated cubic shape boxes. Murphy and coworkers worked on the synthesis of copper oxide nanocubes of edge lengths of 25–200 nm. The reaction was performed using copper(II) salts, ascorbic acid as reducing agent in the presence of polyethylene glycol (PEG), and sodium hydroxide. By variation of the PEG concentration or the sequence of reactants, the control of edge lengths can be achieved [64, 65]. An overview of the shape control for cubical nanoparticles is given by Tao and coworkers [202].

2.2.3 Prisms

By photo-induced transformation, spheres can be turned into prisms in solution. Jin et al. succeeded in the conversion of 8 nm spheres into prisms with 16 nm thickness and 10–60 nm edge lengths (which transform with time to prisms with 100 nm edge length) by irradiating the solution with light of 350–700 nm wavelengths. The shape transformation is combined with changes in the absorption spectrum [101]. A similar work by Mock showed the progression of plasmon shift due to morphological changes. Triangular-shaped particles are transformed to spherical-like particles by heating. Figure 1.2 shows the absorption spectra of the different shapes produced [145]. Due to different modes of the plasmon excitation, three dominant peaks appear from prisms. Hao and colleagues investigated the plasmon bands more intensely using their prisms with 30, 60, 100, and 150 nm edge lengths. They found in-plane dipole excitation as the red-most plasmon band in the spectrum; the middle band is due to in-plane quadrupole excitation and the bluest band to out-of-plane quadrupole excitation at which the first two are mostly affected by the edge lengths. The spectrum is also sensitive to the thickness and to truncation of the tips of the triangles [73].

Truncated triangles of silver were investigated by Chen et al. using a method similar to the seeding growth method for gold nanorods with ascorbic acid as reducing agent and CTAB as stabilizer. The particles produced had an edge size of ca. 68 nm and a thickness of 24 nm. The degree of truncation was 0.35 [29]. As an alternative, a method has been reported using salicylic acid as reducing agent. The reaction yielded a higher concentration of nanoprisms but produced also hexagonal and spherical-like particles. Furthermore, Figlarz's polyol method allows for the preparation of silver nanowires or nanoprisms, respectively, by the reduction of $AgNO_3$ with ethylene glycol in the presence of PVP [198]. The synthesis of silver triangles using the surfactant PSSS was evaluated by Aherne et al. This method results particles with arbitrary localized surface plasmon resonance in visible spectral range [5]. UV-light modulation of triangles to adjusted plasmon peaks was described by Zhang et al. [230].

2.2.4 Tetrahedron/Octahedron

Experiments with silver, platinum, and copper showed the formation of tetrahedral and/or octahedral structures. In the polyol synthesis of silver nanoparticles using trace amounts of sodium chloride and oxygen, tetrahedrons are produced besides truncated cubes. The dimensions of the particles can be controlled from 20 to 80 nm [216, 217]. The same group also investigated the formation of Pt tetrahedron using basically the same way of synthesis. By increasing the molar ratio between $NaNO_3$ and H_2PtCl_6 to 11, the shape of the synthesized particles turned from irregular spheroids to tetrahedra and octahedra with well-defined facets [84]. Already in 1996, a publication showed the shape-controlled synthesis of colloidal Pt nanoparticles. Several shapes (and sizes) were observed: tetrahedra, cubes, irregular prisms, icosahedra, and cubo-octahedra. The morphology was dependent on the ratio of the capping polymer to the platinum concentration [6].

He et al. published the synthesis of octahedrally shaped Cu_2O particles by the reduction of copper nitrate in Triton X-100 water-in-oil (w/o) microemulsions by gamma-irradiation with average edge lengths ranging from 45 to 95 nm as a function of the dose rate [79].

2.2.5 Nanodiscs

Again, the size is controlled by adjusting the involved reagents and the plasmon resonance modes are different from that of spheres [134]. Nanodiscs of 9 nm thickness and 36 nm diameter were synthesized using SiO_2 particles as templates [73].

2.2.6 Multipods and Nanostars

In the next paragraph, more particular structures are discussed as examples for more complex morphologies.

The first structure is called multipods. The particles show a star-like structure but only with three arms. The plasmon bands in the absorption spectra are sensitive to the length and sharpness of the arms. There is less influence of the thickness and total size [73]. Milliron and coworkers published the preparation of even more complex structures. They demonstrated the synthesis of inorganically coupled colloidal quantum dots and rods (CdTe and CdSe), connected epitaxially at branched and linear junctions within single nanocrystals (ZnS). The advantage of this system is the possibility to tune the properties of each component and the nature of their interactions. The arrangement in 3D is achieved in well-defined angles and distances [143].

Nanostars exhibit hot spots on their arm tips with well-defined geometry [74], these structures represent excellent enhancers for Raman spectroscopic applications (SERS) [89].

2.2.7 Nano Dumbells and Dog Bones

Besides the aspect ratio, also the relation between volume and the shape is a dominant factor in the spectral behavior of plasmonic nanoparticles. Outbreaks and overgrowth of rod-like nanoparticles to dumbbells [69] and dog bones results in interesting fine-adjustment of the LSPR band [224].

2.3 *Nanoshells*

Another important development demonstrates the so-called nanoshells. In principle, these structures are prepared by covering dielectric nanoparticles (i.e., SiO_2) with a metal (i.e., gold). Thereafter, the core can be dissolved resulting in a hollow nanoshell.

First calculations were done by Aden and Kerker [4] and continued by Neeves and Birnboim [153]. Besides, Halas and coworkers have been pioneers in practical work in this field. The synthesis of the nanoshells is carried out by decorating the silica particle template with small Au colloids followed by the electroless deposition of gold to build a nearly continuous shell. The minimal shell thickness achievable is ca. 5 nm [71, 215]. The plasmon band in the absorption spectrum shifts quite sensitively as a function of the shell thickness. The plasmon resonance of the nanoshells can easily be positioned in the near IR (800–1,300 nm) by variation of the shell thickness. The absorption of biological matter in the near IR is very low which opens the way to several biological applications [16, 72]. Graf et al. propose silica-Au nanoparticles for in various photonic applications [66]. Investigations concerning the shell showed a pinhole structure which is essential for the etching process. The small holes influence the plasmon band not very strong because their size is quite small (2–5 nm) compared to the total particle size (36 nm) [73].

2.4 Fabrication of Nanoparticle Arrays on Surfaces

E-beam lithography can be applied for the synthesis of nanoparticles on surfaces. This top-down nanofabrication method is expensive and serial (disadvantage) but enables for the preparation of ordered two-dimensional arrays of particles all showing the same size, shape, and defined particle-to-particle distance. The last point represents a key advantage of this approach: The nanostructures can be positioned with a high precision regarding both neighboring particles as well as a technical surrounding such as on a chip surface. Aussenegg and coworkers performed pioneering work in this field [39, 63, 173]. They investigated the influence of particle spacing, polarization direction dependence, shape effects due to variation of the aspect ratio of used ellipses, and the possibility to spectrally store data by a special arrangement of the ellipses.

As already mentioned, spheres lead to one plasmon band in the absorption spectrum independent from the direction of polarization. In contrast, ellipses are reacting differently when irradiated with different polarizations: they show a blue resonance when the polarization is orthogonal, but a red peak when the polarization is parallel to the long axis. When the aspect ratio is increased, the peak for parallel polarization is red-shifted. Because the short axis is not varied in the experiments, there was no effect for the orthogonal polarization-dependent peak [63]. The effects caused by the polarization direction-dependent illumination are proposed to be useful for data storage. The arrangement of one, two, or three elliptically shaped particles with defined orientation to each other shows three different possible answers by interaction with light. By varying the polarization direction, it is possible to direct arrangements only in blue or only in red or even in both regimes, always dependent on the orientation of one or more particles in the arrangement relative to the light [39]. Decreasing the interparticle distance but leaving the aspect ratio constant, the plasmon resonance is shifted to red for a polarization direction parallel to the long axis of the particles. On the other hand, a shift to shorter wavelengths was detected for orthogonal polarization [173, 194].

Experimental and theoretical investigations of 1D arrangements of nanoparticles were also performed. Standard e-beam lithography was used for the formation of such particle chains consisting of elliptically shaped particles with an aspect ratio of 1.4 and 150 nm center-to-center spacing. The polarization direction was chosen parallel to the chains. The plasmon resonance peak is significantly red-shifted compared to the SPR band of a single Au nanoparticle [212].

An alternative method for the production of ordered nanostructures was proposed by Van Duyne and coworkers. Therefore, polystyrene nanosphere monolayers were immobilized on glass substrates serving as a physical masc. Thereafter, gold or silver was deposited onto the substrate followed by the removal of the polystyrene particles leaving behind truncated tetrahedral nanostructures of geometries defined by the size of the polystyrene particles and the amount of deposited metal [77, 100].

Another interesting approach was reported by Alivisatos and colleagues. They produced arrays of nanoparticles and more complex structures such as nanotetrapods by evaporating a nanoparticle solution from the substrate. Hereby, the basis of the assembly mechanism is the interfacial capillary force present during the evaporation [34].

2.5 *Methods for Size Reduction*

Another top-down fabrication approach is based on interaction of light. Gold nanoparticles in the lower nanometer range can be fabricated by laser ablation (wavelength 1,064 nm) of larger particles. Thereby, irradiation at 532 nm (second harmonic) of 8 nm particles resulted in size reduction. The average diameter decreases with laser fluence (6.2, 5.2, and 4.1 nm with 280, 560, and 840 mJ/pulse cm^2). Larger particles are fragmented by irradiation, whereas smaller particles grow by attracting fragments from the solution. By this process, the size distribution is narrowed indicated by narrowing of the width of the plasmon band [133].

The same goal was followed by Takami and colleagues by irradiating 20 or 50 nm nanoparticles with laser light. The size distribution is again narrowed. A minimal diameter of 10 nm was found even at very high radiation. But laser treatment can also cause shape transformation due to particle heating and subsequent melting of the particles resulting in thermodynamically stable spherically shaped particles. Calculations of the particle temperature showed a heating of up to 2,500 K [200].

2.6 *Methods for Size Separation*

Hwang and coworkers suggested capillary electrophoresis for size separation. They showed experimentally the separation of polystyrene (PS) or gold particles of different sizes. It was also possible to distinguish between PS and gold particles of the same size [92]. Alivisatos and coworkers used gel electrophoresis for the separation of gold nanoparticles functionalized with thiol-capped oligonucleotides (see paragraph modifications) [42, 162].

A very precise method for the determination of particle sizes was developed by Galletto and coworkers. They used the hyperpolarizability of first order to detect the size of particles. Small variations in size show a great difference in polarizability. This method could be an effective alternative to the particle size determination with an external reference [57].

3 Bimetallic Particles

Bimetallic particles are classified into two types: first, particles with homogeneous distribution of the two metals are called alloys, and second, particles with heterogeneous arrangement of the two metals, i.e., the so-called core–shell nanoparticles (Fig. 1.3). In principle, the methods for synthesis are derived from the methods for the preparation of monometallic nanoparticles.

3.1 Bimetallic Alloy Nanoparticles

Alloy nanoparticles are synthesized in solution by simultaneous reduction of the two metals of interest (Fig. 1.3). The transformation of core–shell nanoparticles into alloys is another approach that is discussed later.

In general, alloy particles show one characteristic plasmon band. Because of the homogeneous distribution of the metals, alloy particles have homogeneous optical dielectric constants [150]. The position of the LSPR is determined by the ratio of the two metals used. For example, alloy particles consisting of gold and silver

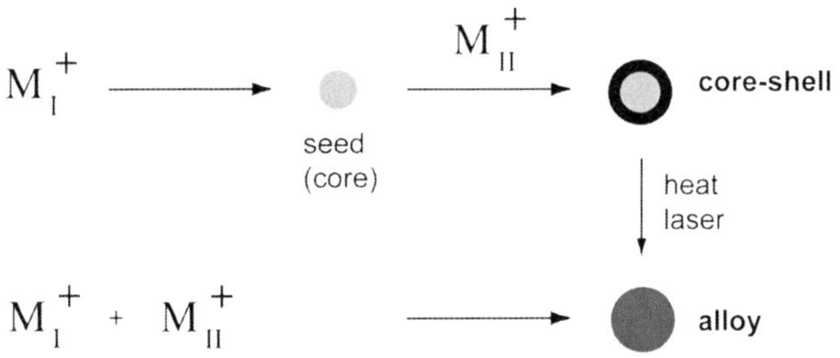

Fig. 1.3 Core–shell particles (*top*) exhibit a defined separation of both metals, whereas alloy particles (*bottom*) consist of a homogeneous mixture. Temperature treatment (e.g., by laser) is known to convert core–shell structures in alloy particles

show a plasmon band ranging between 400 nm (pure silver) and 525 nm (pure gold) [128].

Sànchez-Ramírez et al. synthesized Au/Cu alloys by reducing gold and copper salt solutions with $NaBH_4$ in the presence of PVP as stabilizer. They found a linear law in the position of the LSPR band with the composition of the bimetallic particles as described above [176].

It is possible to prepare a wide variety of alloy particles from several metals. In the early 1990s, scientists found out that alloy nanoparticles provide more catalytic activity than monometallic nanoparticles [205]. Later, the interest for the optical properties has also grown. Alloy nanoparticles of a mixture of gold and silver were also subject matter in many publications. El-Sayed and coworkers synthesized Au/Ag alloys (17–22 nm in size) following a modified protocol by Turkevich [207]. The characterization of the alloys by HRTEM revealed the presence of defects in the homogeneous distribution of the metals such as individual islands of gold and silver within the particles [128]. Alloy synthesis in 2-butanol was applied by Papavassiliou producing 10 nm Au/Ag alloy particles [161].

It is also possible to synthesize alloy particles by laser ablation. Therefore, bulk alloys are treated with laser light in water without any chemical reagents. Lee et al. showed the process for Au–Ag alloy particles with atomic variations of less than 2%. Because of the similar lattice constants, gold and silver are miscible nearly in all proportions. The size distribution could be controlled by the variation of the pulse energy and the time of ablation and the particles could be modified afterwards [125].

3.2 *Bimetallic Core–Shell Nanoparticles*

Core–shell nanoparticles with inhomogeneous distribution of the metals can be synthesized by the successive reduction of two metal salt solutions. The nanoparticles created during the first reduction process are used as seeds for the second reduction (Figs. 1.3 and 1.4b). Absorption measurements have shown two plasmon bands for these particle structures [150, 151]. The core–shell nanoparticles nearly keep their electronic properties with the electronic band structure of the pure metals. The two bands are only slightly shifted due to interactions between the core and the shell material (Fig. 1.4a).

Noble metals prefer the nonsurface region of the particle [43], but with the addition of a suitable reducing agent, noble metals can also be deposited on less noble metals. To the best of our knowledge, core–shell nanoparticles were synthesized by G. Schmid and colleagues for the first time. He prepared particles with 18 nm Au cores and Pt or Pd shell reaching a total particle size of 35 nm [180]. Then, more and more interest was directed to core–shell particles.

Several bimetallic systems with metals which were already known from monometallic nanoparticle synthesis were used. Lu et al. reported the synthesis of large, monodisperse gold–silver core–shell particles with Ag-like optical properties

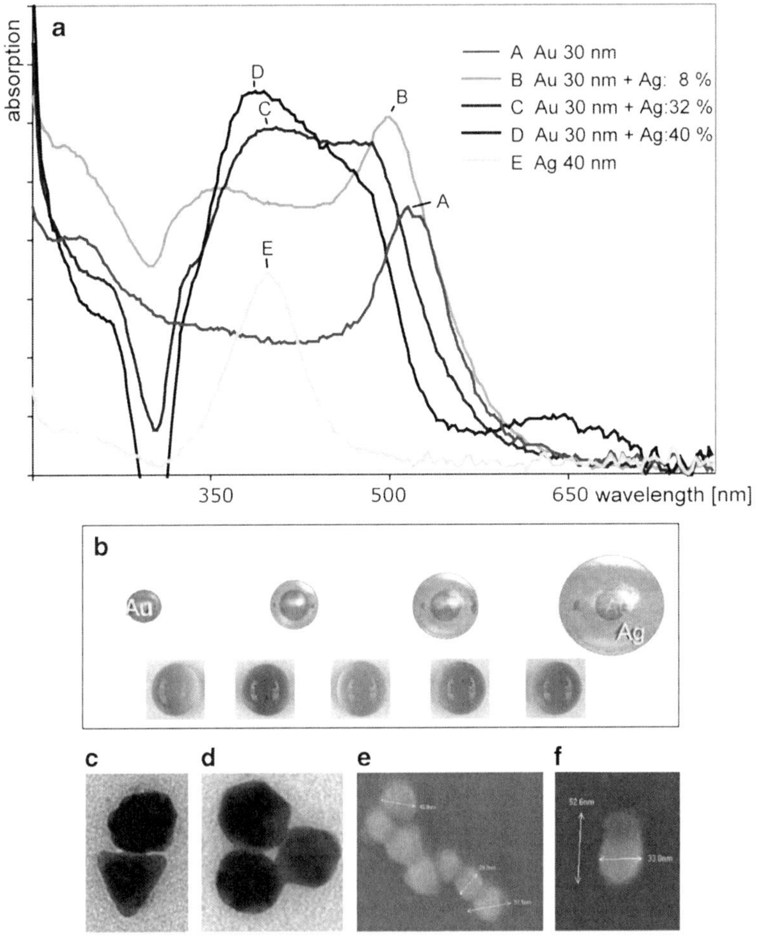

Fig. 1.4 Au-core Ag-shell nanoparticles. (**a**) UV–vis measurements of core–shell particles with increasing silver shell thickness (**b–d**) based on addition of different percentages of silver enhancement solution; for comparison the spectra for pure gold (**a**) and pure silver (**e**) are added; (**b**) scheme of enhancement and photographic pictures of droplets of particle solutions; (**c**, **d**) TEM images and (**e, f**) SEM images of Au-core Ag-shell nanoparticles [189]

and ca. 100 nm in size [51]. Soon, an application was found to utilize the core–shell nanoparticles: SERS. First, monometallic silver nanoparticles were applied to enhance the signal from the sample, but also gold–silver bimetallic nanoparticles could be used. Gold seeds are used to deposit silver shells of varying thicknesses (0–100 nm) [136]. The core–shell character of the particles was always revealed from TEM images where brightness differences indicate the metal arrangement (Fig. 1.4c, d). UV–vis absorption spectra show the absorption peaks that reflect a certain composition of the particles. For example, the silver shell in Au–Ag core–shell particles influences on the position of the LSP band of gold. Increasing the Ag shell thickness results not only in an increase of the absorption signal near 400 nm, but also in a shift of the position of the 520 nm band for gold; i.e., there will occur a blue shift and damping of the LSP signal of gold with increasing Ag content of the particle [3] (Fig. 1.4a).

Besides, the formation of silver core/gold shell particles has been investigated as well. Experiments by Moskovits et al. showed incomplete shell formation only for Au mole fractions below 0.3. Looking more closely to the ultrastructure of the prepared particle, they found a Ag-rich core and a Ag/Au alloy shell whereby the Au fraction grew with increasing Au mole fraction. Furthermore, they theoretically extracted the optical constants from LSP extinction spectra [148]. To suppress the interactions of the two metals in core–shell structures, there is the possibility to separate the core and the shell by an insulating layer of SiO_2. This way, the properties of the core and the shell can be tuned independently and particles with mean total diameters of 120 nm have been synthesized [177].

An interesting method concerning the reducing agent was introduced by Mandal et al. They used UV-switchable so-called "Keggin ions" for the synthesis of Au–Ag core–shell particles [137]. Mirkin's group, the pioneers of nanoparticle modification, also dealt with the modification of bimetallic particles. For alloy structures, the modification was not successful, but the reaction of core–shell particles and thiol-capped DNA resulted in stable particles [24].

Hutter et al. investigated the possibility to connect gold and silver particles together that were synthesized separately in solution. The constructs represent no "real" core–shell particles but it is possible to produce robust gold-encased silver nanoparticles where small (ca. 3 nm) gold particles are attached to ca. 38 nm-sized silver particles. By using larger colloids, chain-like structures are formed [90].

Gold–platinum and gold–palladium are other well-studied systems for bimetallic particle synthesis. Flynn et al. synthesized 11.2 nm gold seeds and deposited Pt with 1–8 mol% [51]. Henglein et al. published the preparation of Pt–Au and Au–Pt nanoparticles by hydrogen reduction and radiolytic techniques. Also, they succeeded in the synthesis of trimetallic particles by the reduction of silver ions by the Au–Pt particles [82]. Moreover, they prepared Pd/Au core–shell particles and trimetallic Pd/Au/Ag particles by Ag deposition on the first mentioned [81].

The experiments of Takatani et al. revealed the importance of the used surfactant for the stabilization of the particles. They tried to synthesize Au/Pt particles with SDS as surfactant, but only monometallic Au (10 nm) and Pt (1 nm) particles were found in the solution. When they used PEG-MS, they achieved a core–shell-like structure. For Au/Pd particles, SDS worked very efficiently as stabilizer reaching 10-nm core–shell particles. The use of PEG-MS resulted also in core–shell formation, but alloys were found in the solution as well and the overall particle size was reduced, whereas the size distribution was broadened [201].

One of the possibilities size distributions to optimize is the use of two-step micro-continuous flow-through method [113]. Such synthesis allows constant residence times and an effective mixing of the components by applying the segmented flow principle. The resulted multi-core–shell nanoparticles offer a narrow size distribution.

3.3 *Transformation of Core–Shell into Alloy Nanoparticles*

Laser treatment of core–shell nanoparticles can result in the formation of alloy nanoparticles (Fig. 1.3). As mentioned earlier, laser irradiation can cause size reduction and shape changes [56, 105, 200]. Particularly, when laser light is used to irradiate nanoparticles with a resonant wavelength, heat is created in the particles. The temperature can exceed 1,000°C [200]. To the best of our knowledge, the alloying of core–shell particles using ns and ps laser equipment was published first by Hodak et al. in 2000. They found alloying and reshaping of particles after the excitation with a ns laser. When the fluence exceeded a power of over 10 mJ/pulse, fragmentation occurred. Due to more efficient heating, ps excitation led to fragmentation processes already at 4 mJ/pulse fluence [87]. Later, Abid et al. also worked in this field. They used Au_{60}/Ag_{40} core–shell particles with absorption peaks at 410 nm (Ag) and 520 nm (Au) and irradiated them at fluences of 16 or 96 mJ/cm^2, respectively. At a high fluence, the absorption spectra showed only one peak at 510 nm after the laser treatment, indicating the removal of the silver shell and leaving monometallic gold particles. At a low fluence, there was also only one peak detectable in the absorption spectrum but regarding its position (455 nm), it seemed to result from Au/Ag alloy particles. Homogeneous mixing of the two metals is only achieved after numerous pulses because of particle interactions with the environment. That means that the particles do not remain heated long enough to mix within a single pulse because heat leaks to the environment. With each pulse, the particle is heated, partial mixing occurs, and then it cools down again. Using nanosecond laser, less mixing takes place compared to picosecond laser treatment [2]. This method offers an opportunity for the well-defined synthesis of alloy particles by the very selective heating to create new materials with a wide range of composition providing unique optical properties [75].

In addition, we want to mention work where the spontaneous (without any laser excitation) alloying of Au–Ag core–shell nanoparticles was investigated. Shibata et al. have found that the interdiffusion of the two metals was limited to the subinterface layer and depended on the core size and the total particle size. In general, parts of the silver shell do remain. Theoretical calculations confirm the practical results that defects or vacancies at the bimetallic interface enhance the radial migration of the metals [185]. Similarly, spontaneous alloying was also reported for Cu [225] and other metals [166].

4 Nanoparticles and Fluorescence

Novel properties arise from the combination of nanoparticles with fluorescence. First, the so-called quenching effect is used whereby the fluorescence signal decreases with decreasing distance between the fluorophore and a metal particle (Fig. 1.5, left). Second, when the metal particle and the fluorophore are brought

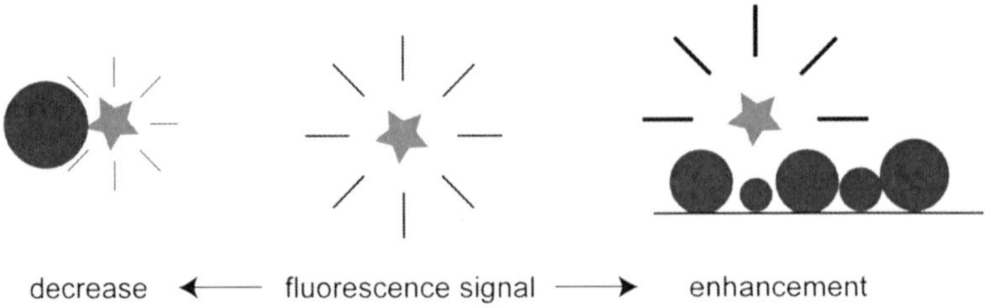

Fig. 1.5 Quenching (*left*) and fluorescence enhancement (*right*) as typical examples for the interactions of fluorescent dyes and metal nanoparticles

together, fluorescence enhancement occurs at certain distances and geometries (Fig. 1.5, right).

4.1 Quenching Effect

Quenching is considered as an energy transfer that occurs when the emission frequency of one molecule (donor) overlaps with the absorption frequency of a second molecule (acceptor). This energy transfer was first observed and investigated by Förster in 1948 [52, 167] and is called Förster resonant energy transfer (FRET). The two interacting molecules can be both fluorophores or a combination of one fluorophore with one metal particle. Here, we only want to discuss fluorophore–nanoparticle systems briefly.

The quenching effect is used in three general approaches: A simple Yes/No answer whether an analyte is present in the test solution, a quantitative analysis of the amount of analyte, or a determination of the distance between donor and acceptor.

Mainly, these tests are DNA-based tests for DNA-analytics or PCR approaches. One approach uses a single-stranded DNA molecule, the so-called molecular beacon that carries both the fluorophore and the metal nanoparticle through covalent attachment [12]. The hairpin loop that is formed to hold the two labels together is opened during the test by hybridization of an analyte to the molecular beacon. Thereby, the distance between the fluorophore and the nanoparticle increases resulting in an increase in fluorescence signal. It is possible to detect SNPs by this principle [42]. The detection of multiple targets in parallel (multiplexing) has been facilitated by the use of several differently colored fluorophores [138, 208]. Furthermore, the identification of retroviruses was shown [209]. Finally, it is possible to monitor nucleic acid detection and PCR quantification in real-time [40].

In addition, the "efficiency" of quenching follows a strict mathematic rule that allows for the determination of the distance between the fluorophore and the metal particle. Förster found that quenching increases with decreasing distance between the two labels following an inverse sixth-order power law. The technique is applied for protein analytics, mainly for monitoring of protein folding/unfolding or the work of the active site of enzymes [127]. FRET in combination with photo-induced electron transfer (PET) even works on the single molecule level.

Instead of "classical" fluorescence molecules, it is possible to use so-called quantum dots. These are nanoparticles made from semiconductor materials (e.g., CdSe) that show high fluorescence and do not suffer from photobleaching, a general problem in fluorescence-labeling techniques. The brightness and photostability of quantum dots allow the detection and observation on the single molecule level [70].

4.2 Fluorescence Enhancement

As described above, quenching normally occurs when a fluorophore and a metal nanoparticle are located very close to each other. But it was described that even fluorescence enhancement can appear when a fluorophore and a metal nanoparticle approach each other [11, 41, 213]. At a distance of ca. 7–10 nm, there exists a local maximum of the fluorescence signal from a fluorophore that is located close to a metal nanoparticle. However, for lower and even higher distances, the fluorescence decreases [103, 121].

The magnitude of the fluorescence enhancement depends on the metal particle, its size, and shape and on the fluorophore type as well. Also the geometry (in the case of multiple metal nanostructures) should influence this effect. There are two pathways upon the interaction of a fluorophore and a metal particle: a radiative and a nonradiative pathway. As mentioned earlier, quenching is due to the absorptive component of the nanoparticle, but fluorescence enhancement, on the other hand, is due to the radiating (scattering) component of the nanoparticle [120]. Therefore, a metallic nanostructure in the vicinity of a fluorophore has great influence on various aspects of the fluorescence emission: the brightness, the lifetime, the absorption and emission spectra, and the effective quantum yield. All these aspect are parts of a causal chain. The starting point is the lifetime of the fluorophore. By decreasing the lifetime, the photostability is increased. Photo-induced destruction can only occur when the molecule is excited. But when the fluorophore is released faster from its excited state, the probability for photobleaching is reduced and the quantum efficiency increases [117, 119, 123].

Lakowicz and coworkers propose the technique for applications in DNA-analytics, immunoassays, for imaging and counting of single molecules, and the control of the flow in microfluidics [121, 122].

Besides these investigations on silvered glass slides, theoretical calculations showed similar facts for fluorophores encased by a metal layer. Enderlein showed theoretically the fluorescence enhancement for polymer beads functionalized with fluorophores that were enclosed by a thin metal layer (5–10 nm of silver or gold) [45].

5 Functionalization of Nanoparticles

Applications of nanoparticles require often a (bio)chemical modification of the particles. This modification can be used to control certain properties of the particle, and/or is needed in order to access the nanoparticle, e.g., for specific binding to substrates or to other binding partners. In addition, a stabilizing effect due to electrostatic and/or steric reasons can be observed, that is even required in many applications involving higher salt concentrations (e.g., bioanalytics).

5.1 Nanoparticles in Microscopy and Analytics

To the best of our knowledge, the first time when nanoparticles were used for a biological approach was in electron microscopy (TEM) by Palade [158] to detect the transport across the endothelium of blood capillaries. Thereafter, Faulk introduced an immunocolloid method for the electron microscopy based on antibody-modified nanoparticles [48]. Several publications followed describing the detection of human serum albumin [22] or immunglobulins [110, 132], respectively. The detection limit achieved with immunglobulins was lower than pM, and quantitative investigations were possible. Several protocols for the preparation of protein–nanoparticle conjugates (cytochrome *C*, protein A, etc.) are available [76, 108].

Alternatively, DNA can be simply adsorbed on nanoparticles [60] or electrostatic interactions can be used for the binding of charged gold nanoparticles to DNA [226].

5.2 Biotin/Streptavidin

One of the most common biological coupling systems today, the biotin–streptavidin system, was discovered by Chaiet and Wolf in 1964 [26]. The protein streptavidin, found in Streptomycetes, specifically recognizes and strongly binds the small molecule biotin (vitamin H). The strength of the bond is comparable to a covalent linkage and the system was used for biological assays. Biotinylated DNA was detected by a (strept)avidin-modified enzyme in a colorimetric assay [124]. Only 5 years later, the biotin–streptavidin interaction was used for the visualization of sites of nascent DNA synthesis by streptavidin–gold nanoparticle binding to biotinylated nucleotides [86]. Shortly after, Henderson and coworkers reported the labeling of oriented linear DNA molecules with 5-nm gold spheres based on biotin–streptavidin recognition [184].

In another approach, biotinylated RNA of bacterial pathogens was detected by streptavidin-modified gold nanoparticles. The detection system was based on the resonant light scattering of the particles that showed significant lower detection limits (10 fM) than conventional fluorescence systems (500 fM) [53]. The system was used for SERS as well. This method is based on the local signal enhancement for molecules adsorbed at roughened metal surfaces, e.g., silver. The effect is due to the local amplification of the electromagnetic field near rough structures in free electron metals [15]. It was predicted that the detection limit is low enough to investigate single molecules (Kneipp 2002). Moreover, biotinylated silver nanoparticles additionally functionalized with SERS markers were used for the efficient molecular sensing on avidin-modified surfaces [111].

5.3 Thiol Ligands

The modification with thiols especially for gold nanoparticles was extensively investigated during the last two decades. It is based on the high affinity of thiol to gold surfaces [156]. Brust used thiols during the particle synthesis as stabilizer [21]. The disadvantage of the system is that the size of the created thiol-modified nanoparticles is limited to 5 nm. The modification with thiols including dithiols, etc. can be carried out after the synthesis resulting in nearly no limitations regarding the sizes as well [168, 227]. The stabilizing capacity differs for the diverse sulfur ligands, e.g., disulfides are not as good stabilizing agents as thiols. The results of the adsorption of disulfides on gold surfaces show that the binding between gold nanoparticles and thiols is considered as covalent [156]. It was reported by Stellaci and coworkers that thiol (and amine) ligands of variable lengths are organized in highly ordered domains on nanoparticles. This effect was not observed on planar substrates. It has been found that curvature (size dependence) was essential for the formation of the domains. It was proposed to create particles with different "poles" by using different ligands [94]. Furthermore, it was found that thiol ligands affect the electronic behavior of gold nanoparticles. As shown by Zhang and Sham thiols strongly bind to the gold nanoparticle surface and cause charge transfer from the gold to the sulfur ligand [229]. By the utilization of dithiols, a covalent linkage of nanoparticles can be created resulting in aggregation [13, 20].

Mulvaney and coworkers used citrate or alkanethiol stabilized nanoparticles, respectively, for the electrophoretical deposition of highly ordered nanoparticle monolayers [61]. Thereby, the interparticle distance is determined by the chain length of the stabilizer.

For the functionalization of gold nanoparticles with thiol-capped oligonucleotides, see Sect. 5.5.

5.4 Other Ligands

The use of several other ligands has been reported for the stabilization of nanoparticles such as amines [126], carboxyls [139], phosphines (i.e., BSPP) [131, 179, 211], and others. Nanoparticles modified with phosphine ligands (nanogold/undecagold) are commercially available from Nanoprobes, Inc. (http://www.nanoprobes.com).

5.5 Oligonucleotide Ligands

Among all possible modifications, the modification of gold nanoparticles with thiol ligands is the most common and widest used. Since DNA can chemically be modified with thiol groups, it shows potential for the modification of nanoparticles.

In 1996, two groups simultaneously developed a strategy to modify gold nanoparticles with thiol-capped noncomplementary oligonucleotides [7, 144].

Mirkin and coworkers applied double-stranded DNA-linkers with complementary "sticky" ends resulting in self-assembly of the nanoparticles. The aggregation is reversible because it is based on hybridization. Here, in contrast to the aggregates based on dithiols, the linkage is noncovalent. The authors suggest the possibility to tailor the optical, electronic, and structural properties of the aggregates by particles of different sizes and materials. In comparison, the main differences of the system of Alivisatos and coworkers are that they use single-stranded linkers to aggregate the DNA–nanoparticle conjugates and that only discrete numbers of nanoparticles are assembled. The system is based on hybridization of complementary strands through Watson-Crick base-pairing that assure the specificity of the recognition too. The authors reported the formation of dimers and trimers upon the addition of a complementary linker and propose the construction of more complex two- and three-dimensional assemblies.

Dozens of publications followed describing applications based on the introduced methods or reporting more detailed investigations in this issue, respectively [24, 37, 44, 78, 102, 140, 163, 165, 174, 188, 192, 193, 203].

For the determination of the number of thiol-capped oligonucleotides bound to nanoparticles, Demers et al. performed fluorescence-based experiments. For example, they found the linkage of 159 DNA-strands (12mer) on a single 15-nm gold particle. This reflects a significantly higher coverage with oligonucleotides for nanoparticles than for planar gold substrates. Analysis of the hybridization efficiency showed that the bases close to the surface of the nanoparticle are less accessive because of steric hindrance. It is recommendable to use C_{12}-linkers between thiol group and oligonucleotide to ensure access to all bases of the strands. So-called "diluent strands" additionally improve the hybridization efficiency. These strands are either noncomplementary DNA-strands or alkanethiols and assure the nearly perpendicular orientation of the oligonucleotides to the nanoparticle surface to achieve maximal accessibility for hybridization (Demers 2000). Alkanethiols such as mercaptohexanol are quite common to "dilute" the oligonucleotide strands on nanoparticles. Both, alkanethiols and noncovalent oligonucleotides, help to destabilize the noncovalent adsorption of DNA via the bases on the nanoparticle surface and they change the conformation of the oligonucleotides as mentioned before (Park 2004). In summary, the coverage with oligonucleotides must be high enough to stabilize the particle and low enough to achieve high hybridization efficiency.

Alivisatos' group mainly dealt with the coupling of defined numbers of oligonucleotides of variable lengths to gold nanoparticles. They used gel electrophoresis to determine the effective diameter of the particles and there from calculated the surface coverage. For low coverage, they yielded perpendicular orientation of the oligonucleotides to the surface. For high coverage, they found perpendicular orientation for short oligonucleotides to the surface and a stretched conformation, whereas longer oligonucleotides showed perpendicular orientation only near to the surface of the particle and a random coil shape for the outer region [162].

The structural properties of oligonucleotide SAM were also investigated on gold surfaces. Like on nanoparticles, the orientation of the strands mainly depends on the steric interactions between neighboring molecules. Thereby, the length influences the conformation of the strands and affects the packing density. Mercaptohexanol can be used to stabilize the SAM film too [171].

Storhoff et al. investigated the optical properties of DNA-linked gold nanoparticle aggregates. The authors found that the oligonucleotide linker kinetically controls the aggregate size. That means that the optical properties are mainly dependent on aggregate size regardless of linker length [193]. Later, some factors that influence the melting properties of the aggregates (hybridization) were investigated. It was found that the melting temperature (T_m) increases with increasing oligonucleotide density on the particles, increasing salt concentration, and increasing interparticle distance. The size of the particles affects the sharpness of the melting profile [102]. These results are also valid for DNA-functionalized Ag/Au core–shell nanoparticles [24].

6 Structures of Defined Geometry

Interparticle interactions depend on the distance and the geometry of the arrangement in case of several involved particles. Molecular tools allow for a definition of these parameters, as shown in the following section.

6.1 Non-DNA-Linkers

Feldheim and coworkers have reported several aspects of the symmetrical assembly of gold and silver nanoparticles using phenylacetylene bridges. They prepared dimer, trimer, and tetramer structures with bridges of variable lengths. The structures have been separated and enriched by ultracentrifugation. The optical properties of spherical particles and dimers differ significantly. Due to the changed aspect ratio for the particle dimmers, an additional peak appears in the absorption spectrum related to the longitudinal plasmon mode. Trimers react similar to spherical nanoparticles [17, 49, 155].

6.2 DNA-Linkers

Based on the principle of the formation of DNA-linked nanoparticle networks introduced by Mirkin and colleagues, Mucic et al. prepared solution-based structures composed of particles of different sizes. They decorated 30 nm-sized with 8 nm-sized particles both made from gold. The authors proposed the "entry into multicomponent nanostructured materials" [149]. Iacopino et al. constructed gold nanoparticle dimers with DNA-linkers of potentially variable lengths based on self-assembly in solution [93]. The use of branched DNA trimers and nanoparticle–DNA conjugates resulted in the formation of nanoparticle trimer and tetramer complexes. Both 5- and 10-nm gold particles were assembled together for the construction of asymmetric structures [31].

The combination of DNA-linkers with different chemistry like thiol or amino on gold or silver particles extends the possibilities for bioconjugation of such nanoparticles [190]. In addition, also core–shell structures can be included in such molecular constructions approaches [191].

Taton et al. reported the formation of supramolecular mono- and multilayered nanoparticle assemblies based on DNA interaction on glass substrates. Using DNA bound to the surface and DNA-modified nanoparticles, it is possible to assemble nanoparticle layers stepwise. In doing so, the DNA controls the interparticle distance. The authors propose the extension of the system to variable nanoparticle sizes and materials [204].

6.3 2D-DNA Patterns

Larger repetitive structures are possibly based on the approaches by Seeman et al. using synthetic DNA with partial overlap in order to create network structures [210, 219]. Such structures allow for highly defined relative distances between positioned nanostructures, and this precision is possible in 2D. This is in contrast to the standard approach of linear DNA mentioned before that is usually 1D and moreover often hampered by the flexibility of the DNA in the case of distances significantly larger than the persistence length (50 nm for standard conditions). These structures were used for positioning of nanoparticles at certain points in the pattern [127].

Current developments are focused on novel 2D-DNA superstructures: the DNA origami [175]. These structures offer a large potential for novel patterning techniques for nanoparticle arrangements in the lower and medium nanometer range with a precision of a few nanometers [38].

7 Applications of DNA-Based Nanoparticle Aggregation

As described above, a nanoparticle-based DNA detection system is several orders of magnitude more selective than current genomic detection systems. Several proteins were also detected in nanoparticle-based assays [76, 110, 132]. Another advantage of nanoparticle-based systems is that the detection can be carried out by a variety of analytical techniques: optical absorption, fluorescence, Raman scattering, atomic and magnetic force, and electrical conductivity. Therefore, nanoparticles are excellent labels for biosensors [95]. Furthermore, nanoparticles have the potential to complement or even replace established labels such as radioactivity, fluorescence, chemiluminescence, and enzymatic/colorimetric detection [182]. They are very stable, nonbleaching, and provide potential for multiplexing because the LSPR peak can be tuned by changing the material, size, or shape.

The first example for applications of DNA-based nanoparticle aggregation immediately followed the introduction of the method for binding oligonucleotides on gold nanoparticles by Mirkin and coworkers. The system was used for a colorimetric assay for the highly selective detection of polynucleotides. Among other parameters, the resonant plasmon peak of nanoparticles is dependent on the interparticle distance. When hybridization of the analyte occurs with the DNA attached to the gold nanoparticles, the aggregation induced by DNA bridges results in a decrease of interparticle distance. That means during hybridization the solution will change its color from red to blue. The reaction could be performed on a solid substrate as well. Then, the color change (signal) is even stronger. The detection limit for an unoptimized process was 10 fM [44]. The same group also investigated the applicability of larger particles (50–100 nm). They modified the nanoparticles with dithiane androsterone-functionalized oligonucleotides to achieve the same coverage as for smaller particles. This resulted also in high detectivity down to 50–100 pM [174].

The very low detection limit that can be achieved with DNA–nanoparticle conjugates allows for the detection of single nucleotide polymorphisms. Willner and coworkers developed a method to detect SNPs by gold nanoparticles that were additionally overgrown by gold to even enhance the signal. They found a detection limit of 1×10^{-15} M for SNP detection [214, 218]. In 2004, Kerman et al. succeeded in the detection and even the identification of SNPs. They used monobase-modified gold nanoparticles that (in case of complementary sequences) bind to the mismatch in the investigated DNA-strand and consequently accumulate on an electrode. By monitoring the changes in the electrochemical signal, the transition and transverse SNP is identified. The process works also in the presence of interfering DNA [109].

Another step was the simultaneous detection and discrimination of multiple DNA or RNA molecule species (multiplexing). For reaching this goal, Mirkin and coworkers published an interesting approach. They combined the advantages of nanoparticle-based assays with the advantages of Raman-active dyes. Raman tags show a very narrow-band spectroscopic fingerprint. A slight modification of a tag will create a new molecule with a different fingerprint spectrum. From this point of view, there is a great pool of nonoverlapping dye molecules available for multiplexing. The differentiation between 6 DNA and 2 RNA molecules was shown with a detection limit of 20 fM [25]. The ultrasensitive detection of DNA (detection limit of 8×10^{-13} M) using SERS was also proved by Kneipp et al. and Smith and colleagues [114, 160].

Another approach uses the possibility to specifically address nanoparticles with light. When illuminated at the resonance wavelength, the nanoparticle heats up and transfers this heat to the environment with local precision (photothermal effect). This behavior can be applied for a drug delivery system where nanoshell-polymer structures were used [183]. On the other hand, induced heating of nanoparticles can be used for nano-localized manipulation. In the case of DNA-conjugated particles, a sequence-specific manipulation of metaphase chromosomes was demonstrated using silver-enhanced gold nanoparticles that were positioned in sequence-specific manner on a metaphase chromosome and irradiated by a laser at the wavelength of the plasmon band [32, 59]. An antenna effect induced heating and thereby damages only at the nanoparticles with a subwavelength radius, as it could be demonstrated using AFM studies before and after irradiation.

Moreover, ultrasensitive assays were reported for single molecule detection based on the observation of individual nanoparticles in dark field microscopy [182] and with optical (reflection/transmission) measurements [33].

Interparticle interaction as observed in the case of aggregation is not required for an assay. It is also possible to monitor the shift of the localized surface plasmon band when the analyte is bound to the particle. The surrounding medium is one critical parameter for the location of the resonant plasmon peak. When the analyte binds to the particle, the surrounding medium is changed followed by a shift of the position of the absorption maximum [46, 150, 157, 223]. This effect can be refined even down to the single particle level [172].

8 Summary

The optical properties of metal nanoparticles depend strongly not only on the metal that is used, but also on the size, the shape, the surrounding medium, and the interparticle distance. By the variation of at least one of these parameters, the optical properties can be tuned within a wide range of wavelengths potentially resulting in broad fields of applications.

In general, it was shown that metal nanoparticles are mainly synthesized by reducing a salt solution with a reducing agent such as sodium citrate, sodium borohydride ($NaBH_4$), ascorbic acid, and others. Especially for the synthesis of silver nanoparticles, laser irradiation of silver salts can be applied. By these synthesis methods, nanoparticles of various sizes and shapes such as spheres, prisms, and cubes are produced by the variation of the ratio of metal salt to reducing agent or the addition of stabilizer(s). Two or even three different metals can be used together resulting in bimetallic or trimetallic nanoparticles, respectively. Alloy particles show a homogeneous distribution of the two metals yielded by the simultaneous reduction of two metal salts. Core–shell particles are produced from seed particles that form the core. In an additional reduction step, the second metal is deposited as a shell around the core.

Novel properties arise from the combination of nanoparticles with fluorescence. First, the so-called quenching effect is used to detect, e.g., SNPs and to monitor nucleic acid detection and PCR quantification in real-time. Second, fluorescence enhancement of a fluorescent dye can occur at certain distances and geometries relative to a metal nanoparticle. The effect is used in DNA-analytics, immunoassays, and for imaging and counting of single molecules.

On the basis of the modification of gold nanoparticles with thiol-capped oligonucleotides, several approaches have been reported such as the construction of structures with defined geometry (nanotools) and applications in DNA-analytics based on hybridization-driven aggregation. For the readout, an optical (colorimetric) detection is used.

9 Outlook

The development of novel or refined synthesis approaches for metal nanoparticles allow for an increased control of the fascinating optical properties of these nanoparticles. The growing knowledge about the interaction of molecular structures with the particles will lead to novel sophisticated bioanalytical assays with high sensitivity in combination with increased robustness.

On the other hand, the potential of molecular construction with the promise of defined structures in the lower nanometer range results in novel opportunities for the realization of nanooptical constructs and devices that will be addressed in future works in the interdisciplinary field of Molecular Plasmonics.

Acknowledgments We thank Robert Möller for helpful discussions, Grit Festag for extensive manuscript corrections and improvements, Franka Jahn for SEM imaging and for help with TEM measurements. This work was funded by grants from the DFG (Fr 1348/12-1).

References

1. Abid J P 2003 Laser Induced Synthesis and Non Linear Optical Properties of Metal Nanoparticles. In: *Laboratorie d'Electrochimie,* (Lausanne: Ecole Polytechnique Federale de Lausanne)
2. Abid J P, Girault H H and Brevet P F 2001 Selective structure changes of core–shell gold–silver nanoparticles by laser irradiation: homogenisation vs. silver removal *Chem. Commun.* 829–30
3. Abid J P, Wark A W, Brevet P F and Girault H H 2002 Preparation of silver nanoparticles in solution from a silver salt by laser irradiation *Chem Commun (Camb)* 792–3
4. Aden A, L. and Kerker M 1951: AIP. pp 1242–6
5. Aherne D, Ledwith D, M., Gara M and Kelly J, M. 2008 Optical Properties and Growth Aspects of Silver Nanoprisms Produced by a Highly Reproducible and Rapid Synthesis at Room Temperature *Advanced Functional Materials* **18** 2005–16
6. Ahmadi and Wang G, Henglein, El-Sayed 1996 Shape-controlled synthesis of colloidal platinum nanoparticles *Science* **272** 1924–7
7. Alivisatos A P, Johnsson K P, Peng X, Wilson T E, Loweth C J, Bruchez M P, Jr. and Schultz P G 1996 Organization of 'nanocrystal molecules' using DNA *Nature* **382** 609–11
8. Anyaogu K C, Fedorov A V and Neckers D C 2008 Synthesis, Characterization, and Antifouling Potential of Functionalized Copper Nanoparticles *Langmuir* **24** 4340–6
9. Aslam M, Gopakumar G, Shoba T L, Mulla I S, Vijayamohanan K, Kulkarni S K, Urban J and Vogel W 2002 Formation of Cu and Cu_2O Nanoparticles by Variation of the Surface Ligand: Preparation, Structure, and Insulating-to-Metallic Transition *Journal of Colloid and Interface Science* **255** 79–90
10. Athawale A A, Katre P P and Majumdar M B 2005 Nonaqueous Phase Synthesis of Copper Nanoparticles *Journal of Nanoscience and Nanotechnology* **5** 991–3

11. Aussenegg F R, Leitner A, Lippitsch M E, Reinisch H and Riegler M 1987 Novel aspects of fluorescence lifetime for molecules positioned close to metal surfaces *Surface Science* **189** 935–45
12. Bernacchi S and Mély Y 2001 Exciton interaction in molecular beacons: a sensitive sensor for short range modifications of the nucleic acid structure *Nucleic Acids Research* **29** e62
13. Bethell D, Brust M, Schiffrin D J and C. K 1996 From monolayers to nanostructured materials: an organic chemist's view of self-assembly *Journal of Electroanalytical Chemistry* **409** 137–43
14. Bigall N C, Hartling T, Klose M, Simon P, Eng L M and Eychmüller A 2008 Monodisperse Platinum Nanospheres with Adjustable Diameters from 10 to 100 nm: Synthesis and Distinct Optical Properties *Nano Letters* **8** 4588–92
15. Brandt E S and Cotton T M 1993 Surface-enhanced Raman scattering *Physical method of chemistry: Vol. IVB. Investigation of surface and interfaces, Part B, 2nd ed. New York: Wiley* 633–718
16. Brongersma 2003 Nanoshells: gifts in a gold wrapper *Nature Materials* **2** 296–7
17. Brousseau L C, Novak J P, Marinakos S M and Feldheim D L 1999 Assembly of Phenylacetylene-Bridged Gold nanocluster Dimers and Trimers *Advanced Materials* **11** 447–9
18. Brown K R, Lyon L A, Fox A P, Reiss B D and Natan M J 2000 Hydroxylamine Seeding of Colloidal Au Nanoparticles. 3. Controlled Formation of Conductive Au Films *Chemistry of Materials* **12** 314–23
19. Brown K R, Walter D G and Natan M J 2000 Seeding of Colloidal Au Nanoparticle Solutions. 2. Improved Control of Particle Size and Shape *Chemistry of Materials* **12** 306–13
20. Brust M, Fink J, Bethell D, Schiffrin D J and Kiely C J 1995 Synthesis and reactions of functionalized gold nanoparticles *Journal of the Chemical Society, Chemical Communications* 1655–6
21. Brust M, Walker M, Bethell D, Schiffrin D J and Whyman R 1994 Synthesis of thiol-derivatized gold nanoparticles in a two-phase liquid-liquid system *Journal of the Chemical Society, Chemical Communications* 801–2
22. Buckle P E, Davies R J, Kinning T, Yeung D, Edwards P R, Pollard-Knight D and Lowe C R 1993 The resonant mirror: a novel optical sensor for direct sensing of biomolecular interactions Part II: applications *Biosensors and Bioelectronics* **8** 355–63
23. Busbee and Obare M 2003 An improved synthesis of high-aspect-ratio gold nanorods *Adv Mat* **15** 414–6
24. Cao Y, Jin R and Mirkin C A 2001 DNA-modified core-shell Ag/Au nanoparticles *J Am Chem Soc* **123** 7961–2
25. Cao Y C, Jin R and Mirkin C A 2002 Nanoparticles with Raman spectroscopic fingerprints for DNA and RNA detection *Science* **297** 1536–40
26. Chaiet L and Wolf F J 1964 The Properties of Streptavidin, a Biotin-Binding Protein Produced by Streptomycetes *Arch Biochem Biophys* **106** 1–5
27. Chen D-H and Hsieh C-H 2002 Synthesis of nickel nanoparticles in aqueous cationic surfactant solutions *Journal of Materials Chemistry* **12** 2412–5
28. Chen L, Zhang D, Chen J, Zhou H and Wan H 2006 The use of CTAB to control the size of copper nanoparticles and the concentration of alkylthiols on their surfaces *Materials Science and Engineering: A* **415** 156–61
29. Chen S and Carroll D L 2002 Synthesis and Characterization of Truncated Triangular Silver Nanoplates *Nano Letters* **2** 1003–7
30. Chen S and Kimura K 1999 Synthesis and Characterization of Carboxylate-Modified Gold Nanoparticle Powders Dispersible in Water *Langmuir* **15** 1075–82
31. Claridge S A, Goh S L, Frechet J M J, Williams S C, Micheel C M and Alivisatos A P 2005 Directed assembly of discrete gold nanoparticle groupings using branched DNA scaffolds *Chemistry of Materials* **17** 1628–35
32. Csaki A, Garwe F, Steinbruck A, Maubach G, Festag G, Weise A, Riemann I, Konig K and Fritzsche W 2007 A Parallel Approach for Subwavelength Molecular Surgery Using Gene-Specific Positioned Metal Nanoparticles as Laser Light Antennas *Nano Lett.* **7** 247–53

33. Csaki A, Kaplanek P, Möller R and Fritzsche W 2003 The optical detection of individual DNA-conjugated gold nanoparticle labels after metal enhancement *Nanotechnology* **14** 1262–8
34. Cui Y, Björk M T, Liddle J A, Sönnichsen S H, Boussert B and Alivisatos A P 2004 Integration of Colloidal Nanocrystals into Lithographically Patterned Devices *Nano Letters* **4** 1093–8
35. Daniel M C and Astruc D 2004 Gold nanoparticles: assembly, supramolecular chemistry, quantum-size-related properties, and applications toward biology, catalysis, and nanotechnology *Chem Rev* **104** 293–346
36. Debye P 1909 Der Lichtdruck auf Kugeln von beliebigem Material *Annalen der Physik* **335** 57–136
37. Demers L M, Mirkin C A, Mucic R C, Reynolds R A, 3 rd, Letsinger R L, Elghanian R and Viswanadham G 2000 A fluorescence-based method for determining the surface coverage and hybridization efficiency of thiol-capped oligonucleotides bound to gold thin films and nanoparticles *Anal Chem* **72** 5535–41
38. Ding B, Deng Z, Yan H, Cabrini S, Zuckermann R N and Bokor J 2010 Gold nanoparticle self-similar chain structure organized by DNA origami *J Am Chem Soc* **132** 3248–9
39. Ditlbacher H, Lamprecht B, Leitner A and Aussenegg F R 2000 Spectrally coded optical data storage by metal nanoparticles *Opt. Lett.* **25** 563–5
40. Drake and Zhao T 2004 Bioconjugated silica particles in bioanalysis - Nanobiotechnology/ Kapitel 27 *Buch : Nanobiotechnology Editors: Niemeyer, Mirkin*
41. Drexhage K H 1974 Interaction of Light with Monomolecular Dye Lasers in Progress in Optics (Wolfe, E., Ed.). North Holland Publishing Company, Amsterdam
42. Dubertret B, Calame M and Libchaber A J 2001 Single-mismatch detection using gold-quenched fluorescent oligonucleotides *Nat Biotechnol* **19** 365–70
43. Edelstein A S and Cammarata R C 1998 *Nanomaterials: synthesis, properties, and applications*: Taylor & Francis
44. Elghanian R, Storhoff J J, Mucic R C, Letsinger R L and Mirkin C A 1997 Selective colorimetric detection of polynucleotides based on the distance-dependent optical properties of gold nanoparticles *Science* **277** 1078–81
45. Enderlein J 2000 A theoretical investigation of single-molecule fluorescence detection on thin metallic layers *Biophys J* **78** 2151–8
46. Englebienne P 1998 Use of colloidal gold surface plasmon resonance peak shift to infer affinity constants from the interactions between protein antigens and antibodies specific for single or multiple epitopes *Analyst* **123** 1599–603
47. Faraday M 1857 Experimental relations of gold (and other metals) to light. *Philos. Trans. R. Soc. London* **147** 145–81
48. Faulk W P and Taylor G M 1971 An immunocolloid method for the electron microscope *Immunochemistry* **8** 1081
49. Feldheim D 2001 Assembly of Metal Nanoparticle Arrays Using Molecular Bridges *The Electrochemical Society Interface* **fall 2001** 22–5
50. Fievet F, Lagier J P, Blin B, Beaudoin B and Figlarz M 1989 Homogeneous and heterogeneous nucleations in the polyol process for the preparation of micron and submicron size metal particles *Solid State Ionics* **32–33** 198–205
51. Flynn N T and Gewirth A A 2002 Attenuation of surface enhanced Raman spectroscopy response in gold–platinum core shell nanoparticles *Journal of Raman Spectroscopy* **33** 243–51
52. Förster T 1948 Zwischenmolekulare Energiewanderung und Fluoreszenz *Annalen der Physik* **437** 55–75
53. Francois P, Bento M, Vaudaux P and Schrenzel J 2003 Comparison of fluorescence and resonance light scattering for highly sensitive microarray detection of bacterial pathogens *Journal of Microbiological Methods* **55** 755–62
54. Frens G 1973 Controlled nucleation for the regulation of the particle size in monodisperse gold suspensions *Nature* **241** 20–2
55. Fritzsche W and Taton T A 2003 Metal Nanoparticles as Labels for Heterogeneous, Chip-Based DNA Detection *Nanotechnology* **14** R63–R73

56. Fujiwara and Yanagida K 1999 Visible laser induced fusion and fragmentation of thionicotinamide-capped gold nanoparticles *J Phys Chem B* **103** 2589–91
57. Galletto and Brevet G, Antoine, Broyer 1999 Size dependence of the surface plasmon enhanced second harmonic response of gold colloids: towards a new calibration method *Chem Commun* 581–2
58. Garg N, Scholl C, Mohanty A and Jin R 2010 The Role of Bromide Ions in Seeding Growth of Au Nanorods *Langmuir*
59. Garwe F, Csaki A, Maubach G, Steinbrück A, Weise A and König K 2005 Laser pulse energy conversion on sequence-specifically bound metal nanoparticles and its application for DNA manipulation *Medical Laser Application* **20** 201–6
60. Gearheart and Ploehn M 2001 Oligonucleotide adsorption to gold nanoparticles: a surface-enhanced Raman spectroscopy study of intrinsically bent DNA *J Phys Chem B* **105** 12609–15
61. Giersig M and Mulvaney P 1993 Preparation of ordered colloid monolayers by electrophoretic deposition *Langmuir* **9** 3408–13
62. Golindano T C, Martínez S I, Delgado O Z and Rivas G P 2005 *Technical Proceedings of the 2005 NSTI Nanotechnology Conference and Trade Show*
63. Gotschy and Vonmetz L, Aussenegg 1996 Optical dichroism of lithographically designed silver nanoparticle films *Opt Lett* **21** 1099–101
64. Gou and Murphy 2004 Controlling the size of Co_2O nanocubes from 200 to 25 nm *J Mater Chem* **14**
65. Gou L and Murphy C J 2003 Solution-Phase Synthesis of Cu_2O Nanocubes *Nano Letters* **3** 231–4
66. Graf C and van Blaaderen A 2002 Metallodielectric Colloidal Core-Shell Particles for Photonic Applications *Langmuir* **18** 524–34
67. Graham T 1861 Liquid Diffusion Applied to Analysis *Phil. Trans. R. Soc. Lond.* **151** 183–224
68. Grochola G, Snook I K and Russo S P 2008 Influence of substrate morphology on the growth of gold nanoparticles *The Journal of Chemical Physics* **129** 154708–14
69. Grzelczak M, Sánchez-Iglesias A, Rodríguez-González B, Alvarez-Puebla R, Pérez-Juste J and Liz-Marzán L, M. 2008 Influence of Iodide Ions on the Growth of Gold Nanorods: Tuning Tip Curvature and Surface Plasmon Resonance *Advanced Functional Materials* **18** 3780–6
70. Gueroui Z and Libchaber A 2004 Single-Molecule Measurements of Gold-Quenched Quantum Dots *Physical Review Letters* **93** 166108
71. Halas N 2002 The Optical Properties of NANOSHELLS *Optics & Photonic News*
72. Halas N J, Lal S, Chang W-S, Link S and Nordlander P 2011 Plasmons in Strongly Coupled Metallic Nanostructures *Chemical Reviews* **111** 3913–61
73. Hao E, Schatz G C and Hupp J T 2004 Synthesis and Optical Properties of Anisotropic Metal Nanoparticles *Journal of Fluorescence* **14** 331–41
74. Hao F, Nehl C L, Hafner J H and Nordlander P 2007 Plasmon Resonances of a Gold Nanostar *Nano Lett.*
75. Hartland and Guillaudeu H 2003 Laser induced alloying in metal nanoparticles - Controlling spectral properties with light *Chapter 9 in ACS Symposium Series No. 844: Molecules as Components in Electronic Devices. M. Liebermann, Editor (2003)*
76. Hayat M H 1989 *Colloidal Gold: Principles, Methods, and Applications* vol 1–3: Academic Press
77. Haynes C L and Van Duyne R P 2001 Nanosphere Lithography: A Versatile Nanofabrication Tool for Studies of Size-Dependent Nanoparticle Optics *The Journal of Physical Chemistry B* **105** 5599–611
78. He L, Musick M D, Nicewarner S R, Salinas F G, Nekovic S J, Natan M J and Keating C D 2000 Colloidal Au-enhanced surface plasmon resonance for ultrasensitive detection of DNA hybridization *J Am Chem Soc* **122** 9071–7

79. He P, Shen X and Gao H 2005 Size-controlled preparation of Cu2O octahedron nanocrystals and studies on their optical absorption *Journal of Colloid and Interface Science* **284** 510–5
80. Henglein A 1993 Physicochemical properties of small metal particles in solution: "micro-electrode" reactions, chemisorption, composite metal particles, and the atom-to-metal transition *Journal of Physical Chemistry* **97** 5457–71
81. Henglein A 2000 Colloidal Palladium Nanoparticles: Reduction of Pd(II) by H_2; PdcoreAushellAgshell Particles *J. Phys. Chem. B* **104** 6683–5
82. Henglein A 2000 Preparation and Optical Absorption Spectra of AucorePtshell and PtcoreAushell Colloidal Nanoparticles in Aqueous Solution *J. Phys. Chem. B* **104** 2201–3
83. Henglein A and Giersig M 2000 Reduction of Pt(II) by H2: Effects of Citrate and NaOH and Reaction Mechanism *The Journal of Physical Chemistry B* **104** 6767–72
84. Herricks T, Chen J and Xia Y 2004 Polyol Synthesis of Platinum Nanoparticles: Control of Morphology with Sodium Nitrate *Nano Letters* **4** 2367–71
85. Hirai H, Nakao Y and Toshima N 1979 *Journal of Macromolecular Science, Part A: Pure and Applied Chemistry* **13** 727
86. Hiriyanna K, Varkey J, Beer M and Benbow R 1988 Electron microscopic visualization of sites of nascent DNA synthesis by streptavidin-gold binding to biotinylated nucleotides incorporated in vivo *J. Cell Biol.* **107** 33–44
87. Hodak and Henglein G, Hartland 2000 Laser-induced inter-diffusion in AuAg core-shell nanoparticles *J Phys Chem B* **104** 11708–18
88. Hostetler M J, Wingate J E, Zhong C J, Harris J E, Vachet R W, Clark M R, Londono J D, Green S J, Stokes J J, Wignall G D, Glish G L, Porter M D, Evans N D and Murray R W 1998 Alkanethiolate Gold Cluster Molecules with Core Diameters from 1.5 to 5.2 nm: Core and Monolayer Properties as a Function of Core Size *Langmuir* **14** 17–30
89. Hrelescu C, Sau T K, Rogach A L, Ja ckel F, Laurent G, Douillard L and Charra F 2011 Selective Excitation of Individual Plasmonic Hotspots at the Tips of Single Gold Nanostars *Nano Letters*
90. Hutter E and Fendler J H 2002 Size quantized formation and self-assembly of gold encased silver nanoparticles *Chem. Commun.* **2002** 378–9
91. Hutter E, Fendler J H and Roy D 2001 Surface Plasmon Resonance Studies of Gold and Silver Nanoparticles Linked to Gold and Silver Substrates by 2-Aminoethanethiol and 1,6-Hexanedithiol *Journal of Physical Chemistry* **105** 11159–68
92. Hwang and Lee B, Choi 2003 Separation of nanoparticles in different sizes and compositions by capillary electrophoresis *Bull. Korean Chem. Soc.* **24** 684–6
93. Iacopino D, Ongaro A, Nagle L, Eritja R and Fitzmaurice D 2003 Imaging the DNA and nanoparticle components of a self-assembled nanoscale architecture *Nanotechnology* **14** 447–52
94. Jacson A M, Myerson J W and Stellacci F 2004 Spontaneous assembly of subnanometreordered domains in the ligand shell of monolayer-protected nanoparticles *Nature Materials* **3** 330–6
95. Jain K K 2003 Nanodiagnostics: application of nanotechnology in molecular diagnostics *Expert Review of Molecular Diagnostics* **3** 153–61
96. Jana and Gearheart O, Johnson, Edler, Mann, Murphy 2002 Liquid crystalline assemblies of ordered gold nanorods *J Mater Chem* **12** 2909–12
97. Jana N R, Gearheart L and Murphy C J 2001 Evidence for Seed-Mediated Nucleation in the Chemical Reduction of Gold Salts to Gold Nanoparticles *Chem. Mater.* **13** 2313–22
98. Jana N R, Gearheart L and Murphy C J 2001 Seed-mediated growth approach for shape-controlled synthesis of spheroidal and rod-like gold nanoparticles using a surfactant template *Advanced Materials (Weinheim, Germany)* **13** 1389–93
99. Jana N R, Gearheart L and Murphy C J 2001 Wet Chemical Synthesis of High Aspect Ratio Cylindrical Gold Nanorods *J. Phys. Chem. B* **105** 4065–7
100. Jensen T, Kelly L, Lazarides A and Schatz G C 1999 Electrodynamics of Noble Metal Nanoparticles and Nanoparticle Clusters *Journal of Cluster Science* **10** 295–317

101. Jin R, Cao Y, Mirkin C A, Kelly K L, Schatz G C and Zheng J G 2001 Photoinduced conversion of silver nanospheres to nanoprisms *Science* **294** 1901–3
102. Jin R, Wu G, Li Z, Mirkin C A and Schatz G C 2003 What Controls the Melting Properties of DNA-Linked Gold Nanoparticle Assemblies? *Journal of the American Chemical Society* **125** 1643–54
103. Johansson P, Xu H and Käll M 2005 Surface-enhanced Raman scattering and fluorescence near metal nanoparticles *Physical Review B* **72** 035427
104. Johnson C J, Dujardin E, Davis S A, Murphy C J and Mann S 2002 Growth and form of gold nanorods prepared by seed-mediated, surfactant-directed synthesis *Journal of Materials Chemistry* **12**
105. Kamat and Flumiani H 1998 Picosecond dynamics of silver nanoclusters. Photoejection of electrons and fragmentation *J Phys Chem B* **102** 3123–8
106. Kannan N, Mukunthan K S and Balaji S 2011 A comparative study of morphology, reactivity and stability of synthesized silver nanoparticles using Bacillus subtilis and Catharanthus roseus (L.) G. Don *Colloids and Surfaces B: Biointerfaces* **86** 378–83
107. Karlsson R 2004 SPR for molecular interaction analysis: a review of emerging application areas *J Mol Recognit* **17** 151–61
108. Keating C D, Kovaleski K M and Natan M J 1998 Protein:Colloid Conjugates for Surface Enhanced Raman Scattering: Stability and Control of Protein Orientation *The Journal of Physical Chemistry B* **102** 9404–13
109. Kerman K, Saito M, Morita Y, Takamura Y, Ozsoz M and Tamiya E 2004 Electrochemical coding of single-nucleotide polymorphisms by monobase-modified gold nanoparticles *Anal Chem* **76** 1877–84
110. Khlebtsov N G, Dykman L A, Bogatyrev V A and Khlebtsov B N 2003 Two-Layer Model of Colloidal Gold Bioconjugates and Its Application to the Optimization of Nanosensors *Colloid Journal* **65** 508–18
111. Kim and Lee K 2003 Isocyanide and biotin-derivatized Ag nanoparticles: an efficient molecular sensing mediator via surface-enhanced Raman spectroscopy *Chem Commun* 724–5
112. Kim M-K, Jeon Y-M, Jeon W-S, Kim H-J, Kim K, Hong S G and Park C G 2001 Novel dendron-stabilized gold nanoparticles with high stability and narrow size distribution *Chem Commun* 667–8
113. Knauer A, Thete A, Li S, Romanus H, Csáki A, Fritzsche W and Köhler J M 2010 Au/Ag/Au double shell nanoparticles with narrow size distribution obtained by continuous micro segmented flow synthesis *Chemical Engineering Journal* **166** 1164–9
114. Kneipp K, Kneipp H, Kartha V B, Manoharan R, Deinum G, Itzkan I, Dasari R R and Feld M S 1998 Detection and identification of a single DNA base molecule using surface-enhanced Raman scattering (SERS) *Physical Review E* **57** 6281–4
115. Korgel B A and Fitzmaurice D 1998 Self-Assembly of Silver Nanocrystals into Two-Dimensional Nanowire Arrays *Advanced Materials* **10** 661–5
116. Kreibig U and Vollmer M 1995 *Optical Properties of Metal Clusters* vol 25 Berlin
117. Kreiter M, Neumann T, Mittler S, Knoll W and Sambles J R 2001 Fluorescent dyes as a probe for the localized field of coupled surface plasmon-related resonances *Physical Review B* **64** 075406
118. Kumar S V and Ganesan S 2011 Preparation and Characterization of Gold Nanoparticles with Different Capping Agents *International Journal of Green Nanotechnology* **3** 47–55
119. Lakowicz J R 2001 Radiative decay engineering: biophysical and biomedical applications *Analytical Biochemistry* **298** 1–24
120. Lakowicz J R 2006 Plasmonics in Biology and Plasmon-Controlled Fluorescence *Plasmonics* **1** 5–33
121. Lakowicz J R, Geddes C D, Gryczynski I, Malicka J, Gryczynski Z, Aslan K, Lukomska J, Matveeva E, Zhang J, Badugu R and Huang J 2004 Advances in Surface-Enhanced Fluorescence *Journal of Fluorescence* **14** 425–41
122. Lakowicz J R, Malicka J and Gryczynski I 2003 Increased intensities of YOYO-1-labeled DNA oligomers near silver particles *Photochem Photobiol* **77** 604–7

123. Lakowicz J R, Ray K, Chowdhury M, Szmacinski H, Fu Y, Zhang J and Nowaczyk K 2008 Plasmon-controlled fluorescence: a new paradigm in fluorescence spectroscopy *Analyst* **133** 1308–46
124. Leary J J, Brigati D J and Ward D C 1983 Rapid and sensitive colorimetric method for visualizing biotin-labeled DNA probes hybridized to DNA or RNA immobilized on nitrocellulose: Bio-blots *Proc Natl Acad Sci USA* **80** 4045–9
125. Lee I, Han S W and Kim K 2001 Production of Au-Ag alloy nanoparticles by laser ablation of bulk alloys *Chem Commun (Camb)* 1782–3
126. Leff D V, Brandt L and Heath J R 1996 Synthesis and Characterization of Hydrophobic, Organically-Soluble Gold Nanocrystals Functionalized with Primary Amines *Langmuir* **12** 4723–30
127. Li H S H P, John H. Reif, Thomas H. LaBean, and Hao Yan 2003 DNA-Templated Self-Assembly of Protein and Nanoparticle Linear Arrays *JACS* **126** 418–9
128. Link and Burda W, El-Sayed 1999 Electron dynamics in gold and gold-silver alloy nanoparticles: the influence of a nonequilibrium electron distribution and the size dependence of the electron-phonon relaxation *J Chem Phys* **111** 1255–64
129. Link S and El-Sayed M A 1999 Spectral Properties and Relaxation Dynamics of Surface Plasmon Electronic Oscillations in Gold and Silver Nanodots and Nanorods *J. Phys. Chem. B* **103** 8410–26
130. Liz-Marzan L M 2004 Nanometals: Formation and color *Materials Today* **7** 26–31
131. Loweth C J, Caldwell W B, Peng X, Alivisatos A P and Schultz P G 1999 DNA als gerüst zur Bildung von Aggregaten aus Gold-Nanokristallen *Angewandte Chemie* **111** 1925–30
132. Lyon L A, Musick M D and Natan M J 1998 Colloidal Au-enhanced surface plasmon resonance immunosensing *Anal Chem* **70** 5177–83
133. Mafuné and Kohno T, Kondow 2001 Dissociation and aggregation of gold nanoparticles under laser irradiation *J Phys Chem* **105** 9050–6
134. Maillard M, Huang P and Brus L 2003 Silver Nanodisk Growth by Surface Plasmon Enhanced Photoreduction of Adsorbed [Ag+] *Nano Letters* **3** 1611–5
135. Mallick K, Wang Z L and Pal T 2001 Seed-mediated successive growth of gold particles accomplished by UV irradiation: a photochemical approach for size-controlled synthesis *Journal of Photochemistry and Photobiology A: Chemistry* **140** 75–80
136. Mandal M, Ranjan Jana N, Kundu S, Kumar Ghosh S, Panigrahi M and Pal T 2004 Synthesis of Au-core–Ag-shell type bimetallic nanoparticles for single molecule detection in solution by SERS method *Journal of Nanoparticle Research* **6** 53–61
137. Mandal S, Selvakannan P R, Pasricha R and Sastry M 2003 Keggin Ions as UV-Switchable Reducing Agents in the Synthesis of Au Core-Ag Shell Nanoparticles *J. Am. Chem. Soc.* **125** 8440–1
138. Marras S A E, Russell Kramer F and Tyagi S 1999 Multiplex detection of single-nucleotide variations using molecular beacons *Genetic Analysis: Biomolecular Engineering* **14** 151–6
139. Mayya K S, Patil V and Sastry M 1996 On the Stability of Carboxylic Acid Derivatized Gold Colloidal Particles: The Role of Colloidal Solution pH Studied by Optical Absorption Spectroscopy *Langmuir* **13** 3944 -7
140. Mbindyo J, Reiss B D, Martin B R, Keating C D, Natan M J and Mallouk T E 2001 DNA-Directed Assembly of Gold Nanowires on Complementary Surfaces *Advanced Materials* **13** 249–54
141. Meltzer S, Resch R, Koel B E, Thompson M E, Madhukar A, Requicha A A G and Will P 2001 Fabrication of Nanostructures by Hydroxylamine Seeding of Gold Nanoparticle Templates *Langmuir* **17** 1713–8
142. Mie G 1908 Beitrage zur Optik truber Medien speziell kolloidaler Metallosungen *Annalen der Physik* **25** 377–445
143. Milliron D J, Hughes S M, Cui Y, Manna L, Li J, Wang L-W and Paul Alivisatos A 2004 Colloidal nanocrystal heterostructures with linear and branched topology *Nature* **430** 190–5
144. Mirkin C A, Letsinger R L, Mucic R C and Storhoff J J 1996 A DNA-based method for rationally assembling nanoparticles into macroscopic materials *Nature* **382** 607–9

145. Mock J J, Barbic M, Smith D R, Schultz D A and Schultz S 2002 Shape effects in plasmon resonance of individual colloidal silver nanoparticles *The Journal of Chemical Physics* **116** 6755–9
146. Mokari T, Rothenberg E, Popov I, Costi R and Banin U 2004 Selective growth of metal tips onto semiconductor quantum rods and tetrapods *Science* **304** 1787–90
147. Monnoyer P, Fonseca A and Nagy J B 1995 Preparation of colloidal AgBr particles from microemulsions *Colloids and Surfaces A: Physicochemical and Engineering Aspects* **100** 233–43
148. Moskovits and Srnová-Sloufová V 2002 Bimetallic Ag-Au nanoparticles: extracting meaningful optical constants from the surface-plasmon extinction spectrum *JChemPhys* **116** 10435–46
149. Mucic R C, Storhoff J J, Mirkin C A and Letsinger R L 1998 DNA-Directed Synthesis of Binary Nanoparticle Network Materials *Journal of the American Chemical Society* **120** 12674–5
150. Mulvaney P 1996 Surface Plasmon Spectroscopy of Nanosized Metal Particles *Langmuir* **12** 788 -800
151. Mulvaney P, Giersig M and Henglein A 1993 Electrochemistry of Multilayer Colloids: Preparation and Absorption Spectrum of Gold-Coated Silver Particles *Journal of Physical Chemistry* **97** 7061–4
152. Murphy and Jana 2002 Controlling the aspect ratio of inorganic nanorods and nanowires *Adv Mat* **14** 80–2
153. Neeves A E and Birnboim M H 1989 Composite structures for the enhancement of nonlinear-optical susceptibility *J. Opt. Soc. Am. B* **6** 787–96
154. Nikoobakht B and El-Sayed M A 2003 Preparation and Growth Mechanism of Gold Nanorods (NRs) Using Seed-Mediated Growth Method *Chem. Mater.* **15** 1957–62
155. Novak J P and Feldheim D L 2000 Assembly of Phenylacetylene-Bridged Silver and Gold Nanoparticle Arrays *JACS* **122** 3979–80
156. Nuzzo R G and Allara D L 1983 Adsorption of Bifunctional Organic Disulfides on Gold Surfaces *Journal of the American Chemical Society* **105** 4481–3
157. Okamoto and Yamaguchi 2000 Local plasmon sensor with gold colloid monolayers deposited upon glass substrates *Optics Letters* **25** 372–4
158. Palade G E 1960 Transport in quanta across the endothelium of blood capillaries *Anat. Rec.* **136** 254
159. Panigrahi M, Kundu S, Ghos S K, Nath S and Pal T 2004 General method of synthesis for metal nanoparticles *Journal of Nanoparticle Research* **6** 411–4
160. Papadopoulou E and Bell S E J 2011 Label Free Detection of Single Base Mismatches in DNA by Surface Enhanced Raman Spectroscopy *Angewandte Chemie International Edition*
161. Papavassiliou G C 1976 Surface plasmons in small Au-Ag alloy particles *Journal of Physics F: Metal Physics* **6** L103
162. Parak W J, Pellegrino T, Micheel C M, Gerion D, Williams S C and Alivisatos A P 2003 Conformation of Oligonucleotides Attached to Gold Nanocrystals Probed by Gel Electrophoresis *Nano Letters* **3** 33–6
163. Park S J, Taton T A and Mirkin C A 2002 Array-based electrical detection of DNA with nanoparticle probes *Science* **295** 1503–6
164. Pastoriza-Santos I and Liz-Marzan L M 2002 Synthesis of Silver Nanoprisms in DMF *Nano Lett.* **2** 903–5
165. Patolsky F, Ranjit K T, Lichtenstein A and Willner I 2000 Dendritic amplification of DNA analysis by oligonucleotide-functionalized Au-nanoparticles *Chemical Communications* 1025–6
166. Pedersen D B, Wang S, Duncan E J S and Liang S H 2007 Adsorbate-induced diffusion of Ag and Au atoms out of the cores of Ag@ Au, Au@ Ag, and Ag@ AgI core-shell nanoparticles *The Journal of Physical Chemistry C* **111** 13665–72
167. Pohl D 2001 *Near-Field Optics and Surface Plasmon Polaritons,* pp 1–13

168. Porter L A, Ji D, Westcott S L, Graupe M, Czernuszewicz R S, Halas N and Lee R T 1998 Gold and Silver Nanoparticles Functionalized by the Adsorption of Dialkyl Disulfides *Langmuir* **14** 7378–86
169. Puntes and Krishnan A 2001 Colloidal nanocrystal shape and size control: the case of cobalt *Science* **291** 2115–7
170. Qi L, Ma J and Shen J 1997 Synthesis of Copper Nanoparticles in Nonionic Water-in-Oil Microemulsions *Journal of Colloid and Interface Science* **186** 498–500
171. Rant U, Arinaga K, Fujita S, Yokoyama N, Abstreiter G and Tornow M 2004 Structural properties of oligonucleotide monolayers on gold surfaces probed by fluorescence investigations *Langmuir* **20** 10086–92
172. Raschke G, Kowarik S, Franzl T, Sonnichsen K, Klar T A, Feldmann J, Nichtl A and Kurzinger K 2003 Biomolecular Recognition Based on Single Gold Nanoparticle Light Scattering *Nano Letters* **3** 935–8
173. Rechenberger W, Hohenau A, Leitner A, Krenn J R, Lamprecht B and Aussenegg F R 2003 Optical properties of two interacting gold nanoparticles *Optics Communications* **220** 137–41
174. Reynolds R A, Mirkin C A and Letsinger R L 2000 Homogeneous, Nanoparticle-Based Quantitative Colorimetric Detection of Oligonucleotides *Journal of the American Chemical Society* **122** 3795–6
175. Rothemund P W 2006 Folding DNA to create nanoscale shapes and patterns *Nature* **440** 297–302
176. Sánchez-Ramírez J F, Vázquez-López C and Pal U 2002 Preparation and optical absorption of colloidal dispersion of Au/Cu nanoparticles *SUPERFICIES y VACIO* **15** 16–8
177. Schierhorn and Liz-Marzan 2002 Synthesis of bimetallic colloids with tailored intermetallic separation *NanoLetters* **2** 13–6
178. Schmid G 2006 *Nanoparticles: from theory to application*: Wiley VCH
179. Schmid G and Lehnert A 1989 The Complexation of Gold Colloids *Angewandte Chemie, International Edition in English* **28** 780–1
180. Schmid G, Lehnert A, Malm J-O and Bovin J-O 1991 Charakterisierung ligandstabilisierter Bimetall-Kolloide durch hochaufgelöste Transmissionselektronenmikroskopie und energiedispersive Röntgenmikroanalyse *Angewandte Chemie* **103** 852–4
181. Schmid G, Pfeil R, Boese R, Bandermann F, Meyer S, Calis G H M and van der Velden J W A 1981 Au55[P(C6H5)3]12Cl6 — ein Goldcluster ungewöhnlicher Größe *Chemische Berichte* **114** 3634–42
182. Schultz S, Smith D R, Mock J J and Schultz D A 2000 Single-target molecule detection with nonbleaching multicolor optical immunolabels *Proc Natl Acad Sci USA* **97** 996–1001
183. Sershen S R, Westcott S L, Halas N J and West J L 2000 Temperature sensitive polymer–nanoshell composites for photothermally modulated drug delivery *Journal of biomedical materials research* **51** 293–8
184. Shaiu W L, Larson D D, Vesenka J and Henderson E 1993 Atomic force microscopy of oriented linear DNA molecules labeled with 5 nm gold spheres *Nucleic Acids Res* **21** 99–103
185. Shibata and Bunker Z, Meisel, Vardeman, Gezelter 2002 Size-dependent spontaneous alloying of Au-Ag nanoparticles *JACS* **124** 11989–96
186. Silvert P-Y, Herrera-Urbina R, Duvauchelle N, Vijayakrishnan V and Elhsissen K T 1996 Preparation of colloidal silver dispersions by the polyol process Part 1 -Synthesis and characterization *J Mater Chem* **6** 573–7
187. Smith D K and Korgel B A 2008 The Importance of the CTAB Surfactant on the Colloidal Seed-Mediated Synthesis of Gold Nanorods *Langmuir* **24** 644–9
188. Souza G R and Miller J H 2001 Oligonucleotide detection using angle-dependent light scattering and fractal dimension analysis of gold-DNA aggregates *J Am Chem Soc* **123** 6734–5
189. Steinbruck A, Csaki A, Festag G and Fritzsche W 2006 Designing plasmonic structures: bi-metallic core-shell nanoparticles. In: *Nanophotonics*, (Strasbourg, France: SPIE) pp 619513–6

190. Steinbrück A, Csaki A, Ritter K, Leich M, Köhler J M and Fritzsche W 2008 Gold-silver and silver-silver nanoparticle constructs based on DNA hybridization of thiol- and amino-functionalized oligonucleotides *Journal of Biophotonics* **1** 104–13
191. Steinbrück A, Csaki A, Ritter K, Leich M, Köhler J M and Fritzsche W 2009 Gold and gold–silver core-shell nanoparticle constructs with defined size based on DNA hybridization *Journal of Nanoparticle Research* **11** 623–33
192. Storhoff J J, Elghanian R, Mucic R C, Mirkin C A and Letsinger R L 1998 One-Pot Colorimetric Differentiation of Polynucleotides with Single Base Imperfections Using Gold Nanoparticle Probes *Journal of the American Chemical Society* **120** 1959–64
193. Storhoff J J, Lazarides A A, Mucic R C, Mirkin C A, Letsinger R L and Schatz G C 2000 What Controls the Optical Properties of DNA-Linked Gold Nanoparticle Assemblies? *Journal of the American Chemical Society* **122** 4640–50
194. Su and Wei Z, Mock, Smith, Schultz 2003 Interparticle coupling effects on plasmon resonances of nanogold particles *NanoLetters* **3** 1087–90
195. Sun S and Murray C B 1999 *Synthesis of monodisperse cobalt nanocrystals and their assembly into magnetic superlattices (invited)* vol 85: AIP
196. Sun Y, Mayers B, Herricks T and Xia Y 2003 Polyol Synthesis of Uniform Silver Nanowires: A Plausible Growth Mechanism and the Supporting Evidence *Nano Letters* **3** 955–60
197. Sun Y and Xia Y 2002 Shape-controlled synthesis of gold and silver nanoparticles *Science* **298** 2176–9
198. Sun Y and Xia Y 2003 Triangular Nanoplates of Silver: Synthesis, Characterization, and Use as Sacrificial Templates For Generating Triangular Nanorings of Gold *Advanced Materials* **15** 695–9
199. Tabakman S M, Chen Z, Casalongue H S, Wang H and Dai H 2011 A New Approach to Solution-Phase Gold Seeding for SERS Substrates *Small* **7** 499–505
200. Takami and Kurita K 1999 Laser-induced size reduction of noble metal particles *JPhysChem B* **103** 1226–32
201. Takatani and Kago N, Kobayashi, Hori, Oshima 2003 Characterization of noble metal alloy nanoparticles prepared by ultrasound irradiation *Rev. Adv. Mater. Sci.* **5** 232–8
202. Tao A R, Habas S and Yang P 2008 Shape control of colloidal metal nanocrystals *Small* **4** 310
203. Taton T A, Mirkin C A and Letsinger R L 2000 Scanometric DNA array detection with nanoparticle probes *Science* **289** 1757–60
204. Taton T A, Mucic R C, Mirkin C A and Letsinger R L 2000 The DNA-mediated formation of supramolecular mono- and multilayered nanoparticle structures *Journal of the American Chemical Society*
205. Toshima 2000 Core/shell-structured bimetallic nanocluster catalysts for visible-light-induced electron transfer *Pure Appl Chem* **72** 317–25
206. Turkevich J, Miner R S and Babenkova L 1986 Further studies on the synthesis of finely divided platinum *Journal of Physical Chemistry* **90** 4765–7
207. Turkevich J, Stevenson P L and Hiller J 1951 Nucleation and growth process in the synthesis of colloidal gold. *Discuss. Faraday Soc.* **11** 55–75
208. Tyagi S, Bratu D P and Kramer F R 1998 Multicolor molecular beacons for allele discrimination *Nature Biotechnology* **16** 49–53
209. Vet J A M, Majithia A R, Marras S A E, Tyagi S, Dube S, Poiesz B J and Kramer F R 1999 Multiplex detection of four pathogenic retroviruses using molecular beacons *Proceedings of the National Academy of Sciences* **96** 6394
210. Wang Y L, Mueller J E, Kemper B and Seeman N C 1991 Assembly and characterization of five-arm and six-arm DNA branched junctions *Biochemistry* **30** 5667–74
211. Weare W W, Reed S M, Warner M G and Hutchison J E 2000 Improved Synthesis of Small (dCORE = 1.5 nm) Phosphine-Stabilized Gold Nanoparticles *J. Am. Chem. Soc.* **122** 12890–1
212. Wei and Zamborini 2004 Directly monitoring the growth of gold nanoparticle seeds into gold nanorods *Langmuir* **20** 11301–4

213. Weitz D A, Garoff S, Hanson C D, Gramila T J and Gersten J I 1982 Fluorescent lifetimes of molecules on silver-island films *Optics Letters* **7** 89–91
214. Weizmann Y, Patolsky F and Willner I 2001 Amplified detection of DNA and analysis of single-base mismatches by the catalyzed deposition of gold on Au-nanoparticles *Analyst* **126** 1502–4
215. Westcott S L, Oldenburg S J, Lee T R and Halas N J 1998 Formation and Adsorption of Clusters of Gold Nanoparticles onto Functionalized Silica Nanoparticle Surfaces *Langmuir* **14** 5396–401
216. Wiley B, Herricks T, Sun Y and Xia Y 2004 Polyol Synthesis of Silver Nanoparticles: Use of Chloride and Oxygen to Promote the Formation of Single-Crystal, Truncated Cubes and Tetrahedrons *Nano Letters* **4** 1733–9
217. Wiley B, Sun Y and Xia Y 2007 Synthesis of Silver Nanostructures with Controlled Shapes and Properties *Accounts of Chemical Research* **40** 1067–76
218. Willner and Patolsky L 2001 Amplified DNA analysis and single-base mismatch detection using DNA-bioelectronic systems *Anal Sci* **17** i351–3
219. Winfree E, Liu F, Wenzler L A and Seeman N C 1998 Design and self-assembly of two-dimensional DNA crystals *Nature* **394** 539–44
220. Wu S-H and Chen D-H 2004 Synthesis of high-concentration Cu nanoparticles in aqueous CTAB solutions *Journal of Colloid and Interface Science* **273** 165–9
221. Xie H, Gu Y and Ploehn H J 2005 Dendrimer-mediated synthesis of platinum nanoparticles: new insights from dialysis and atomic force microscopy measurements *Nanotechnology* **16** S492
222. Xiong Y and Xia Y 2007 Shape-Controlled Synthesis of Metal Nanostructures: The Case of Palladium *Advanced Materials* **19** 3385–91
223. Xu H and Käll M 2002 Modeling the optical response of nanoparticle-based surface plasmon resonance sensors *Sensors and Actuators B: Chemical* **87** 244–9
224. Xu X and Cortie M B 2006 Shape Change and Color Gamut in Gold Nanorods, Dumbbells, and Dog Bones *Advanced Functional Materials* **16** 2170–6
225. Yasuda H, Mori H, Komatsu M and Takeda K 1993 Spontaneous alloying of copper atoms into gold clusters at reduced temperatures *Journal of Applied Physics* **73** 1100–3
226. Yonezawa and Onoue K 2002 Metal coating of DNA molecules by cationic, metastable gold nanoparticles *Chemistry Letters* 1172–3
227. Yonezawa T, Yasui K and Kmizuka N 2001 Controlled formation of smaller gold nanoparticles by the use of four-chained disulfide stabilizer *Langmuir* **17** 271–3
228. Zayats A V and Baron P, Willner 2005 Biocatalytic growth of Au nanoparticles: from mechanistic aspects to biosensors design *Nano Letters* **5** 21–5
229. Zhang and Sham 2002 Tuning the electronic behaviour of Au nanoparticles with capping molecules *Appl Phys Lett* **81** 736–8
230. Zhang Q, Ge J, Pham T, Goebl J, Hu Y, Lu Z and Yin Y 2009 Reconstruction of silver nanoplates by UV irradiation: tailored optical properties and enhanced stability *Angew Chem Int Ed Engl* **48** 3516–9

Chapter 2
Elastic Light Scattering of Biopolymer/Gold Nanoparticles Fractal Aggregates

Glauco R. Souza and J. Houston Miller

1 Introduction

From the interface of biotechnology and nanotechnology are emerging innovative analytical tools for the detection and characterization of nucleic acids, proteins, and protein interactions. This chapter addresses the development and application of a technique that combines angle-dependent light scattering (ADLS), fractal dimension analysis (FD), and gold nanoparticle assembly to detect and characterize nucleic acids and proteins. We will show that specific DNA or protein interactions trigger the assembly of Au-biopolymer fractal aggregates. The angle-resolved light-scattering signal from these aggregates can be used to determine their fractal dimension which is found to be sensitive to concentration, size, shape, and physical properties of both the biopolymers and the nanoparticles forming the aggregates [1, 2].

Researchers in the fields of material and combustion sciences have used ADLS and fractal dimension analysis (ADLS/FD) to study the structure and the formation of colloidal and soot particle aggregates (see Table 2.1) [1, 5, 27, 34–48]. Fractals are self-similar or scale-invariant objects [49, 50]. Regardless of the magnification of an object relative to a given variable, such as length, area, volume, or radius of gyration, the structure of a fractal remains statistically unchanged [48]. The Au colloid aggregate shown in Fig. 2.1 is an example of a fractal object, where the structure of small regions of the aggregate resembles the overall aggregate structure [51].

In this chapter, we will review work performed to analyze elastic light scattering from fractal aggregates and describe some of the work performed in our laboratory to extend this technique to the detection and characterization of biopolymers.

G.R. Souza
Nano3D Biosciences Inc., 7000 Fannin Street, Suite 2140, Houston, TX 77030, USA

J.H. Miller (✉)
Department of Chemistry, The George Washington University, Washington, DC 20052, USA

C.D. Geddes (ed.), *Reviews in Plasmonics 2010*, Reviews in Plasmonics,
DOI 10.1007/978-1-4614-0884-0_2, © Springer Science+Business Media, LLC 2012

Table 2.1 Partial summary of literature on fractal agglomerate and scattering phenomena

Sample type	Technique	Ref.
Colloid	Theoretical calculation	[3]
Latex	Optical microscopy	[4]
Hematite	Light scattering	[5]
Latex	Light scattering	[6]
Au–DNA	Light scattering	[1]
Au colloid	TEM	[7]
Au–protein	Light scattering	[1]
Vaporized metal	Transmission electron microscope (TEM)	[8]
Silica	Light and X-ray scattering	[9]
Au colloid	TEM	[10]
Au colloid	Dynamic light scattering and theoretical calculations	[11]
Au colloid	Dynamic Light scattering, TEM and light scattering	[12]
Silica aerogel	Small-angle X-ray scattering	[13]
Protein-surfactant (BSA-LDS)	Small angle neutron scattering (SANS)	[14]
Silica	Static light scattering	[15]
Au colloid	X-ray scattering	[16]
Silica	SANS	[17]
Silica vapor	Light scattering	[18]
Au colloid	Dynamic light scattering	[19]
Silica	Light scattering	[20]
Au colloid	Dynamic light scattering	[21]
Hematite	Dynamic light scattering	[22]
Au colloid	Dynamic light scattering	[21]
Au colloid	Light scattering and TEM	[23]
Polystyrene latex	Light scattering	[24]
Silica and carbon black	X-ray scattering and SANS	[25]
Silica aerogels	SANS	[26]
Soot aerogels	Light scattering	[27]
Polystyrene	Light scattering	[28]
Soot aerogel	TEM	[29]
Polystyrene latex	Light scattering	[30]
Polystyrene	Light scattering	[31]
Haematite	X-ray scattering	[32]
Ag colloid	TEM	[33]

1.1 Simple Light Scattering Theories: Rayleigh Through Mie

Light scattering is a natural phenomenon that is part of our everyday lives. Rarely is light observed directly from its original source but rather is a result of scattered light [52]. For example, blue skies and red sunsets occur because sunlight is scattered by molecules and particles in the atmosphere, respectively [53]. The observation of blue skies at midday and red skies at sunrise and sunset provide clues about the dependence of scattering signal on not only the size of the particles, but also the angular dependence of the scattered light from the source to the scatterer to the observer [54].

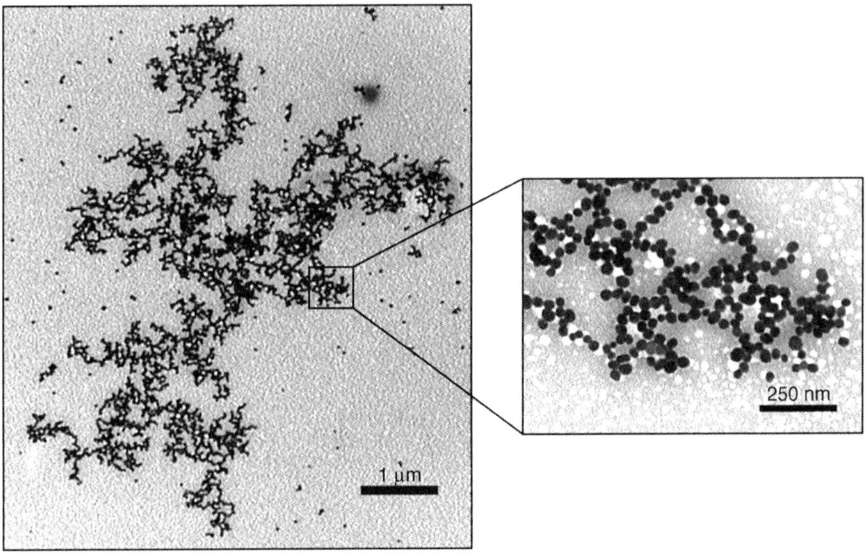

Fig. 2.1 Scanning electron microscope image of Au colloid fractal

It is important to distinguish the meaning of the terms "absorption" and "extinction." The measurement of light extinction (such as that provided in a simple UV/Vis absorption spectrophotometer) reports the sum of scattering and absorption, since both phenomenon attenuate the intensity of a light beam traveling across a medium [55–57]. In this chapter, both phenomenon might occur in the same sample. For example, the observation of surface plasmon absorption in suspensions of gold nanoparticles is an absorption phenomenon as surface plasmon electrons are being excited from one state to another [55]. In the same sample, aggregation of the nanoparticles leads to extinction at both shorter and longer wavelengths that is attributable to scattering from the aggregates particles.

Light scattering can be differentiated as elastic or inelastic light scattering. Inelastic light scattering occurs when the scattered light has a wavelength shift relative to the wavelength of the incident light. This frequency shift can be either higher or lower, depending on if there is a loss or gain of energy of the scattering substance. Examples of inelastic light scattering are Raman and dynamic light scattering. In this chapter, we refer almost exclusively to elastic scattering, where the incident and scattered wave have the same wavelength. However, we will conclude by showing how inelastic scattering might be used to provide complementary information about aggregation of systems of gold nanoparticles induced by biopolymer interactions.

The foundation of light-scattering theory was initially developed near the end of the nineteenth century and in first half of the twentieth century by Rayleigh, Mie, Smoluchowski, Einstein, and Debye motivated in part by the experimental work of Tyndall [58]. In addition to providing an explanation for scattering in the atmosphere, Rayleigh also studied scattering from spheres of arbitrary sizes, where phase relations between the light scattered from different regions of the same particles were regarded as localized, independent, and induced dipoles. In the early 1900s, Debye further developed scattering theory for larger and nonspherical particles.

Matter consists of electrically charged nuclei and electrons that can interact with an oscillating electromagnetic field, e.g., light, and either absorb or scatter light. In either case, there is attenuation in the light intensity, which is described by the exponential decay of the light intensity as it passes through a medium [52, 55, 56, 59]:

$$I = I_0 e^{-kx} \tag{2.1}$$

where I is the intensity of the light transmitted after traveling a distance x across the medium. k changes in definition (and symbol) for different forms of extinction. For molecular absorption, it is the absorption coefficient.[1] For scattering, the extinction coefficient is often given the symbol τ, the turbidity of the medium. In this context, it is related to the number density of particles (n) and their individual extinction cross-section (σ_{ext}),

$$\tau = n \times \sigma_{ext}. \tag{2.2}$$

As noted above, extinction is due to both scattering, which removes light from the incident path by reemitting light in all directions, and absorption, which converts the light into other forms of energy (such as heat). Mathematically,

$$\sigma_{ext} = \sigma_{abs} + \sigma_{scatt}, \tag{2.3}$$

where σ_{abs} and σ_{scatt} are the absorption and scattering cross sections, respectively. This relationship is valid for all particle and molecular systems [56].

As noted, since the late nineteenth century, the relative size of the scatterer and the wavelength of light are critical dependencies of the scattering intensity. For spheres, it is convenient to define the size parameter, α, as a relative measure of the magnitude of the particle radius (a) and the wavelength of light (λ).

$$\alpha = \frac{2\pi a}{\lambda}. \tag{2.4}$$

Rayleigh scattering is the best and simplest mathematical description for the scattering from particles that are small, relative to the wavelength of light. Generally, for Rayleigh scattering we require [52, 54–57, 60]:

$$\alpha << 1 \quad \text{and} \quad m\alpha << 1, \tag{2.5}$$

[1] Chemists remember the extinction coefficient as the product of the molar extinction coefficient and the molar concentration. Gas phase spectroscopists cast this as the product of the gas density, the absorbing species' line strength, and a line shape factor that accounts for how the line is broadened from molecular collisions and molecular motion.

where m is the relative refractive index of the scattering particle. A quantity that is often used to describe a scattering object is its differential scattering cross section,

$$\frac{\mathrm{d}\sigma}{\mathrm{d}\Omega} = \frac{16\pi^4 a^6}{\lambda^4}\left|\frac{m^2-1}{m^2+2}\right|^2 \tag{2.6}$$

which is related to the power scattered (P_{scat}) per unit solid angle (Ω) through [55, 56]:

$$\frac{P_{\mathrm{scat}}}{\Omega} = \frac{\mathrm{d}\sigma}{\mathrm{d}\Omega} I_0, \tag{2.7}$$

where the intensity of incident radiation is I_0.

If the differential scattering cross section is integrated over all space, the total scattering cross section is given by

$$\sigma_{\mathrm{scatt}} = \frac{8\pi}{3}\frac{16\pi^4 a^6}{\lambda^4}\left|\frac{m^2-1}{m^2+2}\right|^2. \tag{2.8}$$

Thus, if the quantity $|(m^2-1)/(m^2+2)|^2$ is weakly dependent on λ, the scattered intensity shows the familiar λ^{-4} dependence.

If the incident light and the scattered light have the same polarization (referred to as "parallel" scattering), the angular distribution of the scattered light depends on the polarization of the incident light. Specifically,

$$I_{\perp} = \frac{1}{r^2}\frac{16\pi^4 a^6}{\lambda^4}\left|\frac{m^2-1}{m^2+2}\right|^2 I_0, \tag{2.9}$$

$$I_{\parallel} = \frac{1}{r^2}\frac{16\pi^4 a^6}{\lambda^4}\left|\frac{m^2-1}{m^2+2}\right|^2 \cos^2\theta I_0, \tag{2.10}$$

$$I = \frac{1}{2}\left(I_{\perp} + I_{\parallel}\right) = \frac{1}{r^2}\frac{16\pi^4 a^6}{\lambda^4}\left|\frac{m^2-1}{m^2+2}\right|^2 \left(1+\cos^2\theta\right) I_0, \tag{2.11}$$

where $I_{\perp}$, $I_{\parallel}$ and I are the intensities of vertically polarized light, horizontally polarized light, and unpolarized light, respectively [55].

As noted by Sorensen [56], three features of Rayleigh scattering can be highlighted:

- The scattering of vertically polarized light is independent of θ and it is isotropic in the scattering plane
- I is a function of λ^{-4}, which implies that short wavelength light scatters more efficiently
- There is large size dependence for the scattered intensity through the a^6 term (proportional to the square of the particle volume). This third feature is often referred to as the Tyndall effect [56, 60]

As seen in (2.3), particles and molecules both scatter and absorb light. For molecules, absorption is described in terms of the interaction of two molecular wavefunctions through the dipole moment operator. For particles, the Rayleigh absorption cross-section is given by

$$\sigma_{\mathrm{abs}} = \frac{8\pi^2 a^3}{\lambda} \mathrm{Im}\left(\pi \frac{m^2 - 1}{m^2 + 2} \right). \tag{2.12}$$

If the relative refractive index is real, there is no absorption [52, 55–57].

Because σ_{abs} and σ_{scatt} have different wavelength, λ, and size, a, dependencies and the scattering cross section depends on both polarization and angle, it is clear that significant information about a particle's morphology is available from interrogation of the scattering signal. While these relationships have been shown explicitly for Rayleigh scatterers, we will show below that similar relationships hold for other scattering regimes [56].

Rayleigh scattering theory is derived with the assumption that the phase of the incident electromagnetic wave does not change across the particle. This requirement [$\alpha \ll 1$ in (2.5)] is relaxed in the Rayleigh–Gans–Debye (RGD) scattering theory [52, 55–57, 60] if

$$|m - 1| \ll 1, \tag{2.13}$$

$$2\alpha |m - 1| \ll 1. \tag{2.14}$$

The RGD scattering cross section for perpendicular detection of light [56], is

$$\frac{\mathrm{d}\sigma}{\mathrm{d}\Omega_{\mathrm{RGD}}} = \frac{\mathrm{d}\sigma}{\mathrm{d}\Omega_{\mathrm{R}}} \left[\frac{9}{u^6} \left(\sin(u) - u \cdot \cos(u) \right)^2 \right], \tag{2.15}$$

where

$$u = qa, \tag{2.16}$$

and

$$q = \frac{4\pi}{\lambda} \sin\left(\frac{\theta}{2} \right). \tag{2.17}$$

q is the scattering wave vector which will be further described below.

Two important characteristics of RGD scattering are [56]:

- For $\theta = 0$, the cross-section reduces to the Rayleigh cross-section.
- In contrast to Rayleigh scattering, the scattering intensity is larger in the forward direction, and this anisotropy increases with increasing particle size. This increase in anisotropy with particle size allows us to use a simple analysis method, the dissymmetry ratio [54, 55, 60], to detect changes in the size of an

aggregate or coalescing particles. Here we calculate the ratio of the intensity of forward scattering to the intensity of back scattering. Larger ratios indicate a larger size for the scattering particle.

Rayleigh and RGD scattering theories represent rigorous treatments of Maxwell's equations. For larger particles, where the approximations used to solve Maxwell's equations for a small particle are not valid, more complex dependencies on angular scattering intensity are observed. For example, Mie scattering theory is often applied to characterize scattering in biological systems [61]. However, Mie scattering should only be applied to systems of spherical symmetry. Even then, the angular distribution of scattering intensity makes it difficult to extract physical insight, particularly when the size distribution is polydispersed.

1.2 Fractal Structures and Scattering

The definition of fractals was first introduced in 1975 by Mandelbrot's in his work [46] *Les Objets Fractals: forme, hazard et dimension* [50] to describe objects that are self-similar or scale-invariant [49, 50]. Regardless of the magnification of an object relative to a given variable, such as length, mass, area, volume, or radius of gyration, the structure of a fractal remains statistically unchanged [48]. Fractal structures and fractal events manifest themselves in many aspects of nature, such as in the distribution of celestial bodies, structure of snowflakes, organization of tree branches, and in the structure of proteins. Fractal events are often identified in phase transition, growth, and diffusion phenomena [48].

Because scattered waves from individual primary particles interfere with one another [62], "classical" light-scattering theories fail to describe the ADLS pattern from an aggregate of noncoalescing particles (Fig. 2.2). New theories have been developed that characterize scattering from these aggregates as a function of the fractal dimension [3, 9, 41, 46, 56, 63–76]. As will be shown below, under certain conditions, the ADLS signal $I(\theta)$ is related to the fractal dimension (D_f) of an aggregate through [38, 46, 56, 63] .

$$I(\theta) \propto q(\theta)^{-Df}. \tag{2.18}$$

Thus a simple data-processing step can be used to quantify this important physical parameter of the scattering system. In the biological systems in focus here, this parameter can be related to the chemistry that forms the particle and is thus of analytical value.

Scattering models based on RGD theory are often used to describe the scattering signal from fractal aggregates [3, 46, 56, 67, 77, 78]. The assumption implicit in this use is that multiple scattering is negligible, and that the individual scatterers are small enough to behave like Rayleigh scatterers. Although the assumption that the

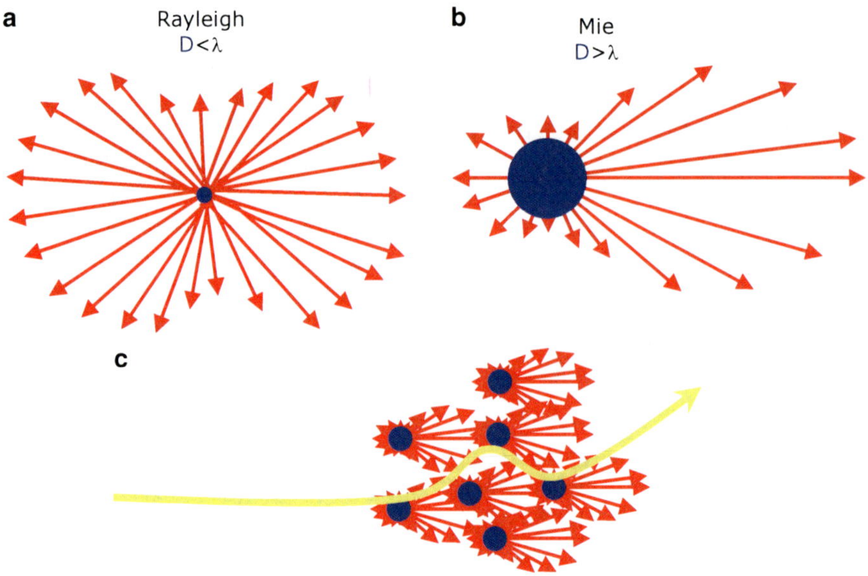

Fig. 2.2 Representation of parallel scattering patterns for (**a**) Rayleigh, (**b**) Mie, and (**c**) RGD-FA particles

individual particles scatter independently is not always accurate, RGD theory provides a reliable approximation when the following conditions exist [3, 46]:

$$|m-1| \ll 1 \quad \text{and} \quad \frac{2\pi a}{\lambda}|m-1| \ll 1, \tag{2.19}$$

where, a is now the radius of the primary particles and m is the relative refractive index of the primary particles. Farias and coworkers [3] have shown that, within 10% accuracy, RGD theory can predict the scattering signal with more relaxed constraints:

$$|m-1| \le 1 \quad \text{and} \quad \frac{2\pi a}{\lambda}|m-1| \le 0.6. \tag{2.20}$$

This method is referred to as the RGD approximation for the optical cross-section of fractal aggregates, RGD-FA.

In addition to the fractal dimension, another important parameter often used to characterize morphology of fractal aggregates is the Radius of Gyration, R_g which describes how mass is distributed around an arbitrary rotation axis. In mechanics, it is the square root of the moment of inertia divided by mass. Thus, R_g is functionally

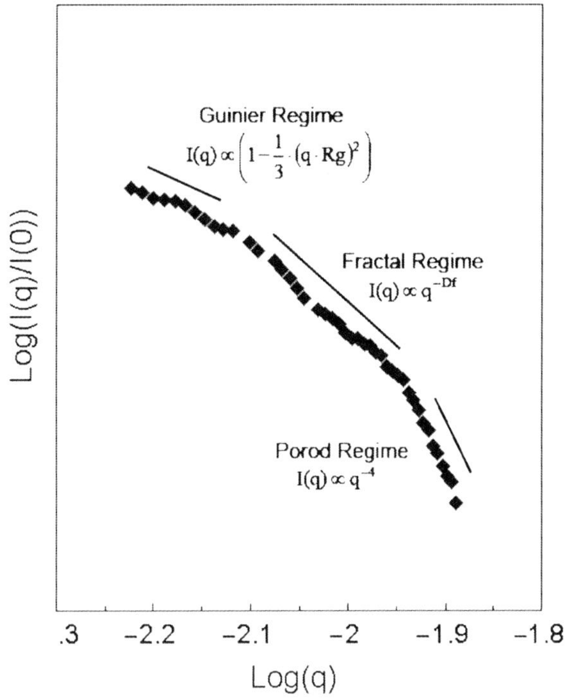

Fig. 2.3 Fractal dimension analysis of scattering signal from 100 nm Au aggregated with a protein showing aggregate light scattering regimes: Guinier, fractal and Porod regimes

a root mean square radius for the aggregate. Further, it is related to the number of primary particles in the aggregate and the fractal dimension through

$$N_p = k_p \left(\frac{R_g}{a} \right)^{Df}. \tag{2.21}$$

In practical terms, the inverse of q (which has units of length) is the size resolution limit of the scattering measurement. A scattering measurement is only sensitive to the structures of size greater than q^{-1}; thus, a scattering measurement is "blind" to structures smaller than q^{-1}. The scattering from a fractal aggregate can be separated into three regimes as a function of q: the fractal regime, the Guinier Regime and the Porod regime, as illustrated in Fig. 2.3. The scattering signal dependence as a function of q in the Guinier regime results from the overall size of the aggregate, where the size of the aggregates is smaller than the inverse of the wave vector ($q < 1/R_g$). In the drop-off region at high values of q, q^{-1} is larger than the size of the individual particles (or $q > 1/a$). In this regime, referred to as the Porod regime, the scattering is attributed to dispersed particles. The intermediate regime is the fractal regime. Here, the scattering intensity is sensitive to the morphology and overall structure of the scattering aggregate, where q^{-1} is smaller than the R_g of the aggregate but larger than size of the individual particles making up the aggregate [56].

In most experiments, polydispersity and heterogeneity (for both the primary particle size distribution and that of the aggregates) always exist and the transition regions from one regime to the next are generally broader and less well defined than the conditions defined above [75, 76, 79]. Even so, the qualitative behavior predicted for light-scattering intensity as a function of angle (through q) is often observed. For example, Fig. 2.3 shows the scattering signal from gold-protein aggregates (see below), where the Guinier, fractal, and Porod regions are clearly evident.

To summarize, fractal aggregate-scattering theory predicts several diagnostic regimes in which the light scattering shows different dependencies on fundamental physical characteristics of the aggregate, notably the radius of gyration and the fractal dimension. Further, manipulation of the angular distribution data can be used to readily extract these quantities. In the Guinier regime, the scattering signal can be rewritten as

$$\frac{I(0)}{I(q)} \propto \frac{1}{3} q^2 R_g^{\ 2}. \tag{2.22}$$

Therefore a plot of $I(0)/I(q)$ vs. q^2, should yield a linear curve with slope of $\frac{1}{3} R_g^{\ 2}$. In the fractal regime, scattering signal is related to the fractal dimension through

$$I(q) \propto q(\theta)^{-\mathrm{Df}}. \tag{2.23}$$

Here the slope of a log–log plot of $I(q)$ vs. q gives the fractal dimension.

2 Experimental Considerations

Traditionally, light-scattering measurements have been performed either by collecting light at a single angle [80–84] or by placing a detector on a rotating stage and collecting the light as the stage rotates across different angles [27, 41, 74, 85]. Both detection schemes have limitations. The first offers fast detection, but, generally, single angle detection cannot be used to differentiate between multiple particle sizes without being coupled to a particle separation technique, such as gel filtration or chromatography [86–89]. The second scheme can provide good angle resolution and it can be used to differentiate between particles of various sizes without the need of a separation technique; however, this scheme is time-consuming because it is limited by the speed of the rotating stage and the detector integration time.

In our work, two apparatuses have been constructed. In the first, a rotating detection optics rail was used to resolve the angular intensity distribution. This can detect scattering angles ranging from 20° to 155° with angular resolution of approximately 6°. The second apparatus used a more elegant design, where an ellipsoidal mirror coupled to a CCD detector was used to detect and resolve the full scattering angle range in a single frame [90]. This apparatus provides fast detection, superior angular resolution, and wider scattering angle detection range compared to traditional light scattering

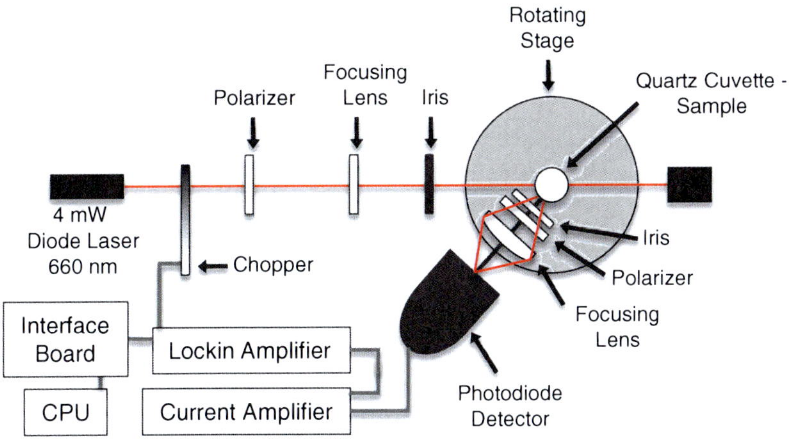

Fig. 2.4 Experimental arrangement used for collecting angle-resolved light scattering using a rotation detection optics arm

detection [54]. We found that the CCD-based apparatus could detect scattering signal in the range of 15–160° in a fraction of a second with angular resolution as low as 0.2°. Further details on these systems are provided below.

2.1 *Rotation Stage Apparatus*

The general schematic of the optical setup for the rotation stage apparatus is shown in Fig. 2.4. The light source was a 4 mW Fabry–Perot diode laser (equipped with a 3-Element Glass Lens, Edmund) with an output wavelength of 660 nm. The incident beam is modulated by a mechanical chopper (running at 1 KHz) that triggered a lock-in amplifier for phase-sensitive detection. The polarization of light can be selected by placing a polarizer before the 10-cm focal point focusing lens. The beam was focused into the center of a 1-cm diameter quartz-scattering cell. All light collection components are mounted on a manual rotation stage. The scattered light passes through a second polarizer that isolates the polarization of the scattered light, and then it is focused onto a silicon photodiode detector. An iris positioned in front of the detector provided approximately 6° angular resolution.

The scattering signal was collected as the detector arm rotated stepwise across the desired angles defined by the rotation stage and imaging iris. The photodiode signal is fed into a current amplifier with 1×10^6 gain. The amplified signal was sent to the lock-in amplifier for first harmonic detection. Custom LabView software on a Windows-based personal computer controlled the data collection and data analysis.

An inherent challenge for all light-scattering measurements is to reduce the background generated by stray light from reflection, diffraction, and scattering from optical components [91] and particulate matter. Stray light can result from imperfections

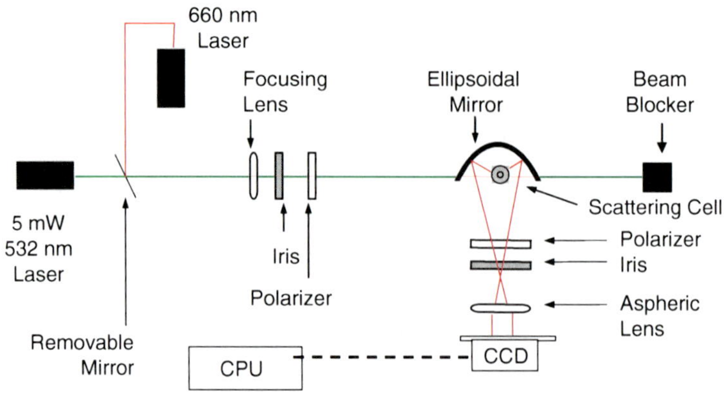

Fig. 2.5 Experimental arrangement used for collecting angle-resolved light scattering using the ellipsoidal mirror/CCD apparatus

in the glass wall of the scattering cell and from changes in the refractive index at the cell air/glass and glass/liquid interfaces as the laser beam enters/exits the scattering cell. In this apparatus, forward-scattered light from the walls of the cuvette compromised the measurements for scattering angles less than 20°. In order to minimize the impact of stray light, background subtraction in a suitable "blank" system can be used. In our measurements, we used either buffer solutions or pure water as a background that could be subtracted from the sample signal.

2.2 *CCD-Based Apparatus*

Figure 2.5 is a schematic diagram of the ellipsoidal mirror and CCD apparatus. The laser consists of either a 5 mW, 532-nm laser (a diode laser-pumped Nd:YVO_4 crystal coupled with KTP as a frequency doubler) or a 632 nm, 10 mW power Fabry–Perot laser. The laser used in a given experiment could be selected using a flip mirror. Either beam first goes through a polarizer and then it is focused into the center of a custom scattering cell through a 100 mm focal length lens. The scattering cell sits on a translation stage located approximately at one of the two focal points of an ellipsoidal mirror [90]. The ellipsoidal mirror is tilted approximately 35° downward relative to the scattering plane that minimized scattered light from reaching the detector.

The scattered light passed through a polarizer, which isolates the scattered light with the same polarization as the incident light, and then through an iris. Finally, a 1″ aspherical lens (with a focal length of 21 mm) focuses the scattered light onto the CCD array detector. The camera used had a 19.4 mm × 19.4 mm cooled CCD chip (1,024 × 1,024 pixels, EG&G PARC CCD detector model 1530-P-1024I). A mechanical shutter controls the light exposure of the CCD detector.

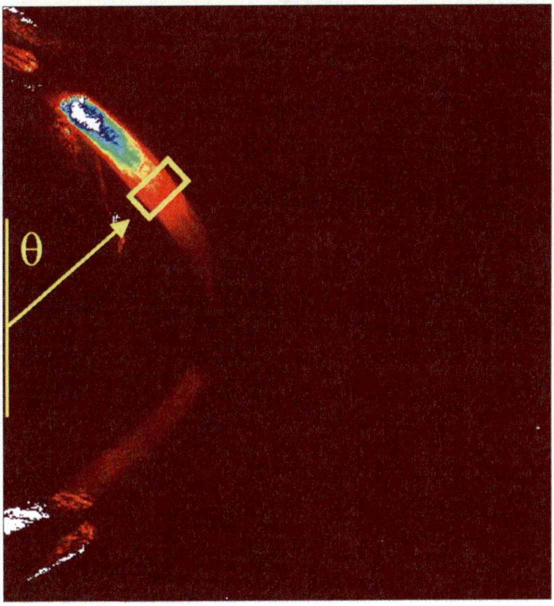

Fig. 2.6 CCD image illustrating angular distribution of scattering

The resulting CCD image is an arch, as shown in Fig. 2.6, from which a plot of scattered light intensity as a function of scattering angle can be extracted. In principle, the angular resolution is defined by the width of the of the CCD pixels (here, $19 \times 19\ \mu m^2$/pixel). In contrast to the rotation stage apparatus, this optical configuration can detect a full scattering angle profile in a single shutter shot with an angle resolution of approximately 0.2° per pixel. Binning of adjacent pixels reduces this resolution but can increase experimental throughput and lower noise.

A computer controls data storage and data analysis. The HIDRIS image software provided with the CCD detector controls scattering data collection. Image files are further analyzed using custom software that converts the image files into a two-dimensional array, subtracts a background image of scattering, and extracts and plots the angular scattering intensity.

To calibrate the angular distribution of scattering intensity, a mixture of a fluorescent dye and 100 nm Au nanoparticles was placed in the sample holder. This solution provides an isotropic scattering signal. The fluorescent dye is Cy3, which emits at 558 nm with 530 nm excitation [92]. A slit mounted on a rotation stage could be used to mask a portion of the scattered intensity onto the detector. The two dimensional location of the resulting spot on the CCD image was then correlated to angle.

As already noted, an inherent challenge for all light scattering measurements is to reduce the background generated by changes in the refractive index as light travels through different media [91]. In this apparatus, the scattering cell used a outer jacket of toluene as an index matching medium between the two reservoirs [90] (Toluene's refractive index is 1.43 which is approximately the same as glass, 1.40 [55]).

2.3 Au Nanoparticles

The most common technique for the synthesis of Au nanoparticles is reduction of Au salts, often using sodium citrate as the reducing agent [93–98]. This method consists of dissolving auchloric acid trihydrated in high purity boiling water, and then reducing the Au salt with specific amounts of trisodium citrate. The concentration of Au salt relative to the concentration of sodium citrate determines the rate of nucleation, which, consequently, determines the final size of the Au nanoparticles. The reduction of Au ions by citrate is accompanied by change in color from the pale yellow of Au salt to dark red of metallic Au colloids. The high purity water is necessary to avoid unwanted Au nucleation induced by impurities. The width of the size distribution of the Au colloids prepared with this method is approximately 10% relative to the mean particle size. These nanoparticles are usually purified by centrifugation or dialyzes, depending on the size of the nanoparticle being synthesized.

3 Studies of DNA–Au Nanoparticle Aggregates

In this section, we will discuss an application of the angle resolve fractal aggregate scattering technique to DNA detection. In summary, thiol-modified oligonucleotides are covalently attached to gold nanoparticles (Au–DNA probe) which hybridize to a target single-strand oligonucleotide of complementary sequence (oligo-target) in a liquid sample, forming an aggregate network of DNA and Au nanoparticles. Conceptually, we anticipate that a greater concentration of the targets in solution will lead to more compact fractal aggregate structure as cross linking occurs (Fig. 2.7).

3.1 Chemistry and Chemical Methods

The Au–DNA probes consisted of 100 nm gold nanoparticles separately modified with either 3′ end or with 5′ end hexanethiol functionalized oligonucleotide. The concentration of the particles was determined by absorbance measurements at 564 nm using an extinction coefficient of $1.62 \times 10^{11}\ M^{-1}\,cm^{-1}$ [82]. The probes were prepared by mixing 9.0 pM of citrate stabilized 100 nm of gold nanoparticles with 5 nM of either 5′ or 3′ end hexanethiol modified oligonucleotide (respectively) without any salt added and by incubating this solution for 24 h at 4°C. The oligonucleotides sequences used were:

5′ end: $HS(CH_2)_6$–CCC–GCG–CCC–3′
3′ end: 5′–CCC–GCG–CCC–$(CH_2)_6SH$

The hybridization temperature (melting temperature) for these olgonucleotides was approximately 36°C. Au–DNA probes were purified by centrifugation (14,000×*g* for 15 min), where the supernatant was discarded and the precipitate, Au–DNA probes, were resuspended in hybridization buffer.

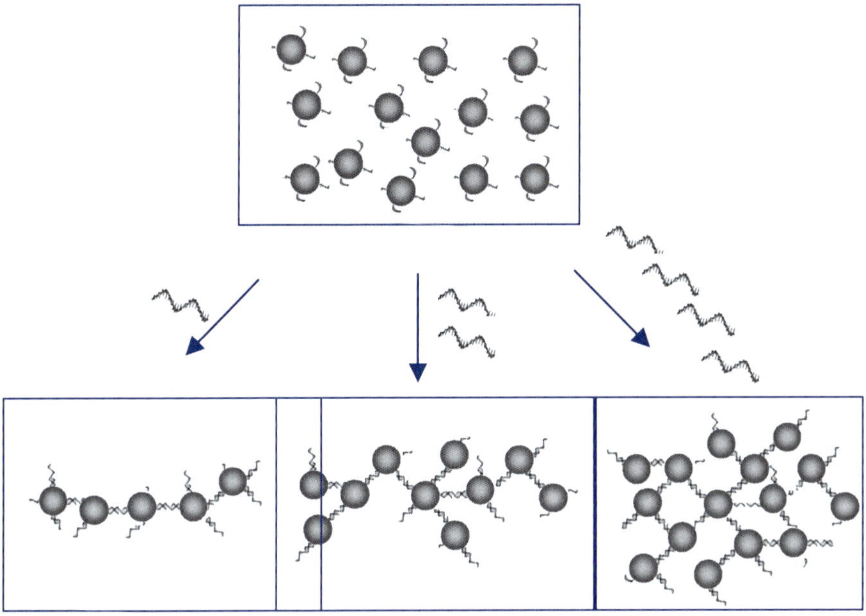

Fig. 2.7 Cartoon illustrating formation of Au–DNA fractal aggregate

The covalent linkage between the Au nanoparticle and the thiol-modified oligonucleotides was done through the thiol moiety and the Au nanoparticle. Although the chemistry of formation of self-assembled monolayers (SAM) of alkanethiols on Au surfaces and colloids has been studied extensively [99–102], there are still questions about the exact mechanism of attachment [103]. In one mechanism, which has been supported by experimental data and theoretical calculations [103–109], thiolate (RS^-) adsorbs on Au:

$$R-S-H+Au_n^0 \rightarrow R-S^-Au^+ \cdot Au_n^0 + \tfrac{1}{2}H_2$$

The thiolate–Au bond to gold surfaces is relatively strong, with a bond strength of approximately 40 kcal/mol [103, 104].

It has also been suggested by Whitesides and coworkers that the chemisorption process can occur with direct formation of a gold–sulfur bond, without the formation of thiolate [99].

$$R-S-H+Au_n^0 \rightarrow (\text{Au nanoparticle})-S-R + \tfrac{1}{2}H_2$$

Whitesides also suggested that Au^0 and Au^{+1} sites coexist, where Au^{+1} are occupied by citrate ions and thiols adsorb on Au^0 sites. The adsorbed citrate ions result from the original synthesis of Au nanoparticle. There is also experimental evidence

[103, 109], that at low coverage, both species, thiolates (R–S$^-$) and intact thiols (R–SH), coexist adsorbed on the Au surface.

Although there are different opinions on the exact mechanism of thiol chemisorption on gold, there is a consensus that chemisorption mechanism, of either intact thiols or thiolates, results in the formation of molecular hydrogen (H_2). The formation of H_2 is energetically necessary to justify the high stability of thiol SAM formation on gold surfaces [99, 100, 103, 104, 110]. The theoretical coverage of these monolayers on Au is approximately 0.77 nmol/cm^2 [99]. However, the extent of HS-DNA primer bound to the Au nanoparticle will be dependent on solution conditions, such as salt concentration and presence of organic solvents [111–115].

3.1.1 Hybridization Procedure

The oligo-targets consisted of synthesized oligonucleotides of either 21 or 30 bases in length where the 9-base sequence on the 3′ and 5′ ends are complementary to the sequences of the Au–DNA probes. The oligonucleotide sequences are:

21 *bases*: 5′–GGG–CGC–GGG–ATA–GGG–CGC–GGG–3′
30 *bases*: 5′–GGG–CGC–GGG–AAA–TAA–AAT–AAA–GGG–CGC–GGG–3′

Samples consisted of 0.1 pM of Au–DNA probe and oligo-target concentrations ranging from 0 to 2,500 nM. The hybridization procedure consisted of three heat and cool cycles where the Au–DNA samples were denatured at 70°C in a water bath for 10 min and then annealed in an ice bath for another 10 min. After the third cycle, the samples were incubated for 24 h at 4°C. Each sample was removed from 4°C just before ADLS measurement; 100 μL extracted and then mixed with 900 μL of hybridization buffer.

All experiments were done using 50% by volume, 50 mM borate buffer at pH 8.6, and 50% 10 mM TE (10 mM Tris + 1 mM EDTA) TE buffer pH 8.0. Modified and bare gold nanoparticles have shown greater stability in sodium borate buffer than in the presence of other salts, such as sodium phosphate and sodium chloride. TE buffer was chosen because it is routinely used by molecular biologists when manipulating DNA.

4 Results and Discussion

Figure 2.8 shows the scattering signals for $30° \leq \theta \leq 150°$ for the Au–DNA hybridization with different concentrations of oligo-target. As the scattering intensity integrated across all angles increases with target concentrations, these results are consistent with literature results [116], where UV–vis measurements have shown that extinction increases as a function of target concentration.

Figure 2.9 further illustrates the sensitivity of the ADLS signal to the formation of Au–DNA aggregates using dissymmetry ratio analysis. Dissymmetry ratio analysis

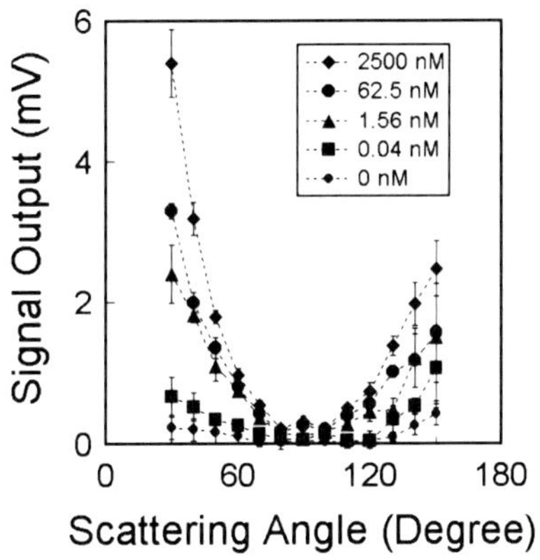

Fig. 2.8 Raw Au–DNA scattering data showing dependence on target concentration: 2,500 nM (*diamond*), 62.5 nM (*circle*), 1.56 nM (*triangle*), 0.04 nM (*square*), and no target oligonucleotide (*small circle*)

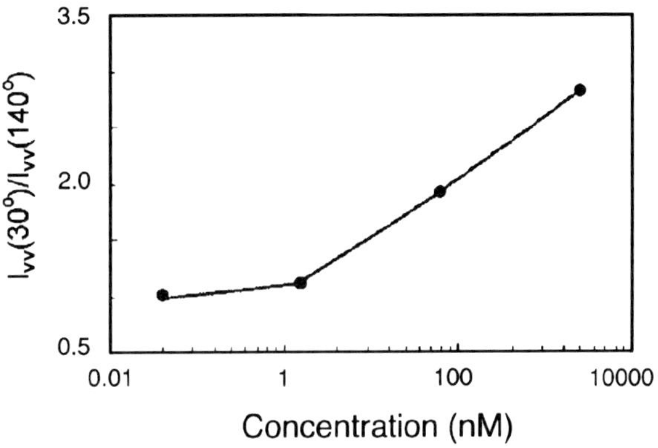

Fig. 2.9 Dissymmetry ratio analysis for raw data of Fig. 2.8

evaluates the ratio between the intensity of forward scattering at 30° and back scattering at 140°. Although this method does not provide quantitative information about the structure of Au–DNA aggregates, it can provide sensitivity to the concentration of the target DNA utilizing a simple instrumental configuration. The increase in the dissymmetry ratio as a function of target concentration results from the increase in volume of the Au–DNA aggregates. The lack of sensitivity at low concentrations of target DNA shown in these results suggests that the low target concentration measurements are near the detection limit of this approach.

Fractal dimension and Guinier analyses can provide quantitative structural information about the Au–DNA aggregates and can do so with greater sensitivity than the dissymmetry ratio analysis. Both Fractal Dimension Analysis (2.23) and Guinier Analysis (2.22) of the raw-scattering data shown in Fig. 2.8 is shown in Fig. 2.10. Numerical results for theses analyses are shown in Table 2.2. These results show that both the radius of gyration and the fractal dimension are found to be sensitive

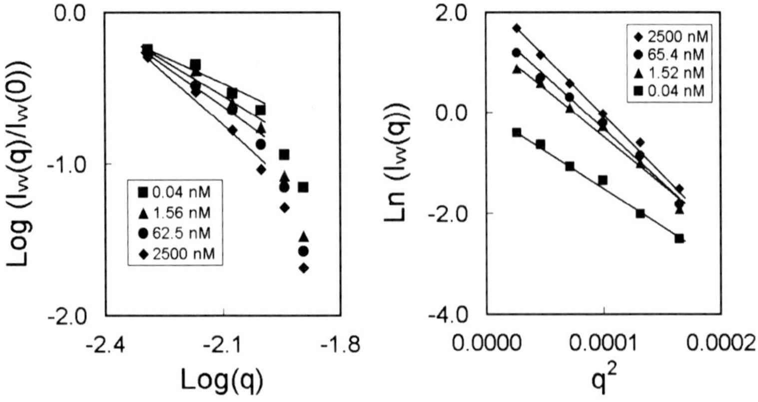

Fig. 2.10 Fractal and Guinier analyses for raw data of Fig. 2.8: 2,500 nM (*diamond*), 62.5 nM (*circle*), 1.56 nM (*triangle*), and 0.04 nM (*square*)

Table 2.2 R_g and Df as a function of oligonucleotide target concentration

Target concentration (nM)	R_g (nm)	Df
0.04	215	1.3
1.56	245	1.6
62.5	251	1.9
2,500	260	2.5

to target concentration. However, the latter shows a much greater sensitivity. As hypothesized, the increase in aggregate density (as shown by the increase in Df with oligo-target concentration) is likely the outcome of increased intraparticle cross-linking at higher oligo-target concentration.

The data of Fig. 2.10 suggest that the morphology of the aggregates can be measurably altered by target concentration. Figure 2.11 shows that the fractal dimension is also dependent on the length of the oligo-target, effectively increasing the spacing between the primary particles. To demonstrate this sensitivity, we compared the fractal dimension values for two systems with different oligo-target lengths: one in which the Au–DNA probe hybridizes to a 21-base long oligo-target, and the other in which the Au–DNA probe hybridizes to a 30-base long oligo-target. Each concentration of target showed lower Df values for the system containing the 30-base long oligo-target, in comparison to the one containing 21-base long oligo-target, indicative of more swollen and better solvated aggregates for the former. Further, the increase in fractal dimension with concentration for both lengths of oligo-target further suggests a transition from string-like aggregates (when Df is near unity) to denser, more compact structures at higher fractal dimension. This exponential dependence of the signal intensity on Df (2.2) makes this method very sensitive and therefore ideal for detecting an oligo-target DNA fragment using Au–DNA probes.

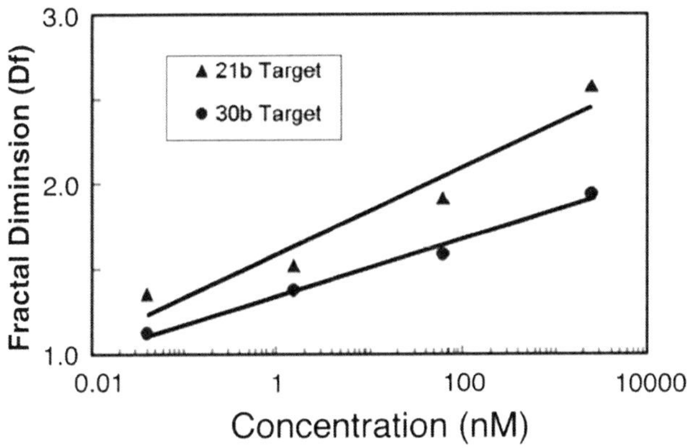

Fig. 2.11 Df as a function of concentration of 21 base- and 30 base-long target oligonucleotides

More quantitative information about the lengths of the targets can be extracted from the ratio between the slopes of the curves shown in Fig. 2.11. This can be accomplished by rewriting (2.21) as:

$$\mathrm{Df} = \frac{\log(N) - \log(kg)}{\log\left(\frac{R_g}{a}\right)} = \frac{\log(N)}{\log\left(\frac{R_g}{a}\right)} - \frac{\log(kg)}{\log\left(\frac{R_g}{a}\right)} \approx \frac{\log(C)}{\log\left(\frac{R_g}{a}\right)} - \frac{\log(kg)}{\log\left(\frac{R_g}{a}\right)}, \quad (2.24)$$

where C is the target concentration, where we assume that N and C are equivalent for this analysis. Then the slope of the curves in Fig. 2.11 is

$$\mathrm{Slope} = \frac{1}{\log\left(\frac{R_g}{a}\right)}. \quad (2.25)$$

Since a is the same for both systems, and if we assume that the two systems (21 bases and 30 bases) have the same number of particles forming an aggregate at each concentration and that the only difference between the two aggregates is the length between the targets, the ratio between the radii of gyrations should reflect the ratio between the length of the two linkers. This latter quantity can be estimated using molecular modeling software as 7.1 nm (21 bases) and 10.2 nm (30 bases). This ratio in lengths of 1.43 is slightly higher than the value extracted from the slopes, 1.26, but is within the uncertainty of the slope determinations.

5 Studies of Protein–Cofactor and Protein–Protein Interactions

In this section, we briefly describe several initial studies that demonstrate how the ADLS can be applied toward the detection and characterization of protein–cofactor and protein–protein interaction through the assembly of nanoparticle aggregates. The general approach for this methodology is to modify gold nanoparticles with a ligand moiety, a cofactor or a protein, that specifically interacts with a target protein, where the interaction induces the assembly of the Au–protein aggregate.

5.1 The Streptavidin–Biotin Complex

An example of a ligand protein system is the streptavidin–biotin complex. Streptavidin is a four-unit 50 kDa recombinant protein [117] that has four binding sites. Biotin (vitamin H) tightly binds to streptavidin with a dissociation constant of 10^{-15} M. The streptavidin–biotin complex has been extensively described in the literature [117–122]. Both biotin and streptavidin can be chemically modified with sulfide (or disulfide) containing linkers and/or other cross-linkers. Finally, the size, shape, and position of binding sites in the protein can be varied by means of a simple one-step conjugation chemistry using a homobifunctional cross-linker (glutaraldehyde) [123, 124] that allows streptavidin to be cross-linked to itself, forming dimers, trimers, and larger protein complexes.

The strong binding between the protein and cofactor occurs through noncovalent interactions consisting of hydrogen bonding between biotin and the serine (Ser) and aspartate (Asp) residues in the streptavidin binding site, and hydrophobic interaction between the biotin tail and the streptavidin side chains along the binding site [118, 119, 125]. Our anticipation was that the interaction through the four available binding sites would allow streptavidin to be used as a cross-linker between biotin-modified gold nanoparticles forming a fractal assembly, as illustrated in Fig. 2.12.

5.1.1 Thiol–Biotin Conjugation

The approach taken to functionalize the Au nanoparticles was to use disulfide moiety to anchor biotin to the surface of the Au nanoparticles. The first step was to chemically modify biotin with L-cystine (cystine), a disulfide-containing molecule. Cystine is an amino acid derived from two cysteine bridged through a disulfide bond. The choice of using the disulfide form instead of plain cysteine or another amine-thiol containing molecule, such as meracaptoethalamine, was because it has been reported that gold SAM originated from di-*n*-alkyl disulfides (RSSR) are more stable than those from alkanethiols (RSH) [99, 103, 104, 109, 110].

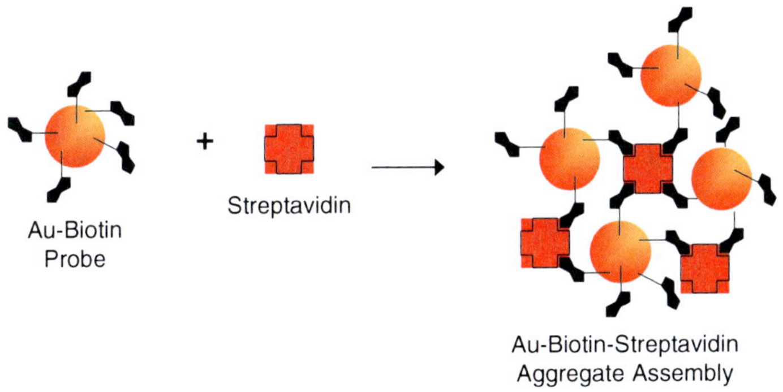

Fig. 2.12 Illustration of cofactor modified biotin (Au-ligand probe) interacting with a target protein to form an Au–protein aggregate

The coupling chemistry used NHS ester modified biotin to generate the disulfide modified biotin. NHS ester cross-linkers are routinely used for modifying proteins and protein cofactors, such as biotin. The NHS ester group reacts with the deprotonated form of the primary amine by nucleophilic attack, which forms a stable amide linkage and *N*-hydroxysuccinimide is released as the by-product of the reaction [124, 126]. Hydrolysis of the NHS ester is a major competing reaction in aqueous solution, and the rate of hydrolysis increases with increasing pH.

5.1.2 Gold–Biotin Coupling

The covalent linkage between the Au nanoparticle and the disulfide-modified biotin (L-cystine-biotin or cyss-biotin) was done through the disulfide bond of the cystine and the Au nanoparticle. The preferred mechanism, which has been supported by experimental data and theoretical calculations [103–109], is the one where the disulfide bond is broken close to the surface with the generation of two thiolates (RS^-) which then adsorbs onto Au [110]:

$$R-S-S-R+Au_n^0 \rightarrow R-S^-Au^+ \cdot Au_n^0$$

Existing literature shows that the rates of adsorption of *n*-alkyl-disulfides is indistinguishable from alkanethiols, but the rate of replacement for thiols is much faster than for disulfides [103, 110, 127]. It also has been suggested that the estimated adsoption energy for dialkyl disulfides is twice as favorable as the adsorption energy for thiols [110, 128]. Determining the exact mechanism of adsorption of thiols and disulfides on gold nanoparticles is not an easy task [103], which explains different opinions on the exact mechanism of thiol and disulfide chemisorption on gold [99, 100, 104, 110]. The theoretical coverage of a SAM on Au is approximately 0.77 nmol/cm^2 [99].

Usually, Au nanoparticles carry an overall negative charge resulting from either citrate groups present on its surfaces resulting from the citrate reduction process during its synthesis, or from hydroxy and chloride groups present in solution which weakly adsorb onto Au nanoparticles [99]. When a thiol or disulfide-containing molecule adsorbs onto the surface of an Au nanoparticle, the overall surface charge of the Au nanoparticle changes. This charge displacement can induce an aggregation of the Au nanopartilcles which can be detected by monitoring the red shift in the plasmon resonance absorption wavelength when aggregation takes place [99, 100, 116].

We determined that cystine concentrations higher than 1.56 mM induce aggregation of 40 nm Au, but that concentrations as high as 12.5 mM cystine do not induce aggregation of 100 nm Au nanoparticles. We also tested the behavior of unmodified particles in the presence of cystine-modified biotin (cys-biotin). In contrast to the cystine experiment, the highest concentration of cyss-biotin available in our stock solution was 0.450 mM, which is approximately three times lower than the cystine concentration that induced aggregation. Both sizes of Au nanoparticles showed negligible aggregation. Unbound cys-biotin was removed through four sequential centrifugation steps (at 14,000 rpm for 15 min), where the supernatant was removed and the sedimented Au is resuspended with nanopure water.

5.1.3 Scattering Signal as a Function of Streptavidin-Complex Concentration

Figure 2.13 shows the angle-resolved scattering signal as a function of streptavidin-complex (target) concentration. The sensitivity of the scattering signal to the concentration of the streptavidin-complex is especially pronounced at forward-scattering angles. Also shown is a control experiment in which unmodified biotin was added to the streptavidn mixture before adding the Au-Biotin probes. The fractal dimension and Guinier analysis for these same experiments are shown in Fig. 2.14. These results indicate that both the radius of gyration and the fractal dimension are sensitive to target streptavidin-complex concentration. However, the latter shows a greater sensitivity. These results also show that the control sample (streptavidin saturated with unmodified biotin) remains nearly unchanged for the various concentrations of protein.

5.1.4 Fractal Dimension Comparison of Streptavidin Monomer and Complex as Target Protein

In this study, we compared the difference in fractal dimensions in aggregates assembled with either streptavidin-complexes or with streptavidin monomers as the protein target. Both 532 and 632 nm excitation light were used, with very similar results. The results of the Guinier and Fractal Dimension Analyses for this system, shown in Fig. 2.14, suggest that the morphology of the aggregates is a function of protein concentration, but also the structure of the target protein. The mixture of larger protein targets, which make up the streptavidin-complex, assembles the Au

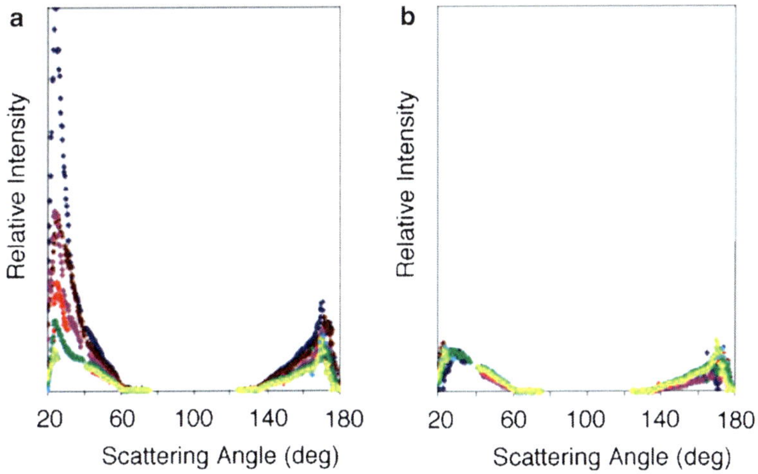

Fig. 2.13 (**a**) Angle-resolved scattering signal for 40 nm Au-biotin probes incubated with various concentrations of streptavidin complexes: 50 nM (*purple*), 25 nM (*brown*), 12.5 nM (*pink*), 6.25 nM (*red*), 3.12 nM (*green*), and no streptavidin (*yellow*; control). (**b**) Control where the streptavidin-complex binding sites from (**a**) were blocked with free biotin (100 μM) prior to mixing with Au-biotin probes. Higher complex concentration showed increase scattering signal (*arrow*)

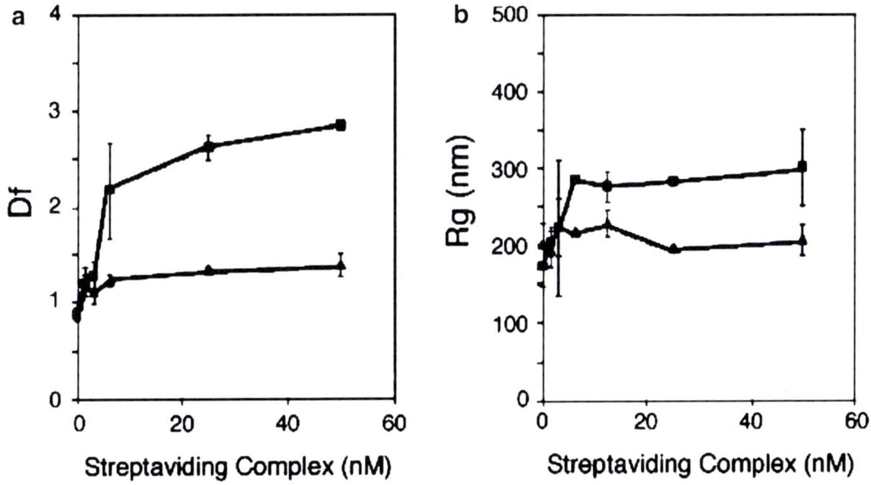

Fig. 2.14 Df of 40 nm Au-biotin as a function of streptavidin complex aggregate (*square*); and R_g of 40 nm Au-biotin as a function of streptavidin complex aggregate (*square*). Control (*triangle*) consists of streptavidin complexes saturated with unmodified biotin prior to introducing Au-biotin

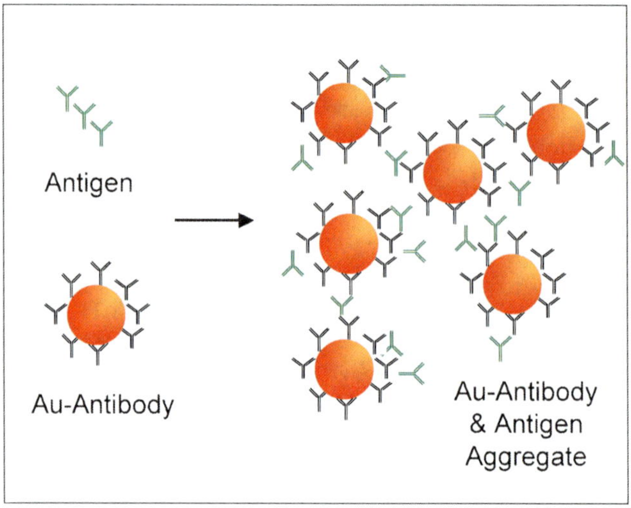

Fig. 2.15 Illustration of antigen-induced aggregation of antibody modified Au nanoparticles (Au-antibody)

nanoparticles into less dense (but probably better solvated) aggregates. However, both systems show an increase in fractal dimension with added protein concentrations, analogous to the behavior observed using the Au–DNA probes.

5.2 Antigen Detection Using Antibody-Modified Au Nanoparticles

To demonstrate an application in an even more complex system, a protein–protein binding event, we applied ADLS and fractal dimension to detect the reaction of 40-nm Au nanoparticles modified with goat antibodies toward a target antigen. The Au nanoparticles were modified with either goat anti-mouse immunoglobulin (Gt anti-Ms IgG) or goat anti-human immunoglobulin (Gt anti-Hu IgG); the target antigen was mouse immunoglobulin G protein (Mouse IgG), which should be reactive mainly with Gt anti-Ms IgG on gold nanoparitcle, as illustrated in Fig. 2.15.

Since an IgG has two binding sites available to interact with an antigen, we anticipated that the presence of a target antigen would trigger the aggregation of the antibody-modified Au nanoparticles, forming an aggregate of nanoparitcle, antibody, and antigen. The results were in agreement with these predictions. Figure 2.16 shows the forward-scattering signal for the two antibody systems and a control. Table 2.3 shows fractal dimensions for the three systems. A system consisting of nanoparticles carrying the Gt anti-Ms IgG antibody showed the highest reactivity toward the target mouse IgG antigen as reflected in the greater forward scattering

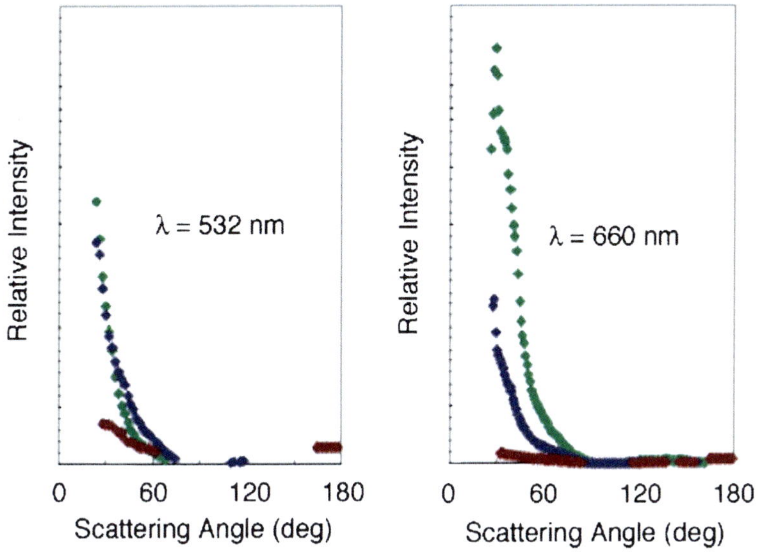

Fig. 2.16 ADLS measurements using 532 nm (*left*) and 660 nm (*right*) incident laser light. Au-anti-Ms IgG (*triangle*), Au-anti-Hu IgG (*square*), and control (Au-anti-Ms IgG; *circle*)

Table 2.3 Df of Au-antibody–antigen aggregates

	Df $\lambda_{ext}=532$ nm	Df $\lambda_{ext}=660$ nm
Au–Gt anti-Ms IgG	2.86±0.05	2.95±0.20
Au–Gt anti-Hu IgG	2.40±0.05	2.32±0.07
Control, Au–Gt anti-Ms IgG	1.65±0.10	1.87±0.03

and the greatest fractal dimension. However, the system in which Au nanoparticles were modified with Gt anti-Hu also showed an increased forward-scattering signal, which indicates cross reactivity between the Gt anti-Hu IgG antibody toward mouse IgG. This is not a surprising result because the Gt anit-Hu bound to the Au nanoparticles are not "mouse serum adsorbed" grade, which means that the Gt anti-Hu IgG which cross react with mouse IgG were not removed during the affinity purification process. Further, because of the common mammalian denomination of humans and mouse, Gt anti-Hu antibodies still show cross reactivity to mouse IgG, as our results suggest.

The use of Au nanoparticles and fractal dimension analysis could be further adapted toward the determination of antibody titer toward a specific antigen. Because antibodies are being routinely developed to be used in drug discovery as drug delivery and diagnostic agents, the determination of the reactivity with target antigen must be accompanied by measurement of its cross reactivity with other proteins. This experiment is a good example, where the Au nanoparticles modified with Gt anti-Hu showed cross reactivity toward the mouse IgG.

6 Future Trends and Possibilities

This chapter presents an innovative approach where ADLS-FA can be used to probe biomolecular assembly processes. Probing these events in a protein's or nucleic acid's native environment is an important step toward the eventual use of biopolymer modified nanoparticles as building blocks for nanostructure assembly. The method provides the flexibility for defining different biosensor formats, by using either the assembly of Au-biopolymer aggregates or the disassembly of these aggregates as the signal transducers. For example (and in contrast to the aggregate assembly demonstrated here), one could take advantage of aggregate disassembly by creating nanoparticle-biopolymer aggregates, where the protein or nucleic acid bridging the nanoparticles undergoes translocation upon contact with a specific protein or molecule. As a result of this translocation or cleaving, the disassembly and/or change in fractal dimension of these aggregates takes place. Many enzymatic-mediated processes are accompanied by translocation, such as, proteolytic digestion of protein by proteases, DNA and RNA cleavage by restriction enzymes and many phosphorylation-mediated processes.

Another direction we are pursuing is the combination of ADLS/FD with inelastic scattering, specially surface-enhanced Raman scattering (SERS), to attempt to extract not only structural information about the Au-biopolymers through ADLS/FD, but also to obtain information about chemical interaction between the biopolymer and Au-probe at the interface between particle and solution. This could be particularly useful for studying peptide–protein interactions, where Au nanoparticles are modified with peptides. There is also considerable interest in understanding the process of SERS, and increased attention has been paid in the relationship between the large surface enhancements and fractal dimension characteristics of metal surfaces and metal nanoparticle assemblies.

References

1. Souza, G.R. and J.H. Miller, *Oligonucleotide detection using angle-dependent light scattering and fractal dimension analysis of gold-DNA aggregates.* Journal of the American Chemical Society, 2001. **123**(27): p. 6734–6735.
2. Souza, G.R. and J.H. Miller, *Fractal Dimension Analysis Of Nanoparticle Aggregates Using Angle Dependent Light Scattering For The Detection And Characterization Of Nucleic Acids And Proteins.* 2003: USPTO Application.
3. Farias, T.L., U.O. Koylu, and M.G. Carvalho, *Range of validity of the Rayleigh-Debye-Gan theory for optics of fractal aggregates.* Appl. Opt., 1996. **35**(33): p. 6560.
4. Johnson, C.P., X. Li, and B.E. Logan, Env. Sci. & Tech., 1996. **30**: p. 1911.
5. Bushell, G.C., R. Amal, and J.A. Raper, *The effect of a bimodal primary particle size distribution on scattering from hematite aggregates.* Physica A: Statistical and Theoretical Physics, 1996. **233**(3–4): p. 859–866.
6. Thill, A., et al., *Structural Interpretations of Static Light Scattering Patterns of Fractal Aggregates; II. Experimental Study.* Journal of Colloid and Interface Science, 2000. **228**(2): p. 386–392.

7. Zenkevich, A.V., et al., *Formation of Au fractal nanoclusters during pulsed laser deposition on highly oriented pyrolitic graphite.* Physical Review B, 2002(65).
8. Forrest, S.R. and T.A. Witten, J. Phys. A, 1979. **12**: p. L109.
9. Schaefer, D.W., et al., *Fractal Geometry of Colloidal Aggregates.* Phys. Rev. Lett., 1984. **52**: p. 2371–2374.
10. Weitz, D.A. and M. Oliveria, *Fractal Structures formed by Kinetic Aggregation of Aqueous Gold Colloids.* Phys. Rev. Lett., 1984. **52**: p. 1433–1436.
11. Weitz, D.A., et al., Phys. Rev. Lett., 1984. **53**: p. 1657.
12. Weitz, D.A., et al., *Limits of the Fractal Dimension for Irreversible Kinetic Aggregation of Gold Colloids.* Phys. Rev. Lett., 1985. **54**(13): p. 1416–1419.
13. Schaefer, D.W. and K.D. Keefer, Phys. Rev. Lett., 1986. **56**: p. 2199.
14. Chen, S.H. and J. Teixeira, Phys. Rev. Lett., 1986. **57**: p. 2583.
15. Aubert, C. and D.S. Cannell, Phys. Rev. Lett., 1986. **56**: p. 738–740.
16. Dimon, P., et al., Phys. Rev. Lett., 1986. **57**: p. 595.
17. Wong, K., B. Cabane, and R.J. Duplessix, J. Colloid and Interface Sci., 1988. **123**: p. 466.
18. Hurd, A.J. and W.L. Flower, J. Colloid and Interface Sci., 1988. **122**: p. 178.
19. Wilcoxon, J.P., J.E. Martin, and D.W. Schaefer, *Aggregation in colloidal gold.* Phys. Rev. A, 1989. **39**: p. 2675–2688.
20. Martin, J.E., et al., Phys. Rev. A, 1990. **41**: p. 4379.
21. Olivier, B.J. and C.M. Sorensen, *Variable aggregation rates in colloidal gold: Kernel homogeneity dependence on aggregant concentration.* Phys. Rev. A, 1990. **41**: p. 2093–2100.
22. Amal, R., et al., *Structure and kinetics of aggregating colloidal haematite.* Colloids and Surfaces, 1990. **46**(1): p. 1–19.
23. Liu, J., et al., *Fractal colloidal aggregates with finite interparticle interactions: Energy dependence of the fractal dimension.* Physical Review A, 1990. **41**(6): p. 3206–3213.
24. Stoll, S., A. Elaissari, and E. Pefferkorn, J. Colloid and Interface Sci., 1990. **140**: p. 98.
25. Schaefer, D.W., et al., Aerosol Sci., 1991. **22**: p. S447.
26. Foret, M., J. Pelous, and R. Vacher, J. de Physique, 1992. **2**: p. 791.
27. Sorensen, C.M., J. Cai, and N. Lu, Appl. Opt., 1992. **31**(30): p. 6547–6557.
28. Oles, V., J. Colloid and Interface Sci., 1992. **154**: p. 351.
29. Cai, J., N. Lu, and C.M. Sorensen, Langmuir, 1993. **9**: p. 2861.
30. Stankiewicz, J., et al., Progr. Colloid & Polymer Sci., 1993. **93**: p. 358.
31. Carpineti, M. and M. Giglio, Adv. Colloid and Interface Sci., 1993. **46**: p. 73.
32. Amal, R., D. Gazeau, and T.D. Waite, Part. & Part. Syst.Charact. II, 1994: p. 315–320.
33. Weber, A.P., J.D. Thorne, and S.K. Friedlander, MRS Symposium Proceedings, 1995. **380**: p. 87.
34. Lin, M.Y., et al., Nature, 1989. **339**: p. 360.
35. Lin, J., et al., *Gold-Coated Iron (Fe@Au) Nanoparticles: Synthesis, Characterization, and Magnetic Field-Induced Self-Assembly.* Journal of Solid State Chemistry, 2001. **159**(1): p. 26–31.
36. Burns, J.L., et al., Langmuir, 1997. **13**: p. 6413–6420.
37. McCauley, J.L., *Chaos, Dynamics, and Fractals.* 1994: Cambridge University Press.
38. Avnir, D., D. Farin, and P. Pfeifer, *Chemistry in noninteger dimensions between two and three. II. Fractal surfaces of adsorbents.* J. Chem. Phys., 1983. **79**(7): p. 3566–3569.
39. Avnir, D. and P. Pfeifer, *Fractal Dimension in Chemistry. An Intensive Characteristic of Surface Irregularity.* Nouveau Journal De Chimie, 1983. **7**(2): p. 71–73.
40. Avnir, D., D. Farin, and P. Pfeifer, Nature, 1984. **308**: p. 261–263.
41. Koylu, U.O. and G.M. Faeth, *Optical Properties of Overfire Soot in Buoyant Turbulent Diffusion Flames at Long Residence Times.* Transactions of the ASME, 1994. **116**: p. 152–159.
42. Sorensen, C.M., N. Lu, and J. Cai, *Fractal Cluster Size Distribution Measurement Using Static Light Scattering.* Journal of Colloid and Interface Science, 1995. **174**(2): p. 456–460.
43. Sorensen, C.M., et al., *Aerogelation in a Flame Soot Aerosol.* Physical Review Letters, 1998. **80**(8): p. 1782.

44. Sorensen, C.M. and C. Oh, *Divine proportion shape preservation and the fractal nature of clster-cluster aggregates*. Physical Review E, 1998. **58**(6): p. 7545.
45. Sorensen, C.M. and G.M. Wang, *Size distribution effect on the power law regime of the structure factor of fractal aggregates*. Physical Review E, 1999. **60**(6): p. 7143.
46. Bushell, G.C., et al., *On techniques for the measurement of the mass fractal dimension of aggregates*. Advances in Colloid and Interface Science, 2002. **95**(1): p. 1–50.
47. Bonczyk, P.A. and R.J. Hall, Langmuir, 1991. **7**: p. 1274–1280.
48. Dewey, T.G., *Fractal in Molecular Biophysics*. 1997, New York: Oxford University Press.
49. Mandelbrot, B.B., *The Fractal Geometry of Nature*. 1982, San Francisco: Freeman.
50. Mandelbrot, B.B., *Les Objets Fractals: Forme, Hasard et Dimension*. 1975, Paris: Flammarion.
51. Schroeder, M., *Fractals, Chaos, and Power Laws*. 1991, New York: W. H. Freeman and Company.
52. van de Hulst, H.C., *Light Scattering by Small Particles*. 1981, New York: Dover.
53. Young, A.T., Physics Today, 1982. **35**(1): p. 42–48.
54. Muller, R.H. and W. Mehnert, *Particle and Surface Characterisation Methods*. 1997: GmbH Scientific Publishers.
55. Bohren, C.F. and D.R. Huffman, *Absorption and Scattering of Light by Small Particles*, ed. D.R. Huffman. 1983, New York: John Wiley and Sons, Inc.
56. Sorensen, C.M., *Scattering and Absorption of Light by Particles and Aggregates*, in *Handbook of Surface and Colloid Chemistry*, K.S. Birdi, Editor. 1997, CRC Press: New York.
57. Kerker, M., *The Scattering of Light and other Electromagnetic Radiation*. 1969, New York: Academic Press.
58. Berne, B. and R. Pecora, *Dynamic light scattering: with applications to chemistry, biology, and physics*. 2nd Edition ed. 2000, New York: Dover.
59. Saleh, B.E. and M.C. Teich, *Fundamentals of Photonics*. 1991, New York: Wiley, John & Sons, Inc.
60. Hunter, R.J., *Introduction to Modern Colloid Science*. 1993, New York: Oxford Science Publications.
61. Gillilanda, K.O., et al., *Distribution, spherical structure and predicted Mie scattering of multilamellar bodies in human age-related nuclear cataracts*. Experimental Eye Research, 2004. **79**: p. 563–576.
62. Mountain, R.D. and G.W. Mulholland, Langmuir, 1998. **4**: p. 1321–1326.
63. Avnir, D. and P.W. Schmidt, *The Fractal Approach to Heterogeneous Chemistry.*, ed. D. Avnir. 1989, New York: John Wiley & Sons, Inc.
64. Pfeifer, P. and D. Avnir, *Chemistry in noninteger dimensions between two and three. I. Fractal theory of heterogeneous surfaces*. J. Chem. Phys., 1983. **79**(7): p. 3558.
65. Kimura, H., *Light-scattering properties of fractal aggregates: numerical calculations by a superposition technique and the discrete-dipole approximation*. Journal of Quantitative Spectroscopy and Radiative Transfer, 2001. **70**(4–6): p. 581–594.
66. Lambert, S., et al., *Structural Interpretations of Static Light Scattering Patterns of Fractal Aggregates; I. Introduction of a Mean Optical Index: Numerical Simulations*. Journal of Colloid and Interface Science, 2000. **228**(2): p. 379–385.
67. Wang, G. and C.M. Sorensen, *Experimental test of the Rayleigh-Debye-Gans theory for light scattering by fractal aggregates*. Applied Optics, 2002. **41**(22): p. 4645–4651.
68. Castellano, A.C., et al., *X-ray small angle scattering of the human transferrin protein aggregates. A fractal study*. Biophysical Journal, 1993. **64**(2): p. 520–524.
69. Seri-Levy, A. and D. Avnir, *Fractal analysis of surface geometry effects on catalytic reactions*. Surface Science, 1991. **248**(1–2): p. 258–270.
70. Pfeifer, P., D. Avnir, and D. Farin, *Ideally irregular surfaces, of dimension greater than two, in theory and practice*. Surface Science, 1983. **126**(1–3): p. 569–572.
71. Shalaev, V.M., M.I. Stockman, and R. Botet, *Resonant excitations and nonlinear optics of fractals*. Physica A: Statistical and Theoretical Physics, 1992. **185**(1–4): p. 181–186.

72. Danilova, Y.E., A.I. Plekhanov, and V.P. Safonov, *Experimental study of polarization-selective holes, burning in absorption spectra of metal fractal clusters.* Physica A: Statistical and Theoretical Physics, 1992. **185**(1–4): p. 61–65.
73. Carl, A., G. Dumpich, and S. Friedrichowski, *Electron diffusion in percolating gold clusters.* Physica A: Statistical and Theoretical Physics, 1992. **191**(1–4): p. 454–457.
74. Koylu, U.O. and G.M. Faeth, Transactions of the ASME, 1994. **116**: p. 971–979.
75. Farias, T.L., U.O. Koylu, and M.G. Carvalho, *Effects of polydispersity of primary particle and aggregate sizes on radiative properties of simulated soot.* J. Quant. Spectrosc. Radiat. Transfer, 1995. **55**: p. 357–371.
76. Bushell, G.C., *Primary particle polydispersity in fractal aggregates*, in *Chemical Enginnering and Industrial Chemistry*. 1998, New South Wales: New South Wales. p. 202.
77. Marsh, P., G. Bushell, and R. Amal, *Scattering Behavior of Restructured Aggregates: A Simulation Study.* Journal of Colloid and Interface Science, 2001. **241**(1): p. 286–288.
78. Selomulya, C., et al., *Evidence of Shear Rate Dependence on Restructuring and Breakup of Latex Aggregates.* Journal of Colloid and Interface Science, 2001. **236**(1): p. 67–77.
79. Khlebtsov, N.G. and A.G. Melnikov, *Structure Factor and Exponent of Scattering by Polydisperse Fractal Colloidal Aggregates.* Journal of Colloid and Interface Science, 1994. **163**(1): p. 145–151.
80. Harding, S.E., D.B. Sattelle, and V.A. Bloomfield, *Laser Light Scattering in Biochemistry*, ed. V.A. Bloomfield. 1992, England: Redwood Press Ltd.
81. Stover, J.C. Optical Scattering Measurements and Analysis. 1995, Bellingham: The Society for Optical Engineering.
82. Yguerabide, J. and E.E. Yguerabide, Anal. Biochemistry, 1998. **262**: p. 137–156.
83. Parkash, J., et al., Biophysical Journal, 1998. **74**: p. 2089–2099.
84. Trulson, M.O., et al., SPIE, 1998. **3259**: p. 234–240.
85. Bryant, G. and J.C. Thomas, Langmuir, 1995. **11**: p. 2480–2485.
86. Gabriel, M.K. and E.T. McGuinness, FEBS Letters, 1984. **175**(2): p. 419–421.
87. Korgel, B.A., J.H. Van Zanten, and H.G. Monbouquette, Biophysical Journal, 1998. **74**: p. 3264–3272.
88. Machtle, W., Biophysical Journal, 1999. **76**: p. 1080–1091.
89. Qian, R.L., R. Mhatre, and I.S. Krull, J. of Chromatography, 1997. **787**: p. 101–109.
90. Tsutsui, K., K. Koya, and T. Kato, *An investigation of continuous-angle laser light scattering.* Review of Scientific Instruments, 1998. **69**(10): p. 3482.
91. Chu, B., *Laser Light Scattering*. 1974, New York: Academic Press.
92. Mujumdar, S.R., et al., Bioconjugate Chem., 1996. **7**: p. 356–362.
93. Pow, D.V. and J.F. Morris, *Membrane routing during exocytosis and endocytosis in neuroendocrine neurons and endocrine cells: Use of colloidal gold particles and immunocytochemical discrimination of membrane compartments.* Cell Tissue Res., 1991. **264**: p. 299–316.
94. Horisberger, M., *Evaluation of colloidal gold as a cytochemical marker for transmission and scanning electron microscope.* Biol. Cell, 1979. **36**: p. 253–258.
95. Horisberger, M. and M.F. Celerc, *Labeling of golloidal gold with protien A.* Histochemistry, 1985. **82**: p. 219.
96. Horisberger, M. and J. Rosset, *Colloidal gold, a useful marker fo transmission and scanning electron microscopy.* J. Histochem. Cytochem., 1977. **25**: p. 295.
97. Horisberger, M., J. Rosset, and H. Bauer, *Colloidal gold granules as markers for cell surface receptors in the scanning electron microscope.* Experientia, 1975. **31**: p. 1147–1149.
98. Horisberger, M. and M. Tacchini-Vonlanthen, *Ultrastructural localization of Kunitz inhibitor on thin sections of Glcine max (soybean) cv. Maple Arrow by the gold method.* Histochemistry, 1983. **77**: p. 37–50.
99. Weisbecker, C.S., M.V. Merritt, and G.M. Whitesides, *Molecular Self-Assembly of Aliphatic Thiols on Gold Colloid.* Langmuir, 1996. **12**(16): p. 3763–3772.
100. Whitesides, G.M., et al., Critical Reviews in Surface Chemistry, 1993. **3**(1): p. 49–65.
101. Bamdad, C., Biophysical Journal, 1998. **75**: p. 1997–2003.

102. Bryant, M.A. and J.E. Pemberton, J. Am. Chem. Soc., 1991. **113**: p. 8284–8293.
103. Gronbeck, H., A. Curioni, and W. Andreoni, *Thiols and Disulfides on the Au(111) Surface: The Headgroup-Gold Interaction*. J. Am. Chem. Soc., 2000. **122**: p. 3839–3842.
104. Ulman, A., et al., *Mixed alkanethiol monolayers on gold surfaces: Wetting and stability studies*. Advances in Colloid and Interface Science, 1992. **39**: p. 175–224.
105. Walczak, M.W., et al., J. Am. Chem. Soc., 1991. **113**: p. 2370.
106. Porter, M.D., et al., J. Am. Chem. Soc., 1987. **109**: p. 3559.
107. Nuzzo, R.G., F.A. Fusco, and D.L. Allara, J. Am. Chem. Soc. **109**: p. 2358.
108. Bain, C.D., H.A. Biebuyck, and G.M. Whitesides, Langmuir, 1989(5): p. 723.
109. Nuzzo, R.G., B.R. Zegarski, and L.H. Dubois, J. Am. Chem. Soc., 1987(109): p. 733.
110. Ulman, A., *Formation and Structure of Self-Assembled Monolayers*. Chem. Rev., 1996. **96**: p. 1533–1554.
111. Herne, T.M. and M.J. Tarlov, J. Am. Chem. Soc., 1998. **119**: p. 8916–8920.
112. Elghanian, R., et al., Science, 1997. **227**(5329): p. 1078–1081.
113. Storhoff, J.J. and C.A. Mirkin, Chem. Rev, 1999. **99**: p. 1849–1862.
114. Lewis, M. and M.J. Tarlov, J. Am. Chem. SOc., 1995. **117**: p. 9574–9575.
115. Peterlinz, K.A., et al., J. Am. Chem. Soc., 1997. **119**: p. 3401–3402.
116. Reynolds, R.A., C.A. Mirkin, and R.L. Letsinger, J. Am. Chem. Soc., 2000. **122**: p. 3795–3796.
117. Savage, D., G. Mattson, and S. Desai, *Avidin-Biotin Chemistry: A Handbook*. 1992, Rockford, IL.: Pierce Chemical.
118. Katz, B., B. Liu, and R. Cass, *Structure-Based Design Tools: Structural and Thermodynamic Comparison with Biotin of a Small Molecule That Binds to Streptavidin with Micromolar Affinity*. J. Am. Chem. Soc., 1996. **118**(34): p. 7914–7920.
119. Hyre, D.E., *Ser45 plays an important role in managing both the equilibrium and transition state energetics of the streptavidin-biotin system*. Protein Science, 2000. **9**: p. 878–885.
120. Badia, A., et al., *Probing the electrochemical deposition and/or desorption of self-assembled and electropolymerizable organic thin films by surface plasmon spectroscopy and atomic force microscopy*. Sensors and Actuators B: Chemical, 1999. **54**(1–2): p. 145–165.
121. Piscevic, D., W. Knoll, and M.J. Tarlov, *Surface plasmon microscopy of biotin-streptavidin binding reactions on UV-photopatterned alkanethiol self-assembled monolayers*. Supramolecular Science, 1995. **2**(2): p. 99–106.
122. Wang, J., et al., *Metal nanoparticle-based electrochemical stripping potentiometric detection of DNA hybridization*. Analytical Chemistry, 2001. **73**(22): p. 5576–5581.
123. Avrameas, S.a.T., T., *The cross-linking of proteins with glutaraldehyde and its use for the preparation of immunosorbents*. Immunochemistry, 1969. **6**: p. 53–66.
124. Wong, S.S. Chemistry of Protein Conjugation and Cross-Linking. 1993: CRC Press.
125. Freitag, S., et al., *Structural studies of the streptavidin binding loop*. Protein Science., 1997. **6**: p. 1157–1166.
126. Hermanson, G.T., *Bioconjugate Techniques*. 1996, San Diego: Academic Press.
127. Biebuyck, H.A. and G.M. Whitesides, Langmuir, 1994. **10**: p. 1825.
128. Schlenoff, j.B., M. Li, and H. Ly, J. Am. Chem. Soc., 1995. **117**: p. 12528.

Chapter 3
Influence of Electron Quantum Confinement on the Electronic Response of Metal/Metal Interfaces

Antonio Politano and Gennaro Chiarello

1 Introduction

The reduced dimensionality may induce the appearance of novel electronic states. In semi-infinite media (surfaces), surface states are formed within energy gaps of the projected bulk band structure due to the broken symmetry at the solid/vacuum interface. Electrons occupying such states are confined at the top layer of the bulk sample and form a two-dimensional nearly free-electron gas. On the other hand, in thin films discrete states are formed, characterized by a notable dependence of their energy on thickness. Such states, called quantum well states (QWS), describe standing electron waves confined within the film .

Recent experimental [1–8, 151] studies have demonstrated the existence of variations with film thickness for properties such as the electronic density of states, electron–phonon coupling, chemical reactivity, superconductivity, magnetism, surface energy, and thermal stability.

Further information on the physical and chemical properties of thin films can be obtained from their electronic collective excitations and, in particular, from the dispersion relation of the surface plasmon (SP) [9–14]. Nevertheless, the influence of electron quantum confinement on the dispersion relation of electronic collective excitations in nanoscale thin films has not been investigated yet. Moreover, the different distribution of occupied and unoccupied electronic states in thin films with respect to surfaces would imply enhanced damping mechanisms for collective excitations by creating electron-hole pairs [15–18].

A. Politano (✉)
Dipartimento di Fisica, Università degli Studi della Calabria, 87036 Rende (Cs), Italy

Universidad Autónoma de Madrid, Departamento de Fisica de la Materia Condensada, 28049 Madrid, Spain
e-mail: antonio.politano@fis.unical.it

G. Chiarello
Dipartimento di Fisica, Università degli Studi della Calabria, 87036 Rende (Cs), Italy

C.D. Geddes (ed.), *Reviews in Plasmonics 2010*, Reviews in Plasmonics,
DOI 10.1007/978-1-4614-0884-0_3,

High-resolution electron energy loss spectroscopy is a suitable technique for such aims as it allows measuring the dispersion curve of the frequency and the line-width of the SP [19–24].

Herein HREELS measurements of systems exhibiting electron quantum confinement, i.e., Ag/Ni(111), Ag/Cu(111), and Na/Cu(111) are reported. Our results should shed light on the influence of QWS on dynamical screening phenomena in thin films.

2 Formation of QWS

The electronic structure of films has been widely investigated by photoemission spectroscopy in recent years. These studies focused particularly on two main spectral features: the occupied Shockley-type surface state resulting from the termination of the crystal [25–42], and the QWS due to the quantum confinement of the *sp* valence electrons in the adlayer [43–56]. The binding energy of the Shockley state depends critically on film thickness.

Figure 3.1 summarizes experimental photoemission data acquired at 300 K for Ag/Cu(111) by Mathias et al. [57] as a function of silver film thickness. Moreover, the film morphology was found to be a very important parameter influencing the electronic structure. Recently, it has been demonstrated that the dispersion of QWS changes upon annealing the adlayer [58]. Angle-resolved photoemission experiments showed that Ag QWS on Au(111) have flat in-plane dispersion in a disordered film and a nearly free-electron-like dispersion in an annealed and well-ordered film. Accordingly, the sp density of states of the film may be tuned by annealing. As an example, Fig. 3.2 shows photoemission data recorded at room temperature for a silver film (of nominally 15 ML) on Cu(111), both before and after the annealing procedure. The peak position of QWS shifted slightly to a higher binding energy. Moreover, annealing reduced the broadness of the peak as a consequence of a higher flatness of the adlayer (Fig. 3.3).

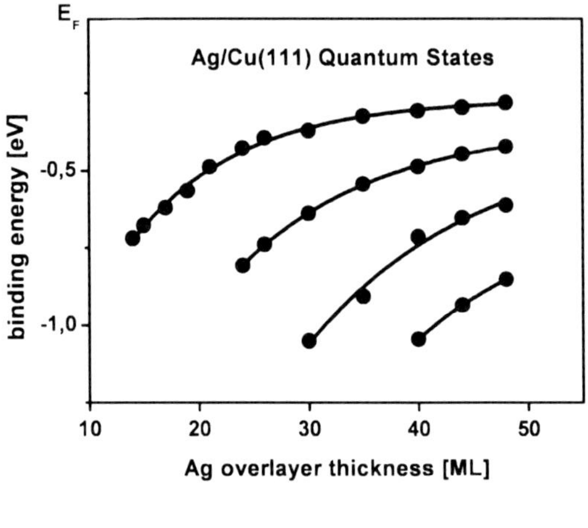

Fig. 3.1 Energy of the $\nu=1$ to $\nu=4$ QWS and resonances as a function of nominal film thickness as determined from photoemission spectra from nonannealed silver films (adapted from [57])

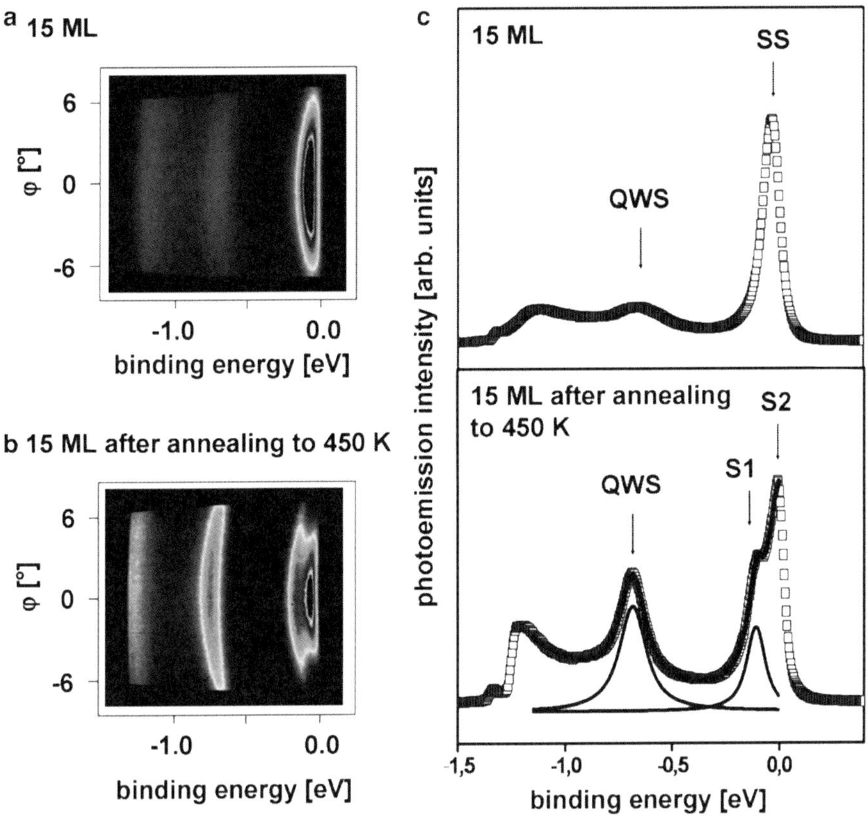

Fig. 3.2 (**a**) Photoemission map of a 15 ML thick silver film (nominal thickness) before (**a**) and after (**b**) annealing at 450 K; (**c**) normal emission spectra taken from (**a**) and (**b**) clearly visible are the modifications in the spectral shape of the quantum wells state (QWS) and in the energy region, where the Shockley surface state (SS) is observed. *S*1 and *S*2 indicate the two distinct Shockley surface states appearing in the spectrum after annealing; the two Lorentzians in the *bottom graph* result from a fit into the QWS-peak and *S*1-peak of the bottom spectrum. Also shown as a *solid line* is the total fit result. A Gaussian was used to reproduce the low energy tail of the spectrum and a Lorentzian to fit the peak *S*2 (Adapted from [57])

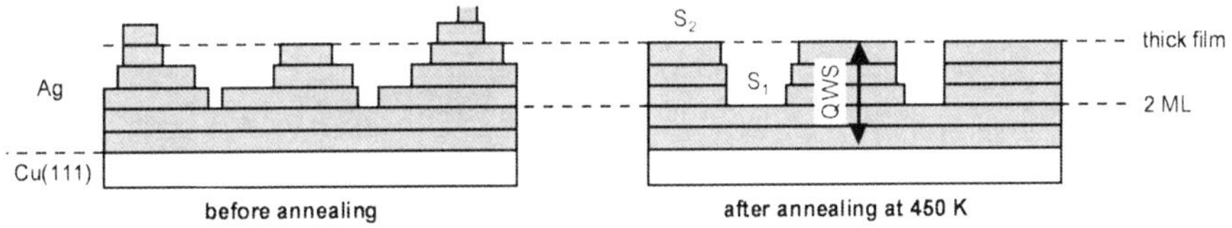

Fig. 3.3 Schematic summary of the morphological changes induced by the heat treatment of the silver film (Adapted from [57])

3 Experimental

Measurements were carried out in a UHV chamber operating at a base pressure of 5×10^{-9} Pa, equipped with standard facilities for surface characterizations, described elsewhere [59–62]. HREEL experiments were performed by using an electron energy loss spectrometer (Delta 0.5, SPECS). The samples were single-crystal surfaces of Ni(111) and Cu(111) with a purity of 99.9999% which were cleaned by repeated cycles of ion sputtering and annealing at 800–900 K. Surface cleanliness and order were checked using Auger electron spectroscopy measurements and low-energy electron diffraction (LEED), respectively. Silver was deposited onto the Ni(111) and Cu(111) surface by evaporating from an Ag wire wrapped on a tungsten filament. Well-ordered Ag films could be obtained at very low deposition rates (≈0.05 ML/min). Alkalis (Na, K) were evaporated in the UHV chamber by means of a well-outgassed dispenser (Saes Getters). Particular care has been dedicated to avoid CO contamination [152–155]. The occurrence of the p(1 × 1)-Ag, (3/2 × 3/2)-Na, and p(2 × 2)-K LEED patterns was used as the calibration point of $\theta_{Ag} = 1.0$ ML, $\theta_{Na} = 0.44$ ML, and $\theta_K = 0.25$ ML, respectively. A constant sticking coefficient was assumed to obtain other desired coverage. Coverage has been also controlled through X-ray photoemission spectroscopy, whose analyzer is described in [156–160]. The energy resolution of the spectrometer was degraded to 7 meV so as to increase the signal-to-noise ratio for off-specular spectra. The angular acceptance α of our electron analyzer was ±0.5°. All depositions and measurements were made at room temperature.

4 Results and Discussion

4.1 Ag Films on Ni(111)

Ag surfaces are characterized by a strong lowering of the SP energy, which follows a positive dispersion as a function of the parallel momentum transfer. Such behavior was ascribed to the presence of filled d bands [12, 20, 64–69]. However, experimental studies on low-dimensional Ag systems, such as ultrathin films on metal [9, 15, 16, 70–74] and semiconductor [18, 75, 76] substrates, nanowires [77], or nanoparticles [78, 79] are less common.

Ag on Ni(111) offers the possibility of investigating the relationship between quantum electron confinement and the damping of collective excitations in ultrathin films as confined electron states in the Ag adlayer exist. As a matter of fact, photoemission measurements [80] revealed the occurrence of Ag QWS for films with thickness from 0 to 15 layers. Ag on Ni(111) is an example of film growth in large (16%) mismatched materials. Such a large misfit determines the silver film to have the crystalline structure of bulk silver even for ultrathin Ag layers [81]. Moreover, the small solubility of Ag into Ni prevents from the dissolution of the silver layer

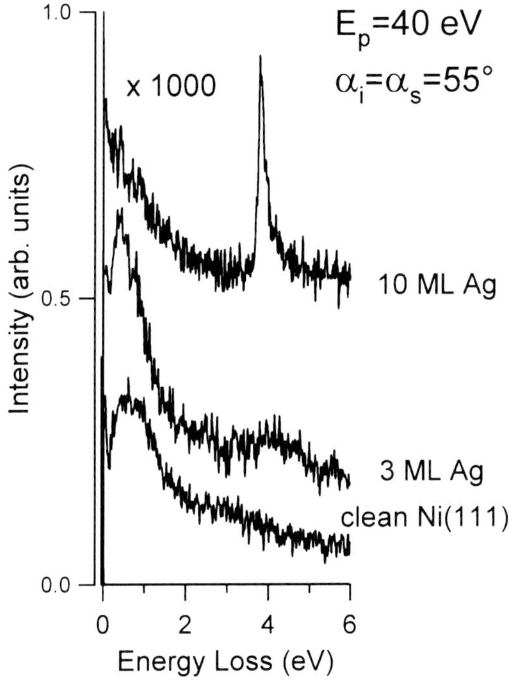

Fig. 3.4 HREEL spectra for Ag/Ni(111) as a function of Ag coverage

into the nickel substrate [82]. As a matter of fact, the formation of a surface alloy was not reported for this system [53, 82, 83]. For coverages higher than three layers, a reconstruction of the silver overlayer was observed [81]. The deposited Ag layer exhibits a 7×7 moiré structure.

Figure 3.4 shows HREEL spectra for thin Ag layers on Ni(111) as a function of Ag coverage. The spectrum of the Ni(111) surface is characterized by a broad peak at 1.0 eV. For less than 2 ML of Ag, the spectrum is extremely broad without a well-defined peak. A broad Ag SP at 4.2 eV arose for Ag coverages above 2 ML. However, its plasma energy was considerably higher than that of the SP of semi-infinite Ag, as also found for Ag/Si(111) [18, 76]. As the Ag coverage is increased, the energy of the plasmonic mode reduced and its line shape became sharper.

Such finding has to be ascribed to the s–d polarization. For thin Ag films, the overall screening of the plasmon via the polarizable d electron medium diminishes and a higher SP frequency occurs [66]. Moreover, the spill-out region not affected by s–d polarization becomes more important, causing a further blue-shift of the SP frequency.

Important information on the screening properties of metal systems is given by the dispersion relation of the SP.

To measure plasmon dispersion, values for the parameters E_p, impinging energy, and θ_i, the incident angle, were chosen so as to obtain the highest signal-to-noise ratio. The primary beam energy used for the dispersion, $E_p = 40$ eV, provided, in fact,

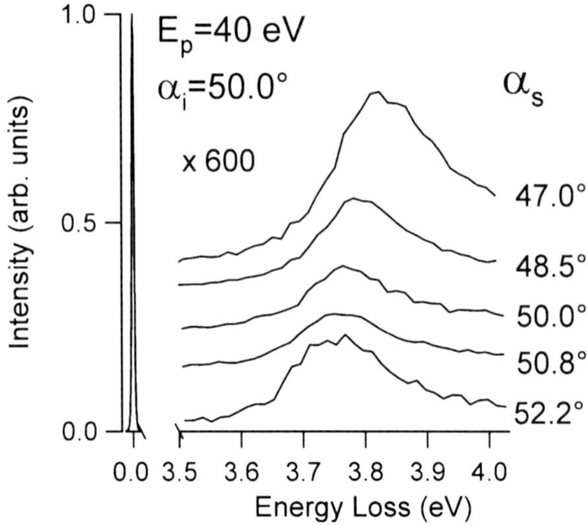

Fig. 3.5 HREEL spectra as a function of the scattering angle for 10 ML Ag/Ni(111) at $T = 300$ K

the best compromise among surface sensitivity, the highest cross-section for the plasmonic excitation, and $q_{\parallel}$ resolution. As,

$$\hbar \vec{q_{\parallel}} = \hbar(\vec{k_i} \sin\theta_i - \vec{k_s} \sin\theta_s) \tag{3.1}$$

the parallel momentum transfer $q_{\parallel}$ is determined by the values of E_p, E_{loss}, θ_i and θ_s:

$$q_{\parallel} = \frac{\sqrt{2mE_p}}{\hbar}\left(\sin\theta_i - \sqrt{1 - \frac{E_{loss}}{E_p}}\sin\theta_s\right), \tag{3.2}$$

where E_{loss} is the energy loss and θ_s is the electron scattering angle [17, 84].

Hence, it is possible to estimate the integration window in reciprocal space:

$$\Delta q_{\parallel} \approx \frac{\sqrt{2mE_p}}{\hbar}\left(\cos\theta_i - \sqrt{1 - \frac{E_{loss}}{E_p}}\cos\theta_s\right)\cdot\alpha, \tag{3.3}$$

where α is the angular acceptance of the apparatus [17, 84]. Under our experimental conditions, $\Delta q_{\parallel} = 0.012$ Å^{-1}, much less than the scanned range in the reciprocal space.

Selected HREEL spectra for 10 ML of Ag as a function of the scattering angle are shown in Fig. 3.5. The Ag SP energy dispersed from 3.751 up to 3.880 eV.

The measured dispersion curve $E_{loss}(q_{\parallel})$ in Fig. 3.6 was fitted by a second-order polynomial given by:

$$E_{loss}(q_{\parallel}) = A + Bq_{\parallel} + Cq_{\parallel}^2, \tag{3.4}$$

where $A = (3.751 \pm 0.002)$ eV, $B = (0.000 \pm 0.004)$ eV Å, and $C = (1.57 \pm 0.04)$ eV Å^2.

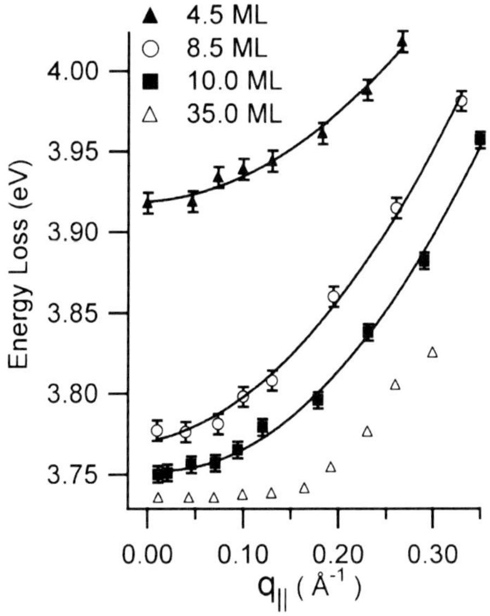

Fig. 3.6 SP dispersion as a function of the parallel transfer momentum for 4.5, 8.5, 10.0, and 35.0 ML Ag/Ni(111) at $T=300$ K. The *solid line* indicates the best-fit curve for experimental data

Hence, the dispersion curve for 10 ML Ag/Ni(111) reported in Fig. 3.5 is purely quadratic, as the linear coefficient B is null. Similar results were obtained for slightly different Ag thicknesses for which QWS were also observed to exist by photoemission spectroscopy [80]. At higher thicknesses at which the discrete electronic structure of QWS [80] is vanishing, the dispersion curve deviates from being purely quadratic. As expected [9, 15, 16, 18, 74, 85, 86], the energy of the SP changed with Ag coverage.

It is worth mentioning that great efforts have been devoted to solve the controversy concerning the quadratic versus linear form of the dispersion curve. The conclusion of this long debate [68, 87–89] is that the dominant coefficient of the SP dispersion for small momenta is always linear. On the other hand, quadratic terms become important at higher values of the parallel transfer momentum $q_{\|}$. However, quadratic terms are nearly absent for Ag(100) [21, 68].

According to Feilbelman's model [90–98] of the SP dispersion, a direct relation between the linear coefficient of the dispersion curve and the position of the centroid of the induced charge, $d_{\perp}$, may be found [9, 99]. In Ag, such centroid lies inside the geometrical surface ($z<0$). The values of the dispersion coefficients and of the calculated position of $d_{\perp}$ for Ag single-crystal surfaces, 10 ML Ag/Ni(111), and 10 ML Ag/Si(111) are reported in Table 3.1. As the linear coefficient B increased, $d_{\perp}$ became more negative. On the contrary, no particular relationship involving the quadratic coefficient C exists.

Feibelman [93, 94] considered the relation between the SP dispersion and the 4d–5s excitations induced by the field of the SP. He argued that the linear coefficient is increased in Ag(100) because the 4d electrons lie closer to the centroid of the oscillating free-electron charge. On the basis of this result, we suggest that for 10 ML

Table 3.1 Best-fit values *A*, *B*, and *C* for different Ag surfaces [21], 10 ML Ag/Ni(111) (our data), and 10 ML Ag/Si(111) [18]

	A	*B*	*C*	$d_\perp$ (Å)
10 ML Ag/Ni(111)	3.751	0.00	1.57	−0.6
Ag(110), <1$\bar{1}$0>	3.683	0.14	3.60	−0.7
10 ML Ag/Si(111)	3.796	0.35	2.56	−0.9
Ag(111)	3.684	0.45	3.40	−1.3
Ag(110), <001>	3.681	0.65	3.10	−1.9
Ag(100)	3.690	1.40	0.06	−2.6

The values of $d_\perp$ were calculated by using (3.4) of [99]. The values for Ag surfaces were reported in the same [99]

Ag/Ni(111) the linear coefficient is null as the centroid of the induced charge of the SP lies much farther with respect to 4d electrons compared to other Ag systems. We ascribe this effect to screening processes [100, 101], enhanced by the existence of Ag QWS [80, 102]. The presence of confined electronic states in the adlayer modifies the shape of the potential barrier that becomes smoother. Very likely, dynamic screening would be more efficient for Ag thin films deposited onto metal substrates with respect to semiconductor surfaces. As a matter of fact, for Ag/Si(111), the linear coefficient of the SP dispersion was decreasing with thickness but it was always different from zero [18]. Moreover, hybridization effects should also be taken into account [103–115]. Their influence should be different for different substrates. Reflection (phase shifts) of the QWS wave functions depends on the interfacial hybridization with the electronic states of the substrate, thus influencing the dynamic screening properties and the electronic response of the Ag/Ni(111) interface.

One advantage of EELS technique is that changing the primary beam energy of the electrons, it is possible to modify their mean free path in the solid and thus their penetration length [116]. This allows direct control of the surface sensitivity. The energy loss position of the Ag SP was found to shift as a function of the primary electron beam energy (Fig. 3.7a). This could be ascribed at the combined effect of the dispersion in the parallel transfer momentum and changes in the penetration length of the primary electrons.

Interestingly, both the Ag SP energy ω_{sp} and the full width at half maximum (FWHM) follow a linear relation as a function of the inverse of the primary beam energy E_{p} (Fig. 3.7b).

$$\omega_{\mathrm{sp}} = a + b*(1/E_{\mathrm{p}}), \tag{3.5}$$

$$\mathrm{FWHM} = c + d*(1/E_{\mathrm{p}}), \tag{3.6}$$

where $a = 3.73$ eV; $b = 4.39$ eV^2; $c = 0.11$ eV; and $d = 3.18$ eV^2.

Increasing the kinetic energy of incident electrons from 20 to 170 eV caused a 200 meV shift of the energy loss position. Furthermore, the FWHM decreased by 140 meV for increasing primary energy. The FWHM of the Ag SP is associated with the damping of the SP, as a result of its decaying into single-particle excitations (Landau damping) through 5sp–5sp or 4d–5sp band transitions.

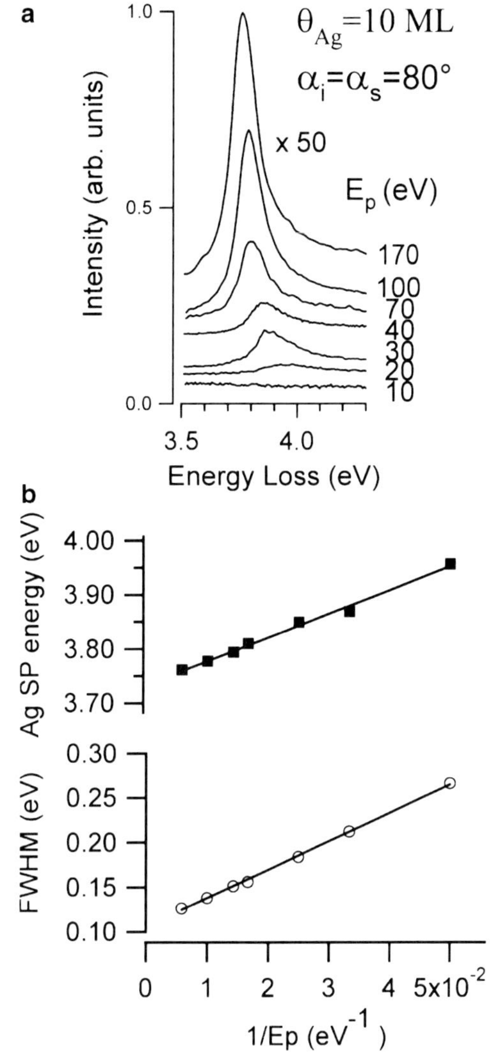

Fig. 3.7 (**a**) HREEL spectra for 10 ML Ag/Ni(111) as a function of the primary electrons beam energy. (**b**) *filled square* Loss energy and *open square* FWHM as a function of the inverse of the kinetic energy of primary electrons

The FWHM of the SP peaks is plotted as a function of Ag SP energy in Fig. 3.8a. The FWHM was found to continuously increase a function of Ag SP energy and no minimum exists (Fig. 3.8a). The loss width notably grows beyond 3.80 eV (empty circles, primary beam energy fixed) indicating the opening of a new efficient decay channel. However, the same plot obtained by changing the impinging beam energy (filled squares) did not present such critical energy. This result was ascribed to effects caused by differences in penetration length of the impinging electrons.

The behavior of the FWHM as a function of the parallel transfer momentum $q_{\parallel}$ (Fig. 3.8b) demonstrated that for $q > 0.2\,\text{Å}^{-1}$, electrons may be promoted from occupied to unoccupied electronic states such that the corresponding plasmon peak would broaden considerably until decaying into the single-particle excitation continuum.

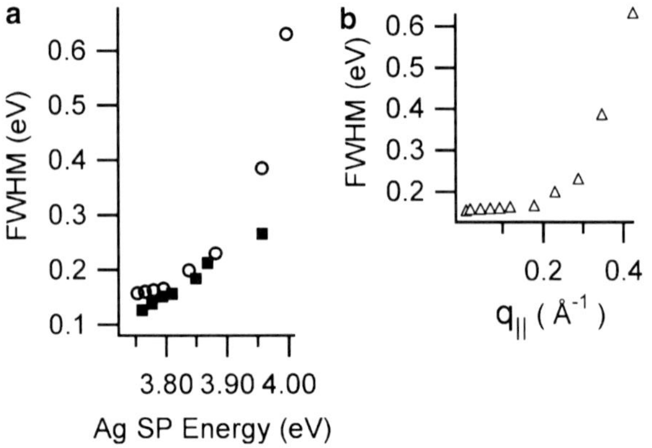

Fig. 3.8 (**a**) FWHM of the Ag SP peak for 10 ML Ag/Ni(111) as a function of the energy of the Ag SP, as obtained in different experimental conditions (*filled square*) $\theta_i=\theta_s=55°$ fixed and E_p varying (*open circle*) $\theta_i=55°$, $E_p=40$ eV fixed and θ_s varying. (**b**) FWHM of the SP peak for 10 ML Ag/Ni(111) as a function of the parallel momentum transfer $q_{\|}$, calculated using with the following experimental parameters: $\theta_i=55°$, $E_p=40$ eV. The abrupt increase of the FWHM beyond $q_{\|}=0.2$ Å^{-1} indicates that an extra decay channel due to indirect single-particle transition between surface electronic states opens up

The width of the SP peak at $q_{\|}=0$ is a sensitive function of the lattice potential, i.e., it is influenced by the periodic potential of the bulk. Such decay mechanisms can be direct or mediated by the exchange of reciprocal lattice vectors or phonons [68]. However, at larger values of $q_{\|}$ the plasmon lifetime becomes less sensitive to the bulk lattice potential and the increasing of the damping is then caused by electron-hole pair excitations in the surface region. In real systems, the lifetime of the plasmon is further limited by the scattering against crystallographic defects. For a thinner film, extra decay channels exist as compared with a thicker film. As a matter of fact, sp–sp interband transitions were found [16, 18, 100] to be more efficient in thinner films rather than in thicker ones.

A comparison of the behavior of the FWHM of the Ag SP in 10 ML Ag/Ni(111), 10 ML Ag/Si(111), and Ag(111) is reported in Fig. 3.9. Panel (a) shows the FWHM dependence on the Ag SP energy, while panel (b) on the parallel transfer momentum. For 10 ML Ag/Si(111), Yu et al. [18] found for different thicknesses, an initial decrease of the FWHM as a function of the Ag SP energy, followed by an abrupt increase. Such an abrupt increase was ascribed to the opening of a new damping channel, i.e., intraband transitions between Ag 5sp-derived QWS. The initial decrease of FWHM was associated with the increased surface barrier due to the ionic pseudopotential of the crystal. On the contrary, the FWHM of Ag SP in 10 ML Ag/Ni(111) has an initial flat dependence on both Ag SP energy and $q_{\|}$ and a stronger dependence beyond a critical value of both Ag SP energy and $q_{\|}$. The notable difference existing in the behavior of the FWHM of Ag SP in 10 ML Ag deposited on Si(111) and on Ni(111) may be tentatively ascribed to substrate effects. The FWHM of the Ag SP in Ag(111) [21] has, in analogy with 10 ML Ag/Ni(111), an initial flat dependence on both Ag SP energy and $q_{\|}$. The critical values of Ag SP

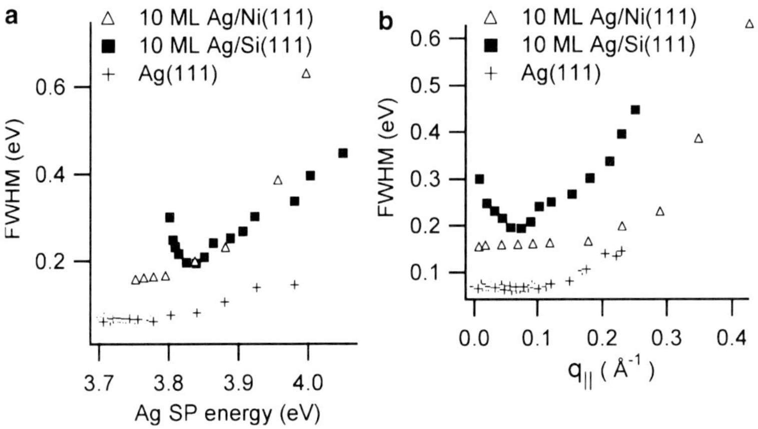

Fig. 3.9 (**a**) FWHM of the Ag SP peak as a function of the energy of the Ag SP for (*open triangle*) 10 ML Ag/Ni(111); (*filled square*) 10 ML Ag/Si(111) (data taken from [18]); and (*plus*) Ag(111) (data taken from [21]). (**b**) FWHM of the SP peak as a function of the parallel momentum transfer $q_{\|}$ for (*open triangle*) 10 ML Ag/Ni(111); (*filled square*) 10 ML Ag/Si(111) (data taken from [18]); and (*plus*) Ag(111) (data taken from [21])

and $q_{\|}$ were found to be 3.78 eV and 0.15 Å^{-1}, respectively. Beyond such values, the FWHM significantly increased. It is worth stressing that the FWHM of the Ag SP in Ag(111) was always notably lower than that of Ag SP in thin films. This finding is ascribed to enhanced damping processes via 5sp–5sp indirect transitions, due to the different electronic properties of thin films with respect to surfaces.

If the surface is exposed to chemically reactive atoms or molecules, the distribution of occupied and unoccupied electronic states changes. Accordingly, differences in damping processes and in the energy of the plasmonic excitation are quite expected. Rather than considering the new overlayer plasmon, we investigated the overlayer-induced modification of the substrate plasmon. Figure 3.10a shows Ag surface excitation spectra for increasing amounts of adsorbed K. At low K coverages, the Ag SP peak is only weakly affected. As the K coverage approaches one monolayer, the FWHM of the SP increased from 0.17 (clean Ag layers) up to 0.70 eV. Moreover, upon K adsorption the plasma energy of the Ag SP shifted from 3.80 down to 3.56 eV. The red-shift of the Ag SP energy may be ascribed to a charge transfer from the Ag substrate to the adsorbates. A reduced Ag SP frequency in the presence of electronegative coadsorbates was reported also for Ag single-crystal surfaces [14, 22]. A similar red-shift of the SP was revealed also for K/Ag(1 1 0) [117] and, hence, we can suggest that it is K-derived.

It is more correct to describe the plasmonic excitation as the K/Ag interface plasmon rather than a red-shifted Ag SP. New adsorbate-induced electronic states arose at the interface. Hence, the significant plasmon broadening is due to new channels for decay into electron-hole pairs at the K/Ag interface. The interband transitions involving the overlayer-induced band below the Fermi level thus influences the SP energy and gives rise to a red-shift of the SP energy.

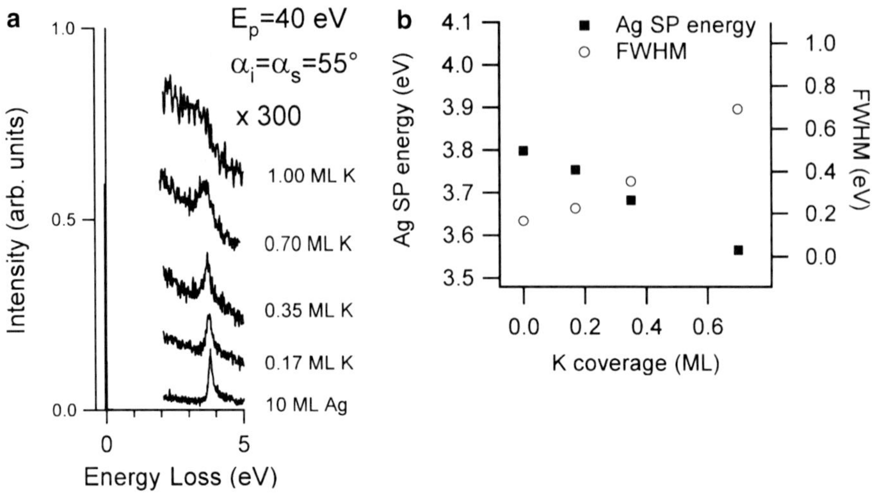

Fig. 3.10 (**a**) HREEL spectra for K/10 ML Ag/Ni(111) as a function of K exposure. (**b**) (*filled square*) Ag SP energy and (*open circle*) FWHM as a function of K coverage

Such SP broadening is ascribed to chemical interface damping. It was observed in absorption spectra of small metal particles embedded in a reactive matrix [118, 119], but it is the first experimental evidence of its occurrence in thin films.

4.2 Ag/Cu(111)

In Ag/Cu(111), silver islands appear for Ag coverages higher than two layers, which instead grow layer-by-layer. Around each spot of the p(1 × 1)-Ag, LEED measurements showed also the occurrence of a (9.5×9.5) reconstruction for the first Ag layer and of a (9×9) reconstruction in correspondence of θAg = 2.0 ML, in excellent agreement with previous structural studies on this system [120, 121].

Hence, this system offers the opportunity to study the dependence of dispersion and damping dispersion of the SP as a function of the growth mode.

Density-functional theory calculations based on s–d polarization model found an initial negative dispersion for two layers of Ag deposited on Al surfaces [66]. By contrast, experiments [18] carried out on Si(111) reported a positive behavior. It would be extremely useful to study dynamic screening processes in the case of a thickness of two layers. In fact, two layers are commonly accepted as a borderline between interface physics and thin-films physics [73, 79, 122–124].

As a matter of fact, such coverage constitutes the minimal thickness necessary to observe in the loss spectra well-distinct features assignable to collective excitations, while the very broad loss features observed from 0 to 2 ML are related to single-particle transitions [125].

However, experimental measurements on Ag nanoscale thin films deposited onto metallic substrates which could verify the effects of s–d screening have not yet been

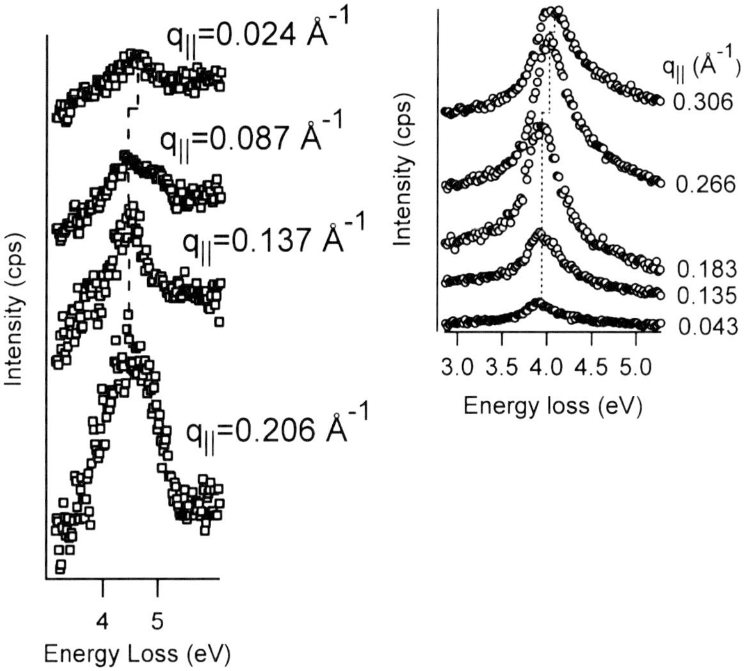

Fig. 3.11 HREEL spectra as a function of the scattering angle for 2 (*left*) and 5 ML Ag/Cu(111) at $T = 300$ K

performed. On the other hand, the collective excitations in two alkali layers on metal substrates were extensively studied [73, 79, 122–124, 126] as significant differences compared to thick films were found [103]. Contradictory results about the existence of the SP at small momenta were reported [81, 122, 123, 127]. The investigation of the behavior of collective excitations also for two Ag layers on metallic substrates should provide a significant advancement in understanding dynamic screening processes at metal surfaces. Moreover, a shifted bulk plasmon (BP) is expected to exist at small momenta, where the SP weight is vanishing [66].

Selected HREEL spectra for 2 ML of Ag as a function of the parallel transfer momentum are shown in Fig. 3.11 (left panel).

In contrast with all previous experimental works on Ag surfaces [20, 64, 66, 88, 99, 104–108] and layer-by-layer Ag films on Si(111) [18], the frequency of the SP did not increase as a function of the parallel transfer momentum (Fig. 3.12).

Present results well agree with theoretical calculations by Liebsch [66]. As the thickness of the Ag film is reduced, the overall screening of the charge associated to the SP via the polarizable d electronic medium diminishes, giving higher plasmon energy. Furthermore, the spill-out region not affected by s–d polarization becomes more important, causing a further blue-shift of the plasmon frequency. Moreover, the differences that we found with respect to measurements on 2.5 ML Ag/Si(111) [18] were ascribed to the enhanced screening properties of metal substrates compared to semiconductor surfaces and to more efficient screening processes due to

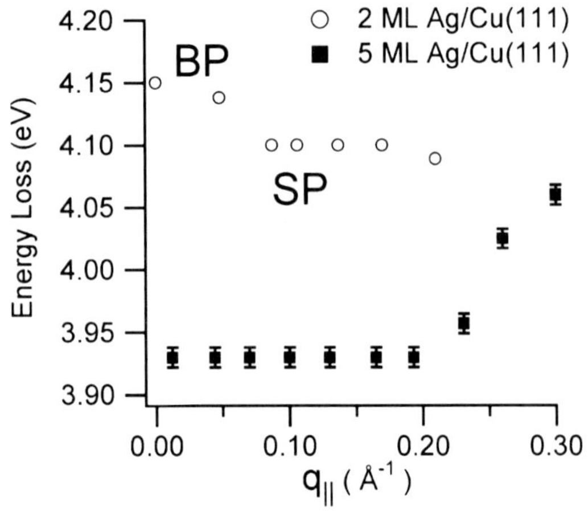

Fig. 3.12 SP dispersion as a function of the parallel transfer momentum for 2 ML and 5 ML of Ag on Cu(111) at $T=300$ K

electron quantum confinement [109, 110]. Interestingly, for two Ag layers, in the limit of small momenta, we observed the excitation of the BP, shifted by s–d screening at a frequency $\omega_p^* = \omega_p / \sqrt{\varepsilon_d}$, where ε_d is the local dielectric function and ω_p is the s–p BP frequency [66]. The real part of the dielectric function decreases as the film thickness is reduced, as a direct consequence of the occurrence of a less sharp onset of transitions involving d states in thin films compared to bulk Ag [66]. As a matter of fact, a significant energy step between small and high momenta exists due to the different nature of the excitation, i.e., BP and SP, respectively. The lack of the SP excitation at $q_{||}=0$ was reported also for ultrathin alkali layers [111, 112, 125]. On the contrary, in ultrathin Ag layers on Si(111) [18] the SP was excited even at small momenta. Once again, such evidence should be taken as a fingerprint of very different screening processes between Ag/Si and Ag/Cu. However, increasing Ag coverage, i.e., 5 ML (right panel of Figs. 3.11 and 3.12), the SP was excited also at small values of $q_{||}$, as generally found for thick alkali layers [113] and Ag/Si(111) [18]. In Fig. 3.13, we compare the SP dispersion for different Ag systems. The behavior found for two Ag layers on metal substrates differs from the behavior found in all other Ag systems exhibiting, instead, a positive and quadratic dispersion. As a matter of fact, the dispersion curve calculated by Liebsch [66] for 2-ML Ag/Al well agree with present results for Ag on Cu(111).

The granularity of the as-deposited film for coverages higher than two layers is argued from the behavior of the dispersion of the collective excitation (Fig. 3.14). The absence of dispersion below a critical wave vector, i.e., 0.15 A^{-1}, indicates that the s electrons oscillate independently in the single (111)-oriented grains. Similar results were reported for Ag/Si(111) [76]. The critical wave vector was suggested [76] to be related to the average island size through the relation

$$\mathrm{Qc} = 2\pi / d. \tag{3.7}$$

From the above equation, we estimate that the grain size in Ag/Cu(111) is about 30–40 Å.

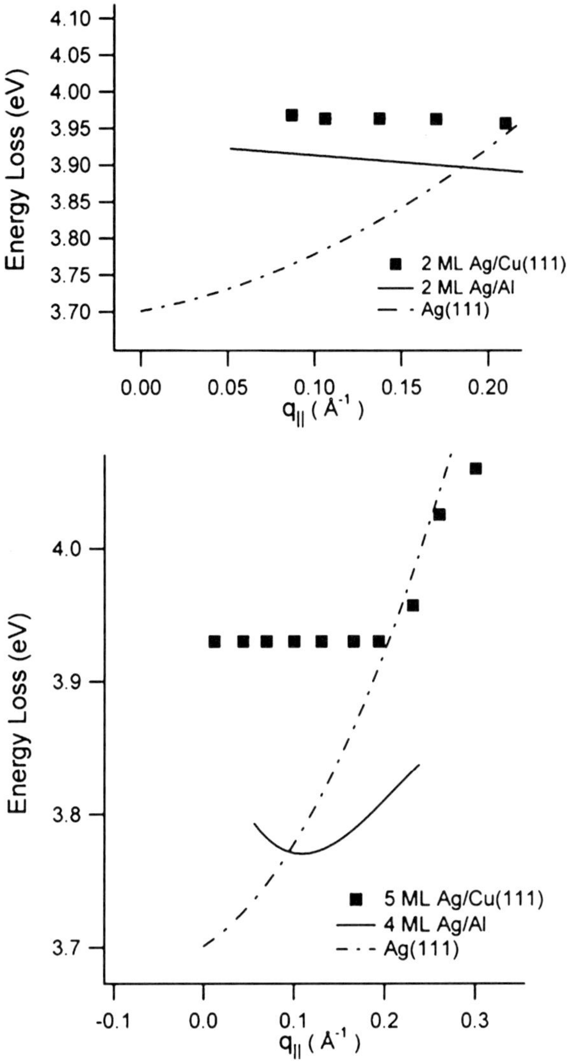

Fig. 3.13 Behavior of the dispersion curve of the Ag SP for Ag(111) (*dashed-dotted line*) [21], 2 ML Ag/Cu(111) (*filled squares* [72]), 2 ML Ag/Al (theory, *continuous line*) [66], and 2.5 ML Ag/Si(111) (*dotted line*) [18]. The values of the SP energy for the various dispersion curves were normalized to their respective SP energies measured at smallest value of $q_{\|}$ in order to put in evidence the different behavior

The propagation of the SP can occur only for modes whose wavelength is smaller than the diameter of the single grain. Interestingly, the Ag grains behave like isolated clusters with respect to the plasmonic excitation.

The behavior of the SP dispersion well agrees with the Stranski–Krastanov growth mode of this system [57]. The increasing strain caused by adsorbed layers destabilizes the film and induces clustering [114]. This behavior arises from the large lattice mismatch between Ag and Cu (13%). LEED experiments showed the occurrence of a 9×9 over-structure of the deposited silver film, as previously reported [115].

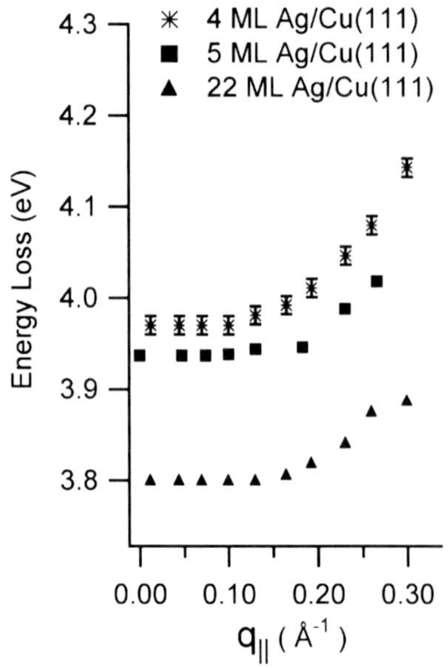

Fig. 3.14 Dispersion of the SP energy for different Ag coverages. The lack of a dispersion before a critical wave-vector indicates that the SP is confined within Ag grains

It is worth noticing that we have no evidences for the existence of Mie plasmons within Ag islands. Two well-distinct Mie plasmons at 3.1 and 3.9 eV were revealed only for Ag deposited on metal-oxide surfaces [78, 79] and not for three-dimensional islands on metal [15, 74] and semiconductor [76] surfaces. As concerns metal/metal interfaces, the occurrence of Mie plasmons was invoked only for Na quantum dots on Cu(111) [4, 128] but only for a very restricted alkali thickness range. Mie plasmon merged into the ordinary SP already for two nominal Na layers.

In order to remove the SP confinement, an annealing of the film at 400 K was performed. Loss spectra in Fig. 3.15 provided evidences for drastic morphological changes in the film and a higher degree of ordering, as suggested by the analysis of the LEED pattern.

Annealing the surface at 400 K caused significant changes in the dispersion curve and, in particular, the loss of the SP confinement. The measured dispersion curve $E_{\mathrm{loss}}(q_{\|})$ of the annealed film, reported in Fig. 3.16, was fitted by a second-order polynomial given by:

$$E_{\mathrm{loss}}(q_{\|}) = A + Bq_{\|} + Cq_{\|}^{2}. \tag{3.8}$$

($A = 3.791 \pm 0.006$ eV; $B = -0.60 \pm 0.09$ eV Å; and $C = 3.4 \pm 0.3$ eV Å^2).

The linear coefficient was found to be slightly negative. This finding was ascribed to the enhanced sp density of states existing in thin Ag films, as a direct consequence of the presence of QWS. Increasing the free-electron character of the QWS by annealing [58] should imply the occurrence of a negative linear term.

Nonetheless, the value of the linear coefficient is still enough higher than the linear coefficient of the SP dispersion curve of alkalis [129], aluminum [23, 130, 131], or alkaline-earth metals [132]. Such finding leads us to suggest that the centroid of the induced charge lies in the close vicinity of the jellium edge [93–95, 98, 129],

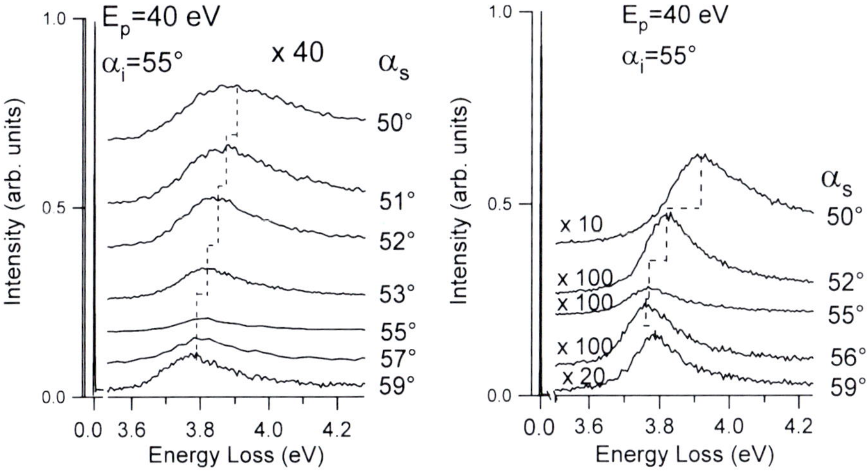

Fig. 3.15 HREEL spectra for as-deposited (*left panel*) and annealed (*right panel*) 22 ML Ag/Cu(111)

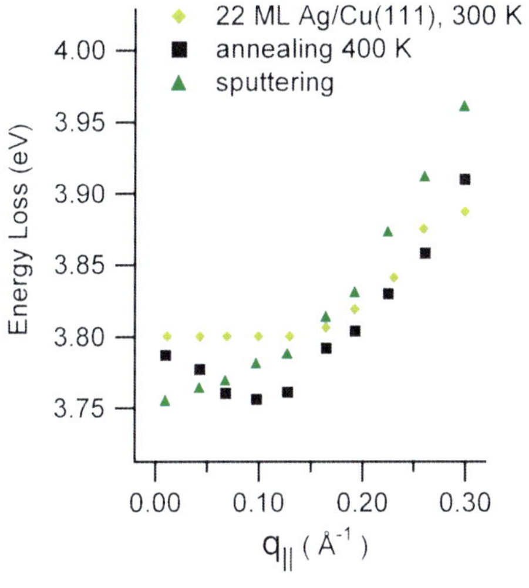

Fig. 3.16 Behavior of the SP dispersion curve

in contrast with all other Ag systems [9, 17, 68, 108, 133], but not outside as for simple metals. Interestingly, the quadratic coefficient coincides with that of SP dispersion in Ag(111) [21], i.e., the surface with the same crystallographic orientation. Significant differences exist between spectra acquired for annealed (Fig. 3.15) and sputtered films (Fig. 3.17).

The dispersion curve measured in a sputtered Ag film reported in Fig. 3.16 shows that the quadratic term is predominant:

($A = 3.760 \pm 0.004$ eV; $B = -0.08 \pm 0.06$ eV Å; and $C = 2.5 \pm 0.2$ eV Å^2).

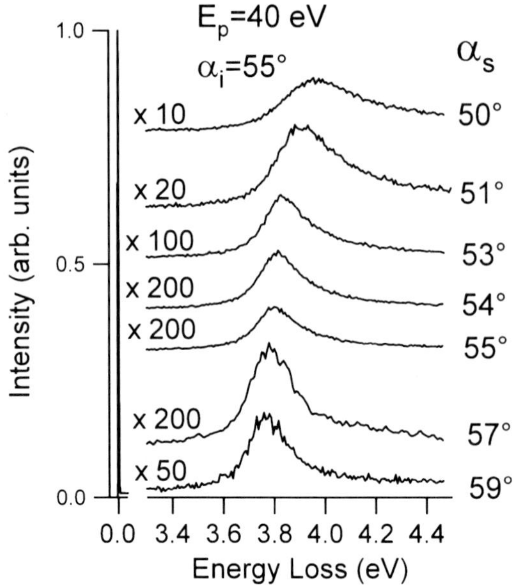

Fig. 3.17 HREEL spectra for a sputtered 22 ML Ag/Cu(111) surface

Accordingly, an increased linear coefficient and a decreased quadratic term were obtained by fit procedure. Contrary to the sputtered Ag(100) surface [19] for the quadratic term, the bulk value of 6 eV Å^2 [122] was not recovered. In our opinion, the link proposed in [19] between the value of the quadratic term of the SP dispersion and that of the bulk plasmon, related to bulk properties, should be revised.

The occurrence of an increased linear coefficient suggests that sputtering induces a significant shift of the position of the centroid of the induced charge associated to the SP compared with that of SP in annealed films.

It is worth mentioning that ion bombardment of a growing film was found to produce both bombardment-induced segregation normal to the film surface and an advancing nanoscale subsurface diffusion zone [123]. Such phenomena should be considered in theoretical studies on the electronic response of sputtered thin films (still lacking). Moreover, our results provide the grounds for angle-resolved photoemission experiments shedding light on the sputtering-induced modifications of the QWS.

4.2.1 Damping Processes of the SP in Ag/Cu(111)

Figure 3.18 (left panel) shows the thickness-dependence of the FWHM as a function of the parallel momentum transfer for Ag/Cu(111). An initial negative behavior of the FWHM was found for ultrathin films, while for higher coverages (22 ML) only a poor dispersion was found (Fig. 3.16). A comparison among the dispersion of the SP line-width for 5 ML Ag/Cu(111), 5 ML Ag/Si(111) [18], and Ag(111) [21] shows (Fig. 3.18, right panel) that for Ag films a critical wave-vector beyond which the dispersion became positive exists, while for the single-crystal surfaces the initial

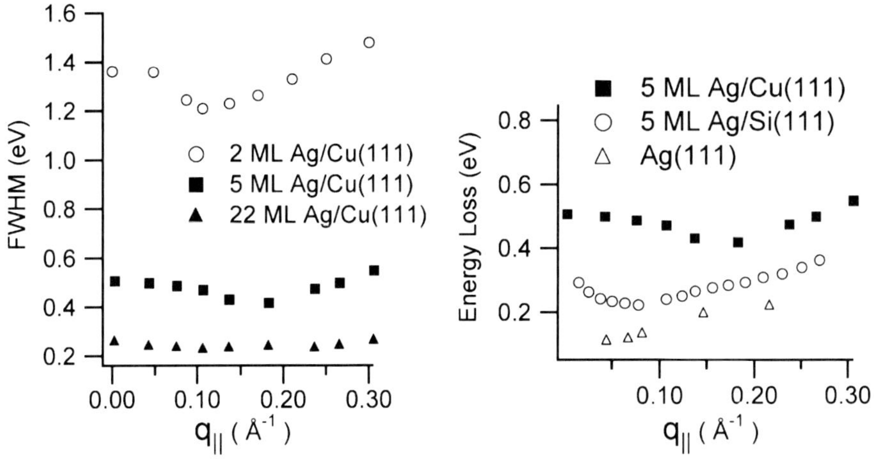

Fig. 3.18 (*left*) Behavior of the FWHM of the Ag SP as a function of Ag thickness on Cu(111). (*right*) Comparison between the dispersion of the FWHM of the Ag SP for different Ag system

dispersion of the FWHM is nearly flat. However, the value of the turning point differs in 5 ML deposited on Cu(111) and Si(111) (0.19 and 0.08 Å^{-1}, respectively).

This behavior of the FHMM is well described by a theoretical model recently proposed [124] on plasmon lifetime in free-standing Ag layers. The behavior of line-width of the Ag SP as a function of the parallel momentum transfer was found to be characterized by a negative behavior of the line-width for small momenta up to a critical wave-vector. This finding was ascribed [124] to the splitting between symmetric and antisymmetric excitation modes and the enhanced electron-hole pair excitation at small $q_{\|}$. For higher values of $q_{\|}$, a linear increase of the line-width was reported.

For the as-deposited Ag film, the initial dispersion of the FWHM is negative (Fig. 3.19). The behavior became positive after a critical wave-vector (0.10 Å^{-1}) This finding is in agreement with results for Ag/Si(111) [18] and recent theoretical calculations for free-standing Ag slabs [124]. However, the behavior of the FHWM for single-crystal Ag surfaces is positive [68] and was recovered by annealing the Ag film. It is worth noticing that in all cases (as-deposited, annealed, and sputtered Ag film), the value of the FWHM is higher than for Ag semi-infinite media (for Ag(111) the FWHM at small momenta is 69 meV [21]). As discussed above, the different distribution of occupied and unoccupied electronic states in thin films with respect to surfaces would imply enhanced damping mechanisms for collective excitations by creating electron-hole pairs.

The different behavior with respect to the case of Ag/Ni(111) should be ascribed to differences in both the growth mode and in the nature of QWS. In fact, due to the absence of a gap in Ni(111), the character of the QWS in such two systems differs substantially [102]. QWSs on Ag/Cu(111) are characterized by standing wave patterns, not supported on Ag/Ni(111). Furthermore, on Ag/Cu(111) the interfacial transmittivity is suppressed, with an enhanced specular reflectivity. The opposite occurs for Ag on Ni(111). Such significant differences in the electronic properties

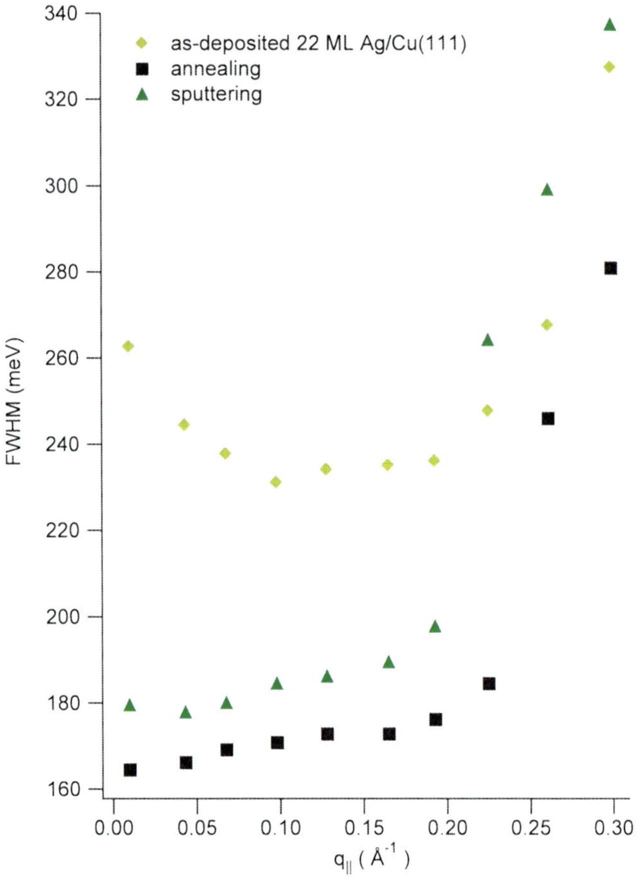

Fig. 3.19 Behavior of the FWHM for as-deposited, annealed, and sputtered Ag films (22 ML) on Cu(111)

between these two bimetallic surfaces should in principle imply quite different Landau damping processes of the plasmonic excitation.

It is worth noticing that sputtering induces a broadening of the SP line-width [74, 126]. The SP peak became progressively asymmetric, as indicated by the increasing of the skewness, i.e., the third standardized momentum [127, 134], upon sputtering.

4.3 Na/Cu(111)

The (111) surface of noble metals (Cu, Ag, Au) exhibits a large confined gap within the projected bulk band structure centered at the $\overline{\Gamma}$ point of the surface Brillouin zone [30, 39, 135, 136]. The adsorption of Na on Cu(111) has been extensively

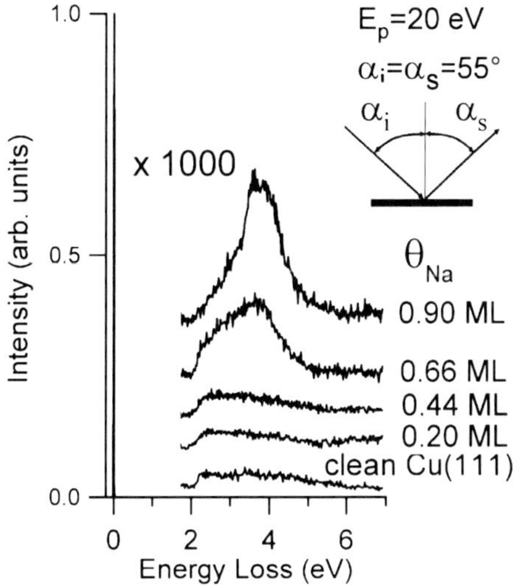

Fig. 3.20 HREEL spectra of Na/Cu(111) as a function of the alkali coverage. The incident beam energy E_p was 20 eV

investigated recently. The existence of Na $3p_z$-derived confined electron states has been well established [120, 140–144]. Multilayers of alkali metals are easily grown on Cu(111) even at room temperature (RT), while in other systems the growth of a second alkali layer is possible only at liquid nitrogen temperature [137]. A study reporting the dispersion curve of the SP in alkali thin layers at RT is still lacking.

HREEL spectra of Na adsorbed at RT on Cu(111) as a function of coverage are shown in Fig. 3.20. For the clean Cu(111) surface, the onset of collective excitations of valence electrons could be detected at 2.1 eV [4, 84, 138]. A broad loss feature, peaked at 3.70 eV and assigned to the Na SP, gradually arose in the spectrum as a function of Na coverage.

Loss spectra in Fig. 3.21 show the SP of two layers of Na/Cu(111) as a function of the scattering angle for $E_p = 20$ eV. The spectrum recorded in the specular geometry is centered at 3.70 eV and it exhibits clear energy dispersion for off-specular angles.

As mentioned above, changing the primary beam energy of the electrons, it is possible to modify their mean free path in the solid and so their penetration length [116]. This allows a direct control of the surface sensitivity. Moreover, the modification of the electron penetration upon changing the impinging energy would imply, according to random-phase approximation calculations [139], a change in the position of the reflection plane at which probing electrons are scattered within the extended electron-density distribution. On the basis of the above result, a strong dependence of dynamic screening processes on impinging energy is expected especially for systems in which electrons are confined into a two-dimensional space. Hence, spectra were acquired also for a higher primary energy, i.e., 100 eV, in order to reveal such behavior.

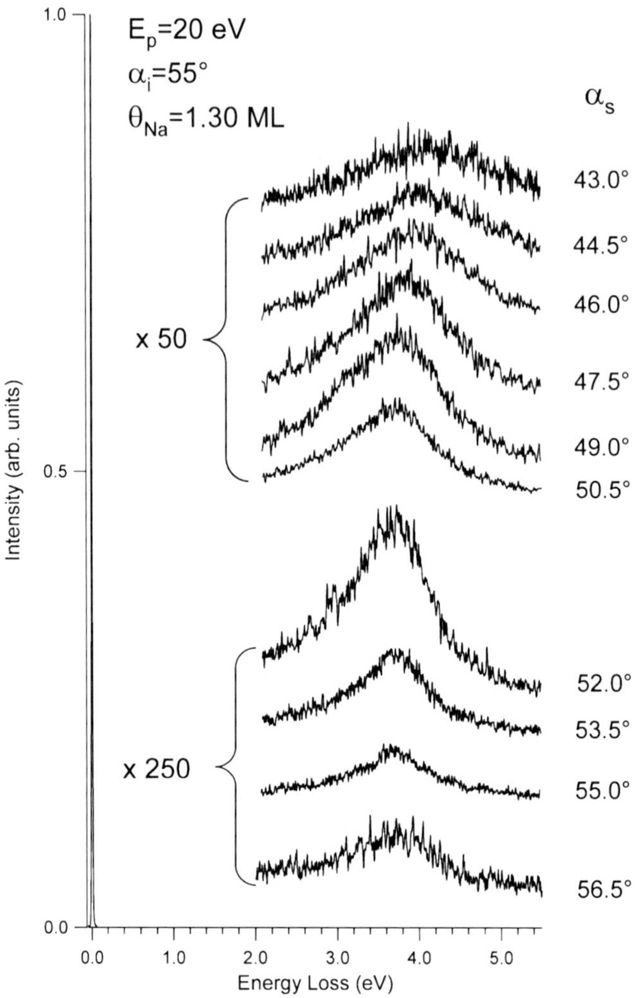

Fig. 3.21 Electron energy loss spectra of 0.90 ML Na/Cu(111) at different scattering angles θ_s. The incident beam energy E_p was held constant at 20 eV and all spectra were recorded at an incident angle of $\theta_i = 55°$ with respect to the sample normal

Multipole SP in Na/Cu(111) became quite evident in spectra taken with $E_p = 100$ eV behind 3 degrees off-specular (Fig. 3.22). Its energy was found to be 4.70 eV, in excellent agreement with the value reported for bulk Na by Tsuei et al. [140], i.e., 4.67 eV. However, the multipole mode was not revealed for lower impinging electron beam energies, thus suggesting the existence of threshold primary beam energy.

The measured dispersion curve $E_{loss}(q_{\|})$ for $E_p = 20$ eV reported in Fig. 3.23 was fitted by a fourth-order polynomial [68] given by:

$$E_{loss}(q_{\|}) = a + bq_{\|} + cq_{\|}^2 + dq_{\|}^3 + eq_{\|}^4, \tag{3.9}$$

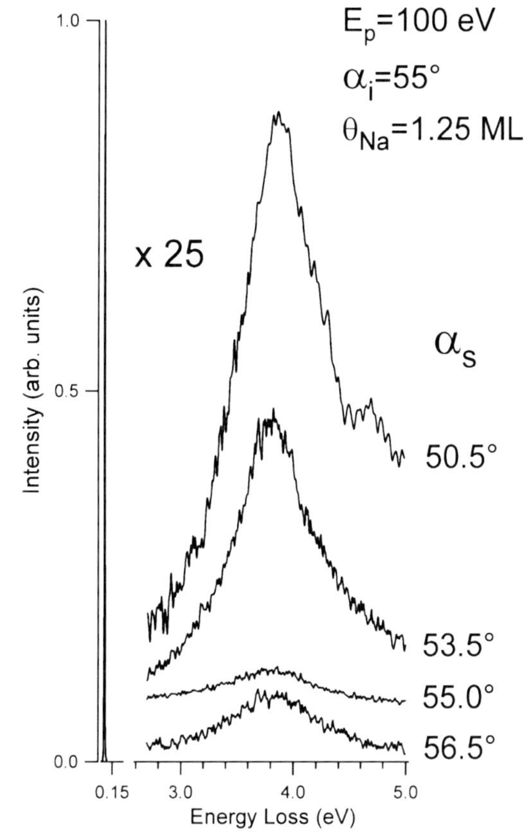

Fig. 3.22 Electron energy loss spectra of 0.90 ML Na/Cu(111) at different scattering angles θ_s, recorded with a primary electron beam energy of 100 eV

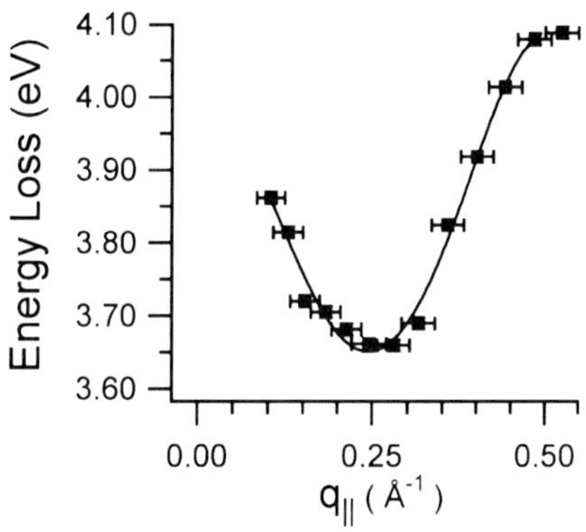

Fig. 3.23 Na SP energy as a function of $q_{\|}$ ($E_p = 20$ eV)

where $a = 3.99$ eV, $b = 0.70$ eV Å, $c = -9.53$ eV Å^2, $d = 113.14$ eV Å^3, and $e = -112.31$ eV Å^4.

The slope of the dispersion curve is negative up to 0.25 Å^{-1}, then the loss energy of the SP increases with increasing $q_{\|}$ and the dispersion becomes definitively

Fig. 3.24 Na SP energy as a function of $q_{\|}$ ($E_p = 100$ eV)

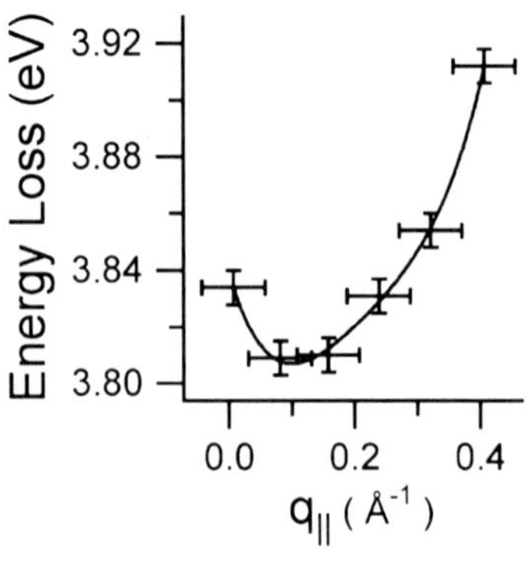

positive. It should be noticed that the SP was not damped into single-particle transitions until $q_{\|} = 0.53$ Å^{-1}, while the critical wave-vector of alkali SPs was found to be around 0.30 Å^{-1} both in thin [111, 125] and thick [129] alkali films.

Furthermore, no loss features were revealed at the lowest values of parallel momentum transfer (below 0.10 Å^{-1}). A similar result was found by Zielasek et al. [112] in Cs/Si(111)-(7 × 7) in which the SP was suppressed for small wave vector. These findings are in agreement with Liebsch's calculations [66] predicting that for two layers of alkali metals the spectral weight of the SP is vanishing at $q_{\|} = 0$. In other words, at $q_{\|} = 0$ only the excitation of the so-called multipole SP is possible. In this regard, we notice that the Na/Cu(111) system behaves differently from K/Ni(111) [125] for which the excitation of the SP was found to be possible.

The energy dispersion of the SP of two Na layers was 430 meV, while it was 130 meV for thick Na films. Such finding could be explained by screening effects that push the position of the induced charge density centroid more outside the substrate jellium edge than in thick Na films.

We propose that electron quantum confinement in Na QWS may be responsible for such result. In fact, the electron confinement of Na electrons on Cu(111) was found [100, 101] to drastically change the dynamical screening properties of this system. The occurrence of electron confinement in Na QWS, moreover, modifies the electron charge-density distribution. Hence, the reflection of external charges at different distances from the surface [139] should affect the electronic response of this system.

The dispersion curve of the SP, measured using a primary electron beam of 100 eV (Fig. 3.24) is quite different from that in Fig. 3.23. The energy dispersion of the SP is lower (70 meV vs. 430 meV). The measured dispersion curve $E_{loss}(q_{\|})$ for $E_p = 100$ eV was fitted by a fourth-order polynomial [68] given by:

$$E_{loss}(q_{\|}) = a + bq_{\|} + cq_{\|}^2 + dq_{\|}^3 + eq_{\|}^4, \tag{3.10}$$

where $a = 3.84$ eV, $b = -0.79$ eV Å, $c = 6.22$ eV Å^2, $d = -17.98$ eV Å^3, and $e = -20.99$ eV Å^4.

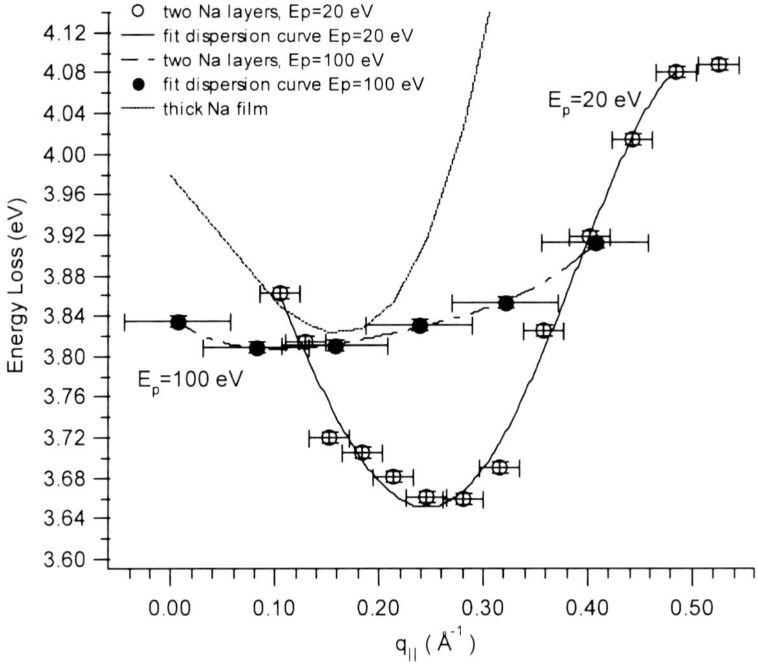

Fig. 3.25 The SP dispersion versus the parallel momentum $q_{\parallel}$ for: (*filled square*) two Na layers, $E_p = 100$ eV; (*open circle*) two Na layers, $E_p = 20$ eV; (continuous line) thick Na film (data taken from [57, 129])

It showed an initial negative behavior that became positive above 0.15 Å^{-1}, in analogy with the dispersion curve of SPs of other simple metals.

It should be noticed that for $E_p = 100$ eV, the SP could be excited even at small momenta. This is a combined effect of the higher dipole/impact scattering ratio of the excitation (see Fig. 3.29 and its discussion) and of the increased integration window [68] in reciprocal space. The latter implies the collection of a finite range of momenta of the outgoing electron.

A dispersion curve not dependent on the primary electron beam energy is expected only for the case of pure sheet plasmons [145–150], i.e., two-dimensional plasmons strictly confined to the surface and with a square root-like dependence on $q_{\parallel}$. However, HREELS measurements revealed that the dispersion curve depends on the impinging energy. When the thickness of the film gets smaller, the tail of the wave function of the QWS can reach and interact with the substrate. The main effect is a hybridization interaction between overlayer and substrate states. We suggest that the occurrence of such interactions between overlayer states and substrate ones may lead to the observed differences in the two dispersion curves.

Figure 3.25 shows a comparison between the dispersion curve obtained for two layers of Na on Cu(111) and the dispersion curve of the SP of a thick Na film [129]. The plasmon energy for a nanoscale thin Na film is lower because screening effects

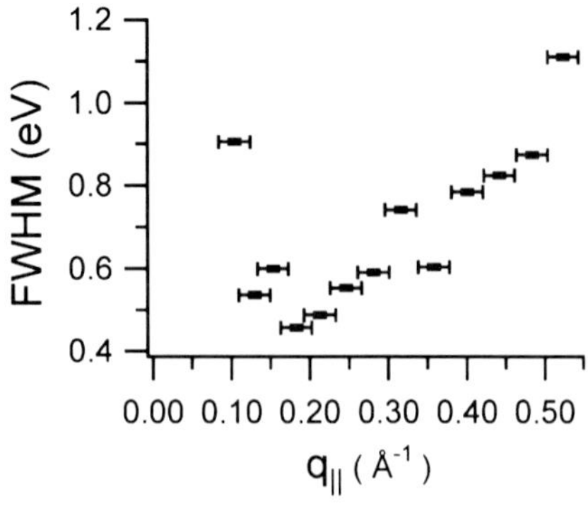

Fig. 3.26 FWHM of the Na SP peak as a function of $q_{\|}$ ($E_p = 20$ eV)

make lower the induced average charge-density and thus the SP energy, as found in two layers of K on Ni(111) [125]. The differences in the critical wave-vector of the SP for each dispersion curve are quite evident, thus suggesting strong differences about damping processes.

The damping of the plasmon excitation is clearly revealed by the trend of the FWHM versus $q_{\|}$, as shown in Fig. 3.26 for $E_p = 20$ eV. The width of the Na plasmon initially decreased, followed by a steep increase as a function of $q_{\|}$. Experimentally, a similar behavior was also observed for a variety of metal surfaces such as Ag surfaces, Mg(0001) [132], Al(111) [23], and Cu(111) [84]. On the contrary, the FWHM of the SP in thick films of Na and K [129] increases for increasing $q_{\|}$, while it was almost constant in two layers of K on Ni(111) [125].

Existing theories [69] predict that, with increasing of the momentum $q_{\|}$ parallel to the surface, the width of the SP rapidly increases due to the decay into electron-hole pairs (Landau damping). The behavior shown in Fig. 3.26 could arise from a more efficient disexcitation channel of the SP by single-particle transitions at small momenta. Such assumption could explain a so high critical wave-vector (0.53 $Å^{-1}$) of the plasmonic excitation too.

The FWHM of the SP for $E_p = 100$ eV was characterized by a negative dispersion vs. $q_{\|}$ (Fig. 3.27). To the best of our knowledge, it is the first time that an FWHM was found to have a similar behavior which disagrees with existing theories. Such finding suggests a strong dependence of the dynamical response of electrons and of screening effects as a function of the impinging energy. Interestingly, the FWHM for thick Na films [129] has the opposite trend as a function of the parallel momentum transfer. This experimental result is a clear evidence that understanding of the broadening mechanisms of an SP requires a careful analysis of the band structure of the system.

The intensity of the SP versus the off-specular angle for $E_p = 20$ eV (Fig. 3.28) clearly demonstrates that such collective mode was excited by impact mechanism because it is peaked at six degrees off-specular [68], while a dipolar loss would have the same behavior of the elastic peak as a function of the off-specular angle. Instead, the intensity of the SP for $E_p = 100$ eV exhibited a maximum around 1.5 degrees

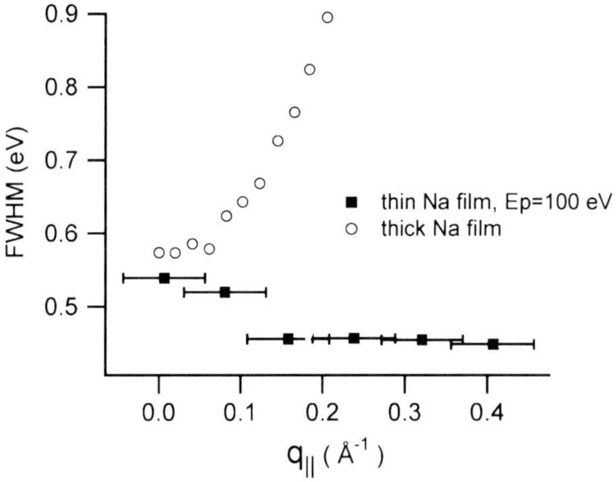

Fig. 3.27 FWHM of the Na SP peak as a function of $q_{\parallel}$ for (*filled square*) 0.90 ML Na/Cu(111), $E_p = 100$ eV and for (*open circle*) thick Na film (data taken from [129])

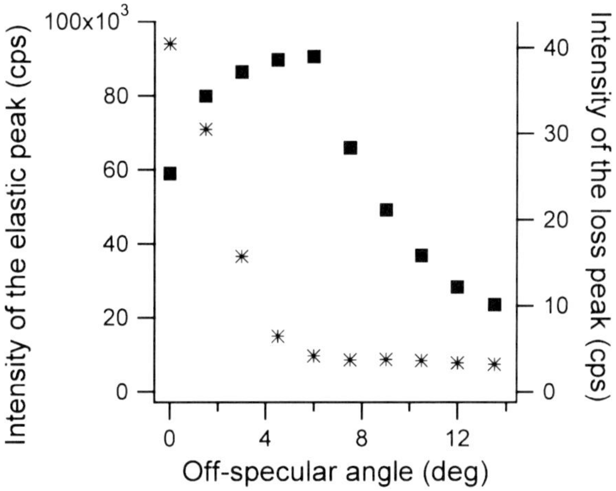

Fig. 3.28 Intensity of the SP (*filled square*) and of the elastic peak (*asterisk*) as a function of the off-specular angle ($E_p = 20$ eV)

off-specular (Fig. 3.29). Thus, the SP excited by a higher primary beam energy has a substantially dipole character while the same mode with a lower impinging energy exhibits a notable impact character. The dependence of the dipole/impact scattering ratio in SP excitation as a function of primary beam energy is still an unexplored research field.

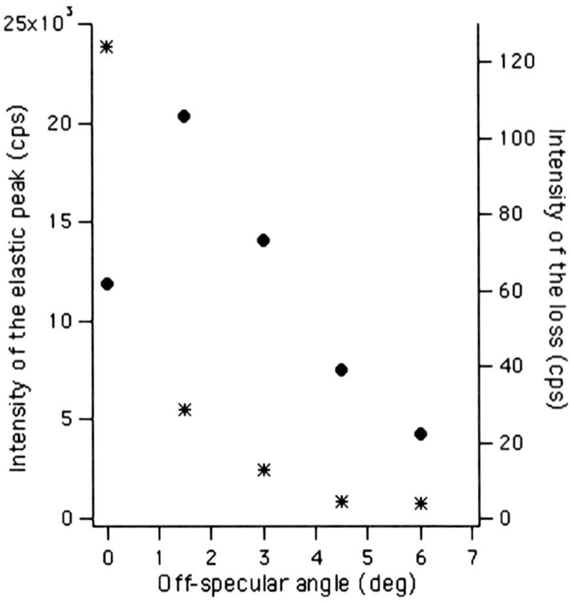

Fig. 3.29 Intensity of the SP (*filled square*) and of the elastic peak (*asterisk*) as a function of the off-specular angle ($E_p = 100$ eV)

5 Conclusions

In conclusion, measurements carried out on Ag layer-by-layer films grown on Ni(111) provided evidences of the relationship between the existence on Ag QWS and the linear term of the dispersion curve, which is null in contrast with all other Ag systems.

Similar loss measurements for Ag/Cu(111), characterized by a Stranski–Krastanow growth mode, revealed SP confinement within islands in as-deposited Ag layers. Annealing caused an enhanced free-electron density of states of the Ag QWS, which renders negative the linear coefficient of the dispersion relation.

The increased value of the FWHM in Ag films compared with Ag single-crystal surfaces suggest the occurrence of enhanced damping mechanism due to the opening of new decay channels of the SP in systems exhibiting QWS.

For two layers of Na on Cu(111), the SP was not excited at small momenta, also found for two layers of Ag on the same substrate. A strong dependence of the dispersion curve on the energy of the primary beam was observed. The FWHM of two Na layers behaves very differently with respect to that reported for a Na thick film. These differences were ascribed to different Landau damping mechanisms of the plasmonic mode in the two cases.

Screening effects enhanced by electron quantum confinement and interactions between overlayer states and substrate states are suggested to be at the basis of these results. Such findings provide the grounds for theoretical studies aimed at characterizing of the nature and dispersion of the excitation modes in nanoscale thin films exhibiting confined electron states.

Acknowledgments We want to thank Dr. Stefan Mathias and Prof. Michael Bauer for having allowed to use their photoemission data, and, moreover, dr. Vincenzo Formoso for many helpful discussions.

References

1. Valla T, Kralj M, Siber A, Milun M, Pervan P, Johnson PD, Woodruff DP (2000) Oscillatory electron-phonon coupling in ultra-thin silver films on V(100). *Journal of Physics: Condensed Matter*, 12 (28), L477–L482
2. Politano A, Formoso V, Chiarello G (2009) Chemical Reactions at Clean and Alkali-Doped Mismatched Metal/Metal Interfaces. *Journal of Physical Chemistry C*, 113 (1), 316–320
3. Wei CM, Chou MY (2002) Theory of quantum size effects in thin Pb(111) films. *Physical Review B*, 66 (23), 233408
4. Politano A, Agostino RG, Colavita E, Formoso V, Chiarello G (2007) High resolution electron energy loss measurements of Na/Cu(111) and H_2O/Na/Cu(111): Dependence of water reactivity as a function of Na coverage. *The Journal of Chemical Physics*, 126 (24), 244712
5. Luh DA, Miller T, Paggel JJ, Chiang TC (2002) Large electron-phonon coupling at an interface. *Physical Review Letters*, 88 (25), 256802
6. Pfennigstorf O, Petkova A, Guenter HL, Henzler M (2002) Conduction mechanism in ultrathin metallic films. *Physical Review B*, 65 (4), 045412
7. Orr BG, Jaeger HM, Goldman AM (1984) Transition-Temperature Oscillations in Thin Superconducting Films. *Physical Review Letters*, 53 (21), 2046
8. Chiang T-C (2004) PHYSICS: Superconductivity in Thin Films. *Science*, 306 (5703), 1900–1901
9. Politano A, Agostino RG, Colavita E, Formoso V, Chiarello G (2008) Purely quadratic dispersion of surface plasmon in Ag/Ni(111): the influence of electron confinement. *Physica Status Solidi-Rapid Research Letters*, 2 (2), 86–88
10. Yu YH, Tang Z, Jiang Y, Wu KH, Wang EG (2006) Thickness dependence of the surface plasmon dispersion in ultrathin aluminum films on silicon. *Surface Science*, 600 (22), 4966–4971
11. Pitarke JM, Nazarov VU, Silkin VM, Chulkov EV, Zaremba E, Echenique PM (2004) Theory of acoustic surface plasmons. *Physical Review B*, 70 (20), 205403
12. Bastidas CL, Liebsch A, Mochan RL (2001) Influence of d electrons on the dispersion relation of Ag surface plasmons for different single-crystal faces. *Physical Review B*, 63 (16), 165407
13. Liebsch A (1998) Prediction of a Ag multipole surface plasmon. *Physical Review B*, 57 (7), 3803–3806
14. Kim JS, Chen LM, Kesmodel LL, GarciaGonzalez P, Liebsch A (1997) Surface plasmon dispersion of Cl/Ag(111). *Physical Review B*, 56 (8), R4402–R4405
15. Politano A, Formoso V, Chiarello G (2009) Dispersion and damping of surface plasmon in Ag thin films grown on Cu(111) and Ni(111). *Superlattices and Microstructures*, 46 (1–2), 137–140
16. Politano A, Formoso V, Chiarello G (2009) Damping of the surface plasmon in clean and K-modified Ag thin films. *Journal of Electron Spectroscopy and related Phenomena*, 173 (1), 12–17
17. Politano A, Formoso V, Chiarello G (2008) Dispersion and Damping of Gold Surface Plasmon. *Plasmonics*, 3 (4), 165–170
18. Yu YH, Jiang Y, Tang Z, Guo QL, Jia JF, Xue QK, Wu KH, Wang EG (2005) Thickness dependence of surface plasmon damping and dispersion in ultrathin Ag films. *Physical Review B*, 72 (20), 205405

19. Savio L, Vattuone L, Rocca M (2003) Surface plasmon dispersion on sputtered and nanostructured Ag(001). *Physical Review B*, 67 (4), 045406
20. Rocca M, Moresco F (1996) HREELS and ELS-LEED studies of surface plasmons on Ag and Pd single crystals. *Progress in Surface Science*, 53 (2–4), 331–340
21. Moresco F, Rocca M, Zielasek V, Hildebrandt T, Henzler M (1997) ELS-LEED study of the surface plasmon dispersion on Ag surfaces. *Surface Science*, 388 (1–3), 1–4
22. Savio L, Vattuone L, Rocca M (2000) Effect of surface interband transitions on surface plasmon dispersion: O/Ag(001). *Physical Review B*, 61 (11), 7324–7327
23. Chiarello G, Formoso V, Santaniello A, Colavita E, Papagno L (2000) Surface-plasmon dispersion and multipole surface plasmons in Al(111). *Physical Review B*, 62 (19), 12676–12679
24. Moresco F, Rocca M, Hildebrandt T, Zielasek V, Henzler M (1998) Influence of surface interband transitions on surface plasmon dispersion: K/Ag(110). *Europhysics Letters*, 43 (4), 433–438
25. Kevan SD (1983) Evidence for a New Broadening Mechanism in Angle-Resolved Photoemission from Cu(111). *Physical Review Letters*, 50 (7), 526
26. Tang SJ, Jeng HT, Ismail, Sprunger PT, Plummer EW (2009) Surface electronic band structure and temperature dependence of the surface state at Ā on Mg (10 $\bar{1}$ 0) surface. *Physical Review B*, 80 (8), 085419,
27. Sklyadneva IY, Heid R, Silkin VM, Melzer A, Bohnen KP, Echenique PM, Fauster T, Chulkov EV (2009) Unusually weak electron-phonon coupling in the Shockley surface state on Pd(111). *Physical Review B*, 80 (4), 045429
28. Scheybal A, Müller K, Bertschinger R, Wahl M, Bendounan A, Aebi P, Jung TA (2009) Modification of the Cu(110) Shockley surface state by an adsorbed pentacene monolayer. *Physical Review B*, 79 (11), 115406
29. Nishimura Y, Kakeya M, Higashiguchi M, Kimura A, Taniguchi M, Narita H, Cui Y, Nakatake M, Shimada K, Namatame H (2009) Surface electronic structures of ferromagnetic Ni(111) studied by STM and angle-resolved photoemission. *Physical Review B*, 79 (24), 245402
30. Mulazzi M, Rossi G, Braun J, Minár J, Ebert H, Panaccione G, Vobornik I, Fujii J (2009) Understanding intensities of angle-resolved photoemission with circularly polarized radiation from a Cu(111) surface state. *Physical Review B*, 79 (16), 165421
31. Kowalczyk PJ (2009) Investigation of STM tip influence on the recorded position of the Shockley surface state on Au(1 1 1). *Surface Science*, 603 (5), 747–751
32. Braun KF, Hla SW (2009) Inelastic quasiparticle lifetimes of the Shockley surface state band on Ni(111). *Applied Physics A: Materials Science and Processing*, 98 (3), 583–588
33. Scheybal A, Müller K, Bertschinger R, Wahl M, Bendounan A, Aebi P, Jung TA (2009) Modification of the Cu(110) Shockley surface state by an adsorbed pentacene monolayer. *Physical Review B*, 79 (11), 115406
34. Nishimura Y, Kakeya M, Higashiguchi M, Kimura A, Taniguchi M, Narita H, Cui Y, Nakatake M, Shimada K, Namatame H (2009) Surface electronic structures of ferromagnetic Ni(111) studied by STM and angle-resolved photoemission. *Physical Review B*, 79 (24), 245402
35. Mulazzi M, Rossi G, Braun J, Minár J, Ebert H, Panaccione G, Vobornik I, Fujii J (2009) Understanding intensities of angle-resolved photoemission with circularly polarized radiation from a Cu(111) surface state. *Physical Review B - Condensed Matter and Materials Physics*, 79 (16), 165421
36. Kowalczyk PJ (2009) Investigation of STM tip influence on the recorded position of the Shockley surface state on Au(111). *Surface Science*, 603 (5), 747–751
37. Nuber A, Higashiguchi M, Forster F, Blaha P, Shimada K, Reinert F (2008) Influence of reconstruction on the surface state of Au(110). *Physical Review B*, 78 (19), 195412
38. Kowalczyk PJ, Puchalski M, Kozłowski W, Dabrowski P, Klusek Z, Olejniczak W (2008) Investigation of the Shockley surface state on clean and air-exposed Au (1 1 1). *Applied Surface Science*, 254 (15), 4572–4576
39. Vergniory MG, Pitarke JM, Echenique PM (2007) Self-energy and lifetime of Shockley and image states on Cu(100) and Cu(111): Beyond the GW approximation of many-body theory. *Physical Review B*, 76 (24), 245416

40. Higashiguchi M, Shimada K, Arita M, Miura Y, Tobita N, Cui X, Aiura Y, Namatame H, Taniguchi M (2007) High-resolution angle-resolved photoemission study of Ni(1 1 1) surface state. *Surface Science*, 601 (18), 4005–4009
41. Schiller F, Laubschat C (2006) Surface states at close-packed surfaces of simple metals. *Physical Review B*, 74 (8), 085109
42. Caravati S, Butti G, Brivio GP, Trioni MI, Pagliara S, Ferrini G, Galimberti G, Pedersoli E, Giannetti C, Parmigiani F (2006) Cu(111) and Cu(001) surface electronic states. Comparison between theory and experiment. *Surface Science*, 600 (18), 3901–3905
43. Zhang X, Liu J, Li B, Wang K, Ming F, Wang J, Xiao X (2010) Effect of substrate doping concentration on quantum well states of Pb island grown on Si(1 1 1). *Surface Science*, 604 (2), 175–180
44. Trontl VM, Pervan P, Milun M (2009) Growth and electronic properties of ultra-thin Ag films on Ni(111). *Surface Science*, 603 (1), 125–130
45. Sawa K, Aoki Y, Hirayama H (2009) Thickness dependence of Shockley-type surface states of Ag(111) ultrathin films on Si (111) 7×7 substrates. *Physical Review B*, 80 (3), 035428
46. Rybkin AG, Shikin AM, Adamchuk VK (2009) Spectra of quantum states in thin metal films and their modification: Al/ W(110) system. *Bulletin of the Russian Academy of Sciences: Physics*, 73 (5), 683–685
47. Pervan P, Milun M (2009) Photoemission from 2D metallic quantum wells. *Surface Science*, 603 (10–12), 1378–1388
48. Okuda T, Takeichi Y, He K, Harasawa A, Kakizaki A, Matsuda I (2009) Substrate dependence of anisotropic electronic structure in Ag(111) quantum film studied by angle-resolved photoelectron spectroscopy. *Physical Review B*, 80 (11), 113409
49. Lin X, Nilius N, Freund HJ, Walter M, Frondelius P, Honkala K, Häkkinen H (2009) Quantum well states in two-dimensional gold clusters on MgO thin films. *Physical Review Letters*, 102 (20), 206801
50. Choi J, Wu J, El Gabaly F, Schmid AK, Hwang C, Qiu ZQ (2009) Quantum well states in Au/ Ru(0001) and their effect on the magnetic properties of a Co overlayer. *New Journal of Physics*, 11, 043016
51. Wang LL, Ma XC, Ji SH, Fu YS, Shen QT, Jia JF, Kelly KF, Xue QK (2008) Epitaxial growth and quantum well states study of Sn thin films on Sn induced Si(111)- (23×23) R30° surface. *Physical Review B*, 77 (20), 205410
52. Tang SJ, Chang WK, Chiu YM, Chen HY, Cheng CM, Tsuei KD, Miller T, Chiang TC (2008) Enhancement of subband effective mass in Ag/Ge(111) thin film quantum wells. *Physical Review B*, 78 (24), 245407
53. Pletikosić I, Trontl VM, Milun M, Okević D, Brako R, Pervan P (2008) D-band quantum well states in Ag(111) monolayer films; Substrate-induced shifts. *Journal of Physics: Condensed Matter*, 20 (35), 355004
54. Miyata N, Horikoshi K, Hirahara T, Hasegawa S, Wei CM, Matsuda I (2008) Electronic transport properties of quantum-well states in ultrathin Pb (111) films. *Physical Review B*, 78 (24), 245405
55. He K, Hirahara T, Okuda T, Hasegawa S, Kakizaki A, Matsuda I (2008) Spin polarization of quantum well states in Ag films induced by the Rashba effect at the surface. *Physical Review Letters*, 101 (10), 107604
56. Algdal J, Balasubramanian T, Breitholtz M, Chis V, Hellsing B, Lindgren SÅ, Walldén L (2008) Sodium and potassium monolayers on Be(0001) investigated by photoemission and electronic structure calculations. *Physical Review B*, 78 (8), 085102
57. Mathias S, Wessendorf M, Passlack S, Aeschlimann M, Bauer M (2006) Morphological modifications of Ag/Cu(111) probed by photoemission spectroscopy of quantum well states and the Shockley surface state. *Applied Physics A: Materials Science and Processing*, 82 (3), 439–445
58. Luh D-A, Cheng C-M, Tsai C-T, Tsuei K-D, Tang J-M (2008) Transition from Disorder to Order in Thin Metallic Films Studied with Angle-Resolved Photoelectron Spectroscopy. *Physical Review Letters*, 100 (2), 027603

59. Politano A, Agostino RG, Formoso V, Chiarello G (2008) Short-range interactions in Na coadsorption with CO and O on Ni(111). *Chemphyschem*, 9 (8), 1189–1194
60. Politano A, Agostino RG, Colavita E, Formoso V, Tenuta L, Chiarello G (2008) Nature of the alkali surface bond at low coverages investigated by vibrational measurements. *Journal of Physical Chemistry C*, 112 (17), 6977–6980
61. Politano A, Formoso V, Chiarello G (2008) Mechanisms Leading to Alkali Oxidation on Metal Surfaces. *Journal of Physical Chemistry C*, 112 (46), 17772–17774
62. Politano A, Formoso V, Chiarello G (2009) Effects of O adsorption on the Na + CO/Ni(111) system. *Superlattices and Microstructures*, 46 (1–2), 10–13
63. Chiarello G, Barberi R, Amoddeo A, Caputi LS, Colavita E (1996) XPS and AFM characterization of a vanadium oxide film on TiO_2(100) surface. *Applied Surface Science*, 99 (1), 15–19
64. Rocca M, Biggio F, Valbusa U (1990) Surface-plasmon spectrum of Ag(001) measured by high-resolution angle-resolved electron-energy-loss spectroscopy. *Physical Review B*, 42 (5), 2835–2841
65. Marini A, Del Sole R, Onida G (2002) First-principles calculation of the plasmon resonance and of the reflectance spectrum of silver in the GW approximation. *Physical Review B*, 66 (11), 1151011
66. Liebsch A (1997) Electronic Excitations at Metal Surfaces, Plenum, New York
67. Li YB, Levi AC, Rocca M (1995) Anisotropy of Surface-Plasmons in Metals. *Surface Science*, 336 (3), 371–376
68. Rocca M (1995) Low-Energy Eels Investigation of Surface Electronic Excitations on Metals. *Surface Science Reports*, 22 (1–2), 1–71
69. Pitarke JM, Silkin VM, Chulkov EV, Echenique PM (2007) Theory of surface plasmons and surface-plasmon polaritons. *Reports on Progress in Physics*, 70, 1–87
70. Politano A, Chiarello G (2009) Tuning the lifetime of the surface plasmon upon sputtering. *Physica Status Solidi-Rapid Research Letters*, 3 (5), 136–138
71. Politano A, Formoso V, Chiarello G (2009) Annealing effects on the plasmonic excitations of metal/metal interfaces. *Applied Surface Science*, 255 (11), 6038–6042
72. Politano A, Formoso V, Chiarello G (2009) Electronic properties of metallic bilayers deposited on Cu(111): A comparative study. *Surface Science*, 603 (6), 933–937
73. Politano A, Formoso V, Chiarello G (2009) Interference effects in the excitation of collective electronic modes in nanoscale thin Ag films. *Superlattices and Microstructures*, 46 (1–2), 166–170
74. Politano A, Formoso V, Colavita E, Chiarello G (2009) Probing collective electronic excitations in as-deposited and modified Ag thin films grown on Cu(111). *Physical Review B*, 79 (4), 045426
75. Fujikawa Y, Sakurai T, Tromp RM (2008) Surface Plasmon Microscopy Using an Energy-Filtered Low Energy Electron Microscope. *Physical Review Letters*, 100 (12), 126803
76. Moresco F, Rocca M, Hildebrandt T, Henzler M (1999) Plasmon confinement in ultrathin continuous Ag films. *Physical Review Letters*, 83 (11), 2238–2241
77. Chelaru LI, Meyer zu Heringdorf FJ (2007) In situ monitoring of surface plasmons in single-crystalline Ag-nanowires. *Surface Science*, 601 (18), 4541–4545
78. Lazzari R, Jupille J, Layet JM (2003) Electron-energy-loss channels and plasmon confinement in supported silver particles. *Physical Review B*, 68 (4), 454281–4542811
79. Nilius N, Ernst N, Freund HJ (2000) Photon Emission Spectroscopy of Individual Oxide-Supported Silver Clusters in a Scanning Tunneling Microscope. *Physical Review Letters*, 84 (17), 3994–3997
80. Varykhalov A, Shikin AM, Gudat W, Moras P, Grazioli C, Carbone C, Rader O (2005) Probing the Ground State Electronic Structure of a Correlated Electron System by Quantum Well States: Ag/Ni(111). *Physical Review Letters*, 95 (24), 247601
81. Mróz S, Jankowski Z (1995) Properties of ultrathin silver layers on the Ni(111) face. *Surface Science*, 322 (1–3), 133–139

82. Mróz S, Jankowski Z, Nowicki M (2000) Growth and isothermal desorption of ultrathin silver layers on the Ni(111) face at the substrate temperature from 180 to 900 K. *Surface Science*, 454 (1), 702–706
83. Mróz S (1995) Directional elastic peak and directional Auger electron spectroscopies - New tools for investigating surface-layer atomic structure. *Progress in Surface Science*, 48 (1–4), 157–166
84. Politano A, Chiarello G, Formoso V, Agostino RG, Colavita E (2006) Plasmon of Shockley surface states in Cu(111) : A high-resolution electron energy loss spectroscopy study. *Physical Review B*, 74 (8), 081401
85. Borensztein Y, Roy M, Alameh R (1995) Threshold and Linear Dispersion of the Plasma Resonance in Thin Ag Films. *EPL (Europhysics Letters)*, 31 (5–6), 311
86. Politano A, Formoso V, Chiarello G (2010) Plasmonic Modes Confined in Nanoscale Thin Silver Films Deposited onto Metallic Substrates *Journal of Nanoscience and Nanotechnology*, 10 (2), 1313–1321
87. Suto S, Tsuei KD, Plummer EW, Burstein E (1989) Surface-plasmon energy and dispersion on Ag single crystals. *Physical Review Letters*, 63 (23), 2590–2593
88. Rocca M, Lazzarino M, Valbusa U (1991) Surface-Plasmon Energy and Dispersion on Ag Single-Crystals - Comment. *Physical Review Letters*, 67 (22), 3197–3197
89. Lee G, Sprunger PT, Plummer EW, Suto S (1991) Lee et al. reply. *Physical Review Letters*, 67 (22), 3198
90. Feibelman PJ (1982) Surface electromagnetic fields. *Progress in Surface Science*, 12 (4), 287–407
91. Feibelman PJ (1973) Sensitivity of surface plasmon dispersion and damping to alkali adsorption. *Surface Science*, 40 (1), 102–108
92. Feibelman PJ (1974) Microscopic calculation of surface-plasmon dispersion and damping. *Physical Review B*, 9 (12), 5077–5098
93. Feibelman PJ (1993) Perturbation of surface plasmon dispersion by "extra" electrons near a surface. *Surface Science Letters*, 282 (1–2), 129–136
94. Feibelman PJ (1994) Comment on Surface plasmon dispersion of Ag. *Physical Review Letters*, 72 (5), 788
95. Feibelman PJ (1989) Interpretation of the linear coefficient of surface-plasmon dispersion. *Physical Review B*, 40 (5), 2752–2756
96. Feibelman PJ (1973) Sensitivity of surface-plasmon dispersion and damping to potential barrier shape. *Physical Review Letters*, 30 (20), 975–978
97. Feibelman PJ (1971) Dependence of the normal modes of plasma oscillation at a bimetallic interface on the electron density profile. *Physical Review B*, 3 (9), 2974–2982
98. Feibelman PJ, Tsuei KD (1990) Negative surface-plasmon dispersion coefficient: A physically illustrative, exact formula. *Physical Review B*, 41 (12), 8519–8521
99. Rocca M, Lazzarino M, Valbusa U (1992) Surface-Plasmon on Ag(110) - Observation of Linear and Positive Dispersion and Strong Azimuthal Anisotropy. *Physical Review Letters*, 69 (14), 2122–2125
100. Silkin VM, Quijada M, Muino RD, Chulkov EV, Echenique PM (2007) Dynamic screening and electron-electron scattering in low-dimensional metallic systems. *Surface Science*, 601 (18), 4546–4552
101. Silkin VM, Quijada M, Vergniory MG, Alducin M, Borisov AG, Muino RD, Juaristi JI, Sanchez-Portal D, Chulkov EV, Echenique PM (2007) Dynamic screening and electron dynamics in low-dimensional metal systems. *Nuclear Instruments & Methods in Physics Research Section B-Beam Interactions with Materials and Atoms*, 258 (1), 72–78
102. Chiang TC (2000) Photoemission studies of quantum well states in thin films. *Surface Science Reports*, 39 (7–8), 181–235
103. Tsuei KD, Plummer EW, Liebsch A, Pehlke E, Kempa K, Bakshi P (1991) The Normal-Modes at the Surface of Simple Metals. *Surface Science*, 247 (2–3), 302–326
104. Rocca M, Lazzarino M, Valbusa U (1992) Plasmon Damping and Surface Interband-Transitions on Ag(001) and (011). *Surface Science*, 270, 560–562

105. Rocca M, Li YB, Demongeot FB, Valbusa U (1995) Surface-Plasmon Dispersion and Damping on Ag(111). *Physical Review B*, 52 (20), 14947–14953
106. Rocca M, Moresco F, Valbusa U (1992) Temperature-Dependence of Surface-Plasmons on Ag(001). *Physical Review B*, 45 (3), 1399–1402
107. Rocca M, Valbusa U (1990) Angular-Dependence of Dipole Scattering Cross-Section - Surface-Plasmon Losses on Ag(100). *Physical Review Letters*, 64 (20), 2398–2401
108. Rocca M, Valbusa U (1993) Electronic Excitations on Silver Single-Crystal Surfaces. *Surface Science*, 287, 770–775
109. Quijada M, Muino RD, Echenique PM (2005) The lifetime of electronic excitations in metal clusters. *Nanotechnology*, 16 (5), S176–S180
110. Quijada M, Borisov AG, Nagy I, Muino RD, Echenique PM (2007) Time-dependent density-functional calculation of the stopping power for protons and antiprotons in metals. *Physical Review A*, 75 (4), 042902
111. Politano A, Agostino RG, Colavita E, Formoso V, Chiarello G (2009) Collective Excitations in Nanoscale Thin Alkali Films: Na/Cu(111). *Journal of Nanoscience and Nanotechnology*, 9 (6), 3932–3937
112. Zielasek V, Ronitz N, Henzler M, Pfnur H (2006) Crossover between monopole and multipole plasmon of Cs monolayers on Si(111) individually resolved in energy and momentum. *Physical Review Letters*, 96 (19), 196801
113. Tsuei KD, Plummer EW, Feibelman PJ (1989) Surface-plasmon dispersion in simple metals. *Physical Review Letters*, 63 (20), 2256–2259
114. Tu KN, Mayer JV, Feldman LC (1992) Electronic Thin Films Science. Macmillan, New York
115. Bendounan A, Fagot Revurat Y, Kierren B, Bertran F, Yurov VY, Malterre D (2002) Surface state in epitaxial Ag ultrathin films on Cu(1 1 1). *Surface Science*, 496 (1–2), L43–L49
116. De Crescenzi M, Piancastelli MN (1996) Electron Scattering and Related Spectroscopies, World Scientific, Singapore
117. Moresco F, Rocca M, Hildebrandt T, Zielasek V, Henzler M (1999) K adsorption on Ag(110), effect on surface structure and surface electronic excitations. *Surface Science*, 424 (1), 62–73
118. Persson BNJ (1993) Polarizability of small spherical metal particles: influence of the matrix environment. *Surface Science*, 281 (1–2), 153–162
119. Hövel H, Fritz S, Hilger A, Kreibig U, Vollmer M (1993) Width of cluster plasmon resonances: Bulk dielectric functions and chemical interface damping. *Physical Review B*, 48 (24), 18178–18188
120. Bendounan A, Forster F, Ziroff J, Schmitt F, Reinert F (2005) Influence of the reconstruction in Ag/Cu (111) on the surface electronic structure: Quantitative analysis of the induced band gap. *Physical Review B*, 72 (7), 075407
121. Schiller F, Cordón J, Vyalikh D, Rubio A, Ortega JE (2005) Fermi Gap Stabilization of an Incommensurate Two-Dimensional Superstructure. *Physical Review Letters*, 94 (1), 016103
122. Zacharias P, Kliewer KL (1976) Dispersion relation for the 3.8 eV volume plasmon of silver. *Solid State Communications*, 18 (1), 23–26
123. He JH, Carosella CA, Hubler GK, Qadri SB, Sprague JA (2006) Bombardment-Induced Tunable Superlattices in the Growth of Au-Ni Films. *Physical Review Letters*, 96 (5), 056105
124. Yuan Z, Gao S (2008) Landau damping and lifetime oscillation of surface plasmons in metallic thin films studied in a jellium slab model. *Surface Science*, 602 (2), 460–464
125. Chiarello G, Cupolillo A, Caputi LS, Papagno L, Colavita E (1997) Collective and single-particle excitations in thin layers of K on Ni(111). *Surface Science*, 377 (1–3), 365–370
126. Politano A, Chiarello G (2010) Sputtering-induced modification of the electronic properties of Ag/Cu(111). *Journal of Physics D: Applied Physics* 43 (8), 085302
127. Stephanov MA (2009) Non-Gaussian Fluctuations near the QCD Critical Point. *Physical Review Letters*, 102 (3), 032301
128. Politano A, Agostino RG, Colavita E, Formoso V, Chiarello G (2007) Electronic properties of self-assembled quantum dots of sodium on Cu(111) and their interaction with water. *Surface Science*, 601 (13), 2656–2659

129. Tsuei K-D, Plummer EW, Feibelman PJ (1989) Surface-plasmon dispersion in simple metals. *Physical Review Letters*, 63 (20), 2256–2259
130. Silkin VM, Chulkov EV (2006) Energy and lifetime of surface plasmon from first-principles calculations. *Vacuum*, 81 (2), 186–191
131. Bagchi A, Duke CB, Feibelman PJ, Porteus JO (1971) Measurement of surface-plasmon dispersion in aluminum by inelastic low-energy electron diffraction. *Physical Review Letters*, 27 (15), 998–1001
132. Sprunger PT, Watson GM, Plummer EW (1992) The normal modes at the surface of Li and Mg. *Surface Science*, 269–270, 551–555
133. Liebsch A, Schaich WL (1995) Influence of a Polarizable Medium on the Nonlocal Optical-Response of a Metal-Surface. *Physical Review B*, 52 (19), 14219–14234
134. Perri S, Lepreti F, Carbone V, Vulpiani A (2007) Position and velocity space diffusion of test particles in stochastic electromagnetic fields. *Europhysics Letters*, 78 (4), 40003
135. Chulkov EV, Silkin VM, Echenique PM (2000) Inverse lifetime of surface states on metals. *Surface Science*, 454, 458–461
136. Steeb F, Mathias S, Fischer A, Wiesenmayer M, Aeschlimann M, Bauer M (2009) The nature of a nonlinear excitation pathway from the Shockley surface state as probed by chirped pulse two photon photoemission. *New Journal of Physics*, 11, 013016
137. Bonzel HP, Bradshaw AM, Ertl G (1989) Alkali Adsorption on Metals and Semiconductors, Elsevier, Amsterdam
138. Palik ED (1985) Handbook of Optical Constants of Solids. Academic Press, New York
139. Nazarov VU (1999) Multipole surface plasmon excitation enhancement in metals. *Physical Review B*, 59 (15), 9866–9869
140. Tsuei KD, Plummer EW, Liebsch A, Kempa K, Bakshi P (1990) Multipole Plasmon Modes at a Metal-Surface. *Physical Review Letters*, 64 (1), 44–47
141. Eremeev SV, Rusina GG, Borisova SD, Chulkov EV (2008) Electron-phonon interaction in the quantum well state of the 1 ML Na/Cu(111) system. *Physics of the Solid State*, 50 (2), 323–329
142. Fuyuki M, Watanabe K, Ino D, Petek H, Matsumoto Y (2007) Electron-phonon coupling at an atomically defined interface: Na quantum well on Cu(111). *Phys Rev B*, 76 (11), 115427
143. Hoffmann G, Berndt R, Johansson P (2003) STM-induced fluorescence from Na monolayers on Cu(111). *Physics of Low-Dimensional Structures*, 3-4, 209–219
144. Politano A, Agostino RG, Colavita E, Formoso V, Chiarello G (2008) Electronic properties of (3/2×3/2)-Na/Cu(111). *J Electron Spectrosc Relat Phenom*, 162 (1), 25–29
145. Langer T, Förster DF, Busse C, Michely T, Pfnür H, Tegenkamp C (2011) Sheet plasmons in modulated graphene on Ir(111). *New J Phys*, 13 (5), 053006
146. Langer T, Baringhaus J, Pfnür H, Schumacher HW, Tegenkamp C (2010) Plasmon damping below the Landau regime: the role of defects in epitaxial graphene. *New J Phys*, 12, 033017
147. Liu Y, Willis RF (2009) The evolution of sheet-plasmon behavior in silver monolayers on Si(111)-($\sqrt{3}\times\sqrt{3}$)-Ag surface. *Surf Sci*, 603 (13), 2115-2119
148. Pfnür H, Langer T, Baringhaus J, Tegenkamp C (2011) Multiple plasmon excitations in adsorbed two-dimensional systems. *J Phys: Condens Matter*, 23 (11), 112204
149. Tegenkamp C, Pfnür H, Langer T, Baringhaus J, Schumacher HW (2011) Plasmon electron–hole resonance in epitaxial graphene. *J Phys: Condens Matter*, 23 (1), 012001
150. Politano A, Marino AR, Formoso V, Farías D, Miranda R, Chiarello G (2011) Evidence for acoustic-like plasmons on epitaxial graphene on Pt(111). *Phys Rev B*, 84 (3), 033401
151. Borca B, Barja S, Garnica M, Minniti M, Politano A, Rodriguez-García JM, Hinarejos JJ, Farías D, Vázquez de Parga AL, Miranda R (2010) Electronic and geometric corrugation of periodically rippled, self-nanostructured graphene epitaxially grown on Ru(0001). *New J Phys*, 12 (9), 093018
152. Politano A, Agostino RG, Colavita E, Formoso V, Tenuta L, Chiarello G (2008) Nature of the alkali surface bond at low coverages investigated by vibrational measurements. *J Phys Chem C*, 112 (17), 6977-6980

153. Politano A, Formoso V, Chiarello G (2008) Temperature effects on alkali-promoted CO dissociation on Ni(111). *Surf Sci*, 602 (12), 2096–2100
154. Politano A, Formoso V, Chiarello G (2008) Alkali adsorption on Ni(111) and their coadsorption with CO and O. *Appl Surf Sci*, 254 (21), 6854–6859
155. Politano A, Formoso V, Chiarello G (2008) Mechanisms Leading to Alkali Oxidation on Metal Surfaces. *J Phys Chem C*, 112 (46), 17772–17774
156. Robba D, Ori DM, Sangalli P, Chiarello G, Depero LE, Parmigiani F (1997) A photoelectron spectroscopy study of sub-monolayer V/TiO2(001) interfaces annealed from 300 up to 623 K. *Surf Sci*, 380 (2–3), 311–323
157. Caputi LS, Chiarello G, Papagno L (1985) Carbonaceous layers on Ni (110) and (100) studied by AES and EELS. *Surf Sci*, 162 (1–3), 259–263
158. De Crescenzi M, Colavita E, Papagno L, Chiarello G, Scarmozzino R, Caputi LS, Rosei R (1983) Electronic properties of Fe80B20 alloys: ordering and disordering effects. *J Phys F Met Phys*, 13 (4), 895–907
159. Chiarello G, Robba D, De Michele G, Parmigiani F (1993) An X-ray Photoelectron-Spectroscopy Study of the Vanadia Titania Catalysts. *Appl Surf Sci*, 64 (2), 91–96
160. Ciambelli P, Bagnasco G, Lisi L, Turco M, Chiarello G, Musci M, Notaro M, Robba D, Ghetti P (1992) Vanadium-Oxide Catalysts Supported on Laser-Synthesized Titania Powders - Characterization and Catalytic Activity in the Selective Reduction of Nitric-Oxide. *Appl Catal B Environ*, 1 (2), 61–77

Chapter 4
Surface Plasmon Resonance Based Fiber Optic Sensors

Banshi D. Gupta

1 Introduction

In recent years, the phenomenon of surface plasmon resonance (SPR) has fascinated a large number of researchers across the world due to its usefulness in various optical devices. Surface plasmons are the electromagnetic excitations generated due to charge density fluctuations at the interface between a metal and a dielectric. These are transverse magnetically (TM) polarized waves that travel along the interface. The field associated with these waves decays exponentially in both the media (metal and dielectric). Because of TM polarized wave, surface plasmons can be excited by a TM or p-polarized light. The resonance between the two occurs when their wave vectors are equal resulting in the transfer of energy to from incident light to surface plasmon wave. The wave vector of surface plasmon wave depends on the dielectric constant of the medium in contact of the metal. Change in the dielectric constant of the medium changes the wave vector of the surface plasmon wave and hence changes the wave vector of the incident p-polarized light at resonance. Knowing the change in the wave vector of the incident beam at resonance, one can determine the dielectric constant of the medium. This technique, called SPR, was first time applied for sensing in 1983 [1]. Since then, numerous SPR-sensing structures for chemical and biochemical sensing have been studied. Optical elements that have been used to excite surface plasmons and hence for the sensing of the dielectric constant of the medium include high index prism, diffraction grating, and waveguide. Prism and waveguide use evanescent wave while the reflection grating uses one of the orders of the diffracted light beam to excite the surface plasmons. The disadvantage of grating-based SPR sensor is that the incident light beam passes through the sample solution (or, the dielectric medium) and hence the flow cell and the sample need to be optically transparent. Although the prism-based SPR sensors have been used

B.D. Gupta (✉)
Department of Physics, Indian Institute of Technology Delhi, New Delhi 110016, India
e-mail: bdgupta@physics.iitd.ernet.in

C.D. Geddes (ed.), *Reviews in Plasmonics 2010*, Reviews in Plasmonics,
DOI 10.1007/978-1-4614-0884-0_4,

widely [2–9], but these have a number of shortcomings such as bulky size and the presence of various optical and mechanical (moving) parts. Further, the prism-based SPR sensing devices cannot be used for remote sensing applications. These shortcomings can be overcome if an optical fiber is used in place of prism. The additional advantage of optical fiber is that the SPR probe can be miniaturized which can be advantageous for samples which are available in minute quantity or are costly. Due to these advantages the SPR-based fiber optic sensors have drawn tremendous attention in recent years [10–28]. Both experimental and theoretical studies have been reported in the literature and the performance of these sensors, in terms of sensitivity, signal-to-noise ratio (SNR) (or, detection accuracy), and operating range has been evaluated.

The present chapter deals on the SPR-based fiber optic sensors including sensing principle, performance parameters and the choice of metals. As examples, the fiber optic sensors for the detection of naringin and pesticide have been presented. To enhance the sensitivity of the sensor, various fiber optic SPR probes that have been analyzed using ray approach and matrix method have also been discussed.

2 Sensing Principle

As mentioned above, metal–dielectric interface supports charge density oscillations along the interface called surface plasma oscillations. The quantum of these oscillations is called surface plasmon. The surface plasmons are accompanied by a longitudinal TM-polarized electric field which decays exponentially in metal as well as in dielectric medium as shown in Fig. 4.1. The maximum of electric field is at the metal–dielectric interface. These characteristics of surface plasmons, TM-polarization, and exponential decay of electric field, are found by solving the Maxwell's equation for semi-infinite media of metal and dielectric with an interface of metal–dielectric. The propagation constant (K_{SP}) of the surface plasmon wave propagating along the metal–dielectric interface is given by

$$K_{\mathrm{SP}} = \frac{\omega}{c}\left(\frac{\varepsilon_{\mathrm{m}}\varepsilon_{\mathrm{s}}}{\varepsilon_{\mathrm{m}}+\varepsilon_{\mathrm{s}}}\right)^{1/2} \tag{4.1}$$

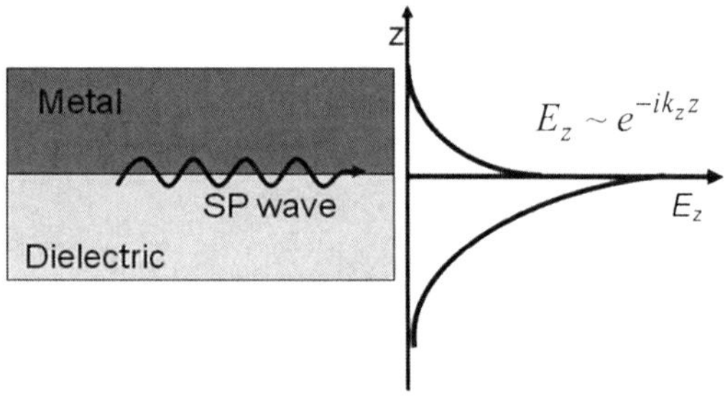

Fig. 4.1 Schematic of surface plasmon wave and the decay of associated electric field in two media

Table 4.1 Values of characteristic parameters associated with surface plasmon wave at metal–water interface at wavelength 630 nm

Parameter	Silver	Gold
Propagation length	20 μm	3 μm
Penetration depth into metal	25 nm	30 nm
Penetration depth into dielectric	220 nm	160 nm
Field in dielectric	90%	85%

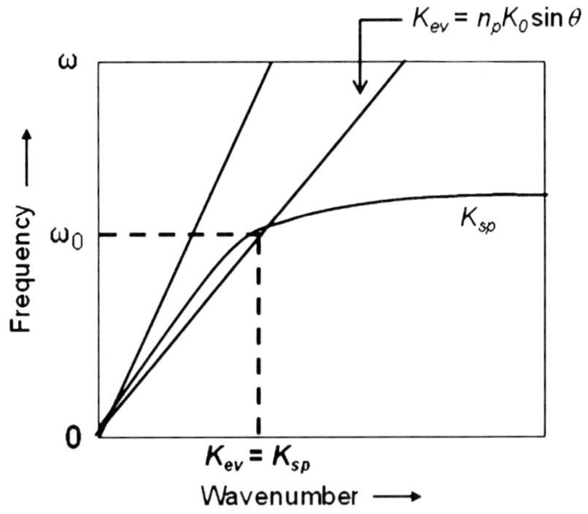

Fig. 4.2 Dispersion curves for direct light, evanescent wave, and the surface plasmon wave

where ε_m and ε_s are the dielectric constants of metal and the dielectric medium respectively; ω is the frequency and c is the velocity of light in vacuum. For gold–water and silver–water interface at wavelength 630 nm the characteristic parameters of surface plasmon wave are tabulated in Table 4.1. It may be noted that the propagation length of the surface plasmon wave is very small which implies that the sensing action takes place directly in the region where the surface plasmon wave is excited by the light. Since the surface plasmons are TM polarized, these are excited by TM or p-polarized light. The propagation constant (K_s) of the light wave with frequency ω propagating through the dielectric medium with dielectric constant ε_s is given by

$$K_s = \frac{\omega}{c}\sqrt{\varepsilon_s}. \tag{4.2}$$

Comparison of (4.1) and (4.2) implies that $K_s < K_{SP}$ because $\varepsilon_m < 0$ (for metal) and $\varepsilon_s > 0$ (for dielectric). Since for the excitation of surface plasmons two propagation wave-vectors should be equal, the inequality of two wave vectors above implies that the direct light cannot excite surface plasmons at a metal–dielectric interface. This is also shown in Fig. 4.2 where frequency has been plotted as a function of propagation constant. The two curves, one for direct light and the other for surface plasmon

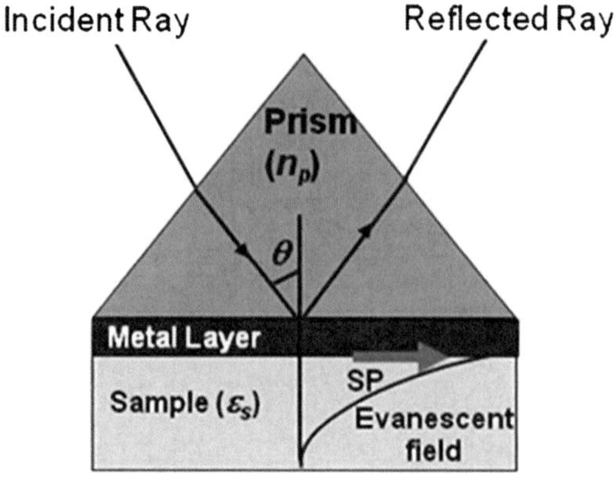

Fig. 4.3 Kretschmann configuration for the excitation of surface plasmon at metal–dielectric interface

wave, do not cross and hence the resonance condition is not satisfied. The surface plasmons can be excited only if the wave vector of the exciting light in dielectric medium is increased. This can be achieved by using evanescent wave propagating along the interface instead of direct light. To obtain the evanescent wave for the excitation of surface plasmons, a high index prism is used.

Kretschmann and Reather [29] devised a prism-based configuration, shown in Fig. 4.3, to excite surface plasmons using evanescent wave. In this configuration, the base of the glass prism is coated with a thin layer of metal. The other side of the metal layer is kept in contact with the dielectric medium of lower refractive index (such as air or some other dielectric sample). When a p-polarized light beam is incident through the prism on the prism–metal interface at an angle θ equal to or greater than the critical angle, the evanescent wave is produced at the prism–metal interface. It propagates along the interface and its electric field decays exponentially in the rarer medium. The propagation constant of the evanescent wave is given by

$$K_{\mathrm{ev}} = \frac{\omega}{c}\sqrt{\varepsilon_{\mathrm{p}}}\sin\theta \tag{4.3}$$

where ε_{p} is the dielectric constant of the material of the prism. It may be noted from (4.3) that an increase in the dielectric constant of the material of the prism or the angle of incidence of the light beam increases the propagation constant of the evanescent wave and hence it can be made equal to the propagation constant of the surface plasmon wave by the proper choice of the dielectric constant of the prism and the angle of incidence of the beam. For the excitation of surface plasmons, the following resonance condition should be satisfied

$$\frac{\omega}{c}\sqrt{\varepsilon_{\mathrm{p}}}\sin\theta_{\mathrm{res}} = \frac{\omega}{c}\left(\frac{\varepsilon_{\mathrm{m}}\varepsilon_{\mathrm{s}}}{\varepsilon_{\mathrm{m}}+\varepsilon_{\mathrm{s}}}\right)^{1/2}. \tag{4.4}$$

For a prism of fixed refractive index, the condition is satisfied at a particular angle of incidence called resonance angle and is represented by θ_{res}. At resonance, the

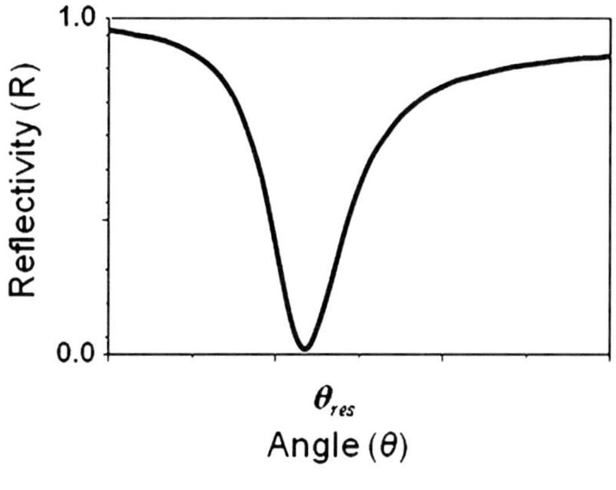

Fig. 4.4 SPR spectrum

transfer of energy of incident light to surface plasmons takes place which reduces the intensity of the reflected light. At resonance angle, the dispersion curve of evanescent wave crosses the dispersion curve of the surface plasmon wave as shown in Fig. 4.2.

In the sensing device based on prism and angular interrogation method, the intensity of the reflected light is measured as a function of angle of incidence θ for fixed values of frequency, metal layer thickness and dielectric constant of the sensing medium. At resonance angle, a sharp dip is observed due to an efficient transfer of energy to surface plasmons as shown in Fig. 4.4. The plot can also be obtained with the help of Fresnel's equations for the three-layer system. For a given frequency of the light source and the dielectric constant of the metal film, one can determine the dielectric constant (ε_s) of the sensing medium by using (4.4) if the value of the resonance angle (θ_{res}) determined experimentally is substituted. The resonance angle is very sensitive to the variation in the refractive index of the sensing layer. A graph between the resonance angle and the refractive index of the sensing medium serves as the calibration curve of the sensor. Any increase in the refractive index of the sensing medium increases the resonance angle.

3 Sensitivity and Detection Accuracy

Sensitivity and detection accuracy are the two important parameters that are used to analyze the performance of an SPR-based sensor. For the best performance, both the parameters of the sensor should be as high as possible. Sensitivity of a SPR-based sensor utilizing angular interrogation method depends on the amount of shift of the resonance angle with the change in the refractive index of the sensing medium. Figure 4.5 shows a plot of reflectance as a function of angle of incidence of the light beam for two different refractive indices of the sensing medium n_s and $n_s + \delta n_s$ in the Kretschmann configuration (Fig. 4.3). Increase in refractive index of the sensing

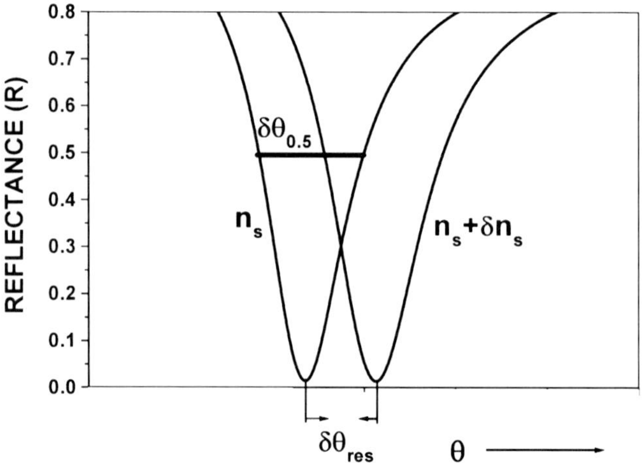

Fig. 4.5 SPR spectra for two different refractive indices of the sensing medium [23]. Reprinted with permission IEEE

medium by δn_s shifts the resonance angle by $\delta\theta_{res}$. The sensitivity of an SPR sensor utilizing angular interrogation method is, thus, defined as

$$S_n = \frac{\delta\theta_{res}}{\delta n_s}. \tag{4.5}$$

The detection accuracy of an SPR-based sensor depends on how accurately and precisely the resonance angle can be determined. The accuracy of resonance angle will decide the accuracy of the measurement of the refractive index of the sensing medium. The detection accuracy depends on the width of the SPR spectrum. Narrower the width, higher is the detection accuracy. The angular width of SPR spectrum is generally taken corresponding to reflectance 0.5 for the determination of the detection accuracy. In some of the cases, the value of reflectance for the determination of detection accuracy is taken greater than 0.5 if the minimum reflectance is high. If $\delta\theta_{0.5}$ is the angular width of the SPR curve corresponding to reflectance 0.5, the detection accuracy of the sensor is assumed to be inversely proportional to $\delta\theta_{0.5}$ (Fig. 4.5). The parameter which combines the two sensitivity and detection accuracy is the SNR. For SPR sensor with angular interrogation, it is defined as [30]

$$\mathrm{SNR} = \frac{\delta\theta_{res}}{\delta\theta_{0.5}}. \tag{4.6}$$

This should be maximized to achieve best performance of the SPR-based sensor.

In the angular interrogation method, the light source is monochromatic and the angle of incidence is varied to obtain the SPR spectrum and hence the resonance angle. If the angle of incidence of the light beam is fixed then wavelength is varied to obtain the SPR spectrum. Since the dielectric constants of the material of the prism and the metal depend on wavelength, the two propagation constants become equal at one particular wavelength and the resonance condition, (4.4), is satisfied. The wavelength corresponding to minimum reflectance is called resonance wavelength and the method is called wavelength interrogation. This method is rarely used in the prism-based SPR sensors but is common in optical fiber-based SPR sensors which are described in the next section. The sensitivity and the detection accuracy are determined in the same way as determined in the case of angular interrogation. The angles are replaced by wavelengths in the definitions of sensitivity and detection accuracy.

4 Fiber Optic Sensors

In the prism-based SPR sensors, evanescent wave required to excite surface plasmons is resulted due to the total internal reflection of the incident beam at the prism–metal interface when the angle of incidence is greater than the critical angle. In an optical fiber, light guidance occurs due to the total internal reflection of the guided rays at the core–cladding interface. The evanescent wave present in the cladding region of the optical fiber propagates along the core–cladding interface. Therefore, to design an SPR-based fiber optic sensor, the prism can be replaced by the core of an optical fiber. To fabricate an SPR-based fiber optic sensor, the cladding from a small portion of the fiber, preferably from the middle, is removed and the unclad core is coated with a metal layer. The metal layer is further, surrounded by a dielectric sensing medium as shown in Fig. 4.6. Since in an optical fiber all the rays with angles greater than critical angle are guided hence, instead of angular interrogation, spectral interrogation method is used in the case of SPR-based fiber optic SPR sensors. Thus the light from a polychromatic source is launched into one end of the optical fiber. The evanescent field produced by the guided rays excites the surface plasmons at the metal–dielectric sensing medium interface. The coupling of evanescent field with surface plasmons strongly depends on wavelength, fiber parameters, probe geometry, and the metal layer properties. Unlike prism-based SPR sensor, the number of reflections for most of the guided rays is greater than one in SPR-based fiber optic sensor. Smaller the angle of incidence of the ray at the core–metal interface, larger is the number of reflections per unit length of the fiber. In addition,

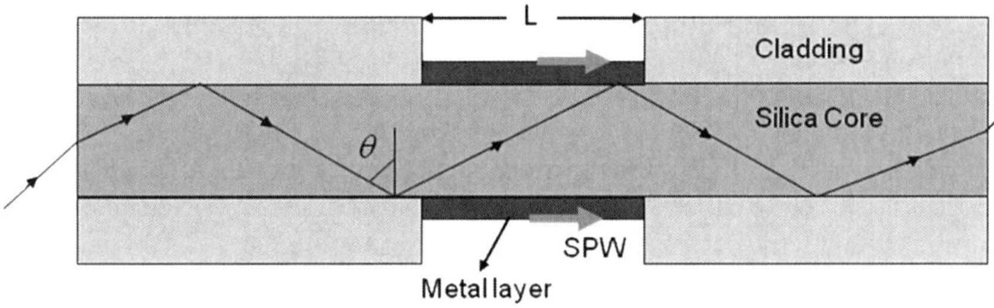

Fig. 4.6 SPR-based fiber optic probe

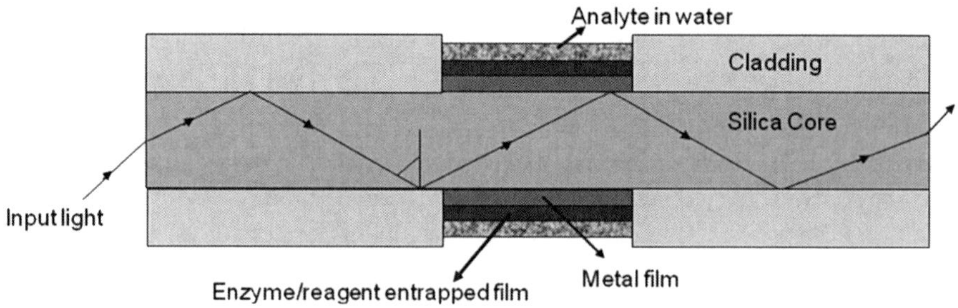

Fig. 4.7 Enzyme/reagent mediated SPR-based fiber optic probe

the number of reflections for any ray also depends on fiber core diameter. The number of reflections affects the width of the SPR curve. Larger the number of reflections, larger is the width. The intensity of the light transmitted through the fiber is detected at the other end as a function of wavelength. The SPR spectrum thus obtained is similar in shape to that shown in Fig. 4.4. The sensing is accomplished by knowing the resonance wavelength that corresponds to the dip in the SPR spectrum. A plot of resonance wavelength with the refractive index of the sensing medium is the calibration curve of the SPR sensor.

SPR-based fiber optic sensors can be used for quantitative detection of various chemical and biological species. These include food quality, medical diagnostics, and environmental monitoring. To detect chemicals, generally, a film containing reagent or enzyme is deposited on the metal layer as shown in Fig. 4.7. When analyte, chemical to be sensed, interacts with the reagent, a change in the refractive index of the film takes place. The change in refractive index increases with the increase in the concentration of the analyte. In this section, two SPR-based fiber optic sensors based on this method will be discussed. One is the detection of naringin [21] and the other is the detection of pesticide [31]. To detect naringin, an enzyme naringinase is deposited on the silver-coated fiber using gel entrapment technique. The SPR spectra are obtained for different concentrations of naringin in water around the probe. One such SPR spectrum is shown in Fig. 4.8 for 150 μg/ml concentration of naringin [21]. The resonance wavelength for this concentration of naringin is 525 nm. The calibration curve of the SPR-based fiber optic sensor obtained for the detection of naringin is shown in Fig. 4.9 [21]. As the concentration of naringin increases, the resonance wavelength increases. The increase in resonance wavelength implies increase in the refractive index of the immobilized layer. The reason of increase in refractive index is due to the formation of a complex between naringin and naringinase in the film. Figure 4.10 shows the variation of sensitivity with the concentration of naringin [21]. The sensitivity, in this case, has been defined as the change in SPR wavelength per unit change in the concentration of naringin ($\delta\lambda_{res}/\delta c$). Sensitivity increases as the concentration of naringin increases. Figure 4.11 shows the variation of SNR with the concentration of naringin [21]. It decreases with the increase in the concentration of naringin. This occurs due to the increase in the width of the SPR spectrum with the increase in the concentration of naringin or the refractive index of the sensing film.

Fig. 4.8 SPR spectrum for 0.15 mg/ml concentration of naringin [21]. Reprinted with permission Elsevier

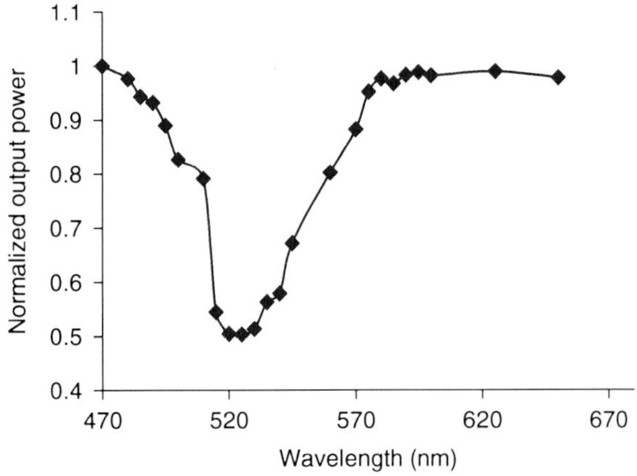

Fig. 4.9 Calibration curve of the naringin sensor

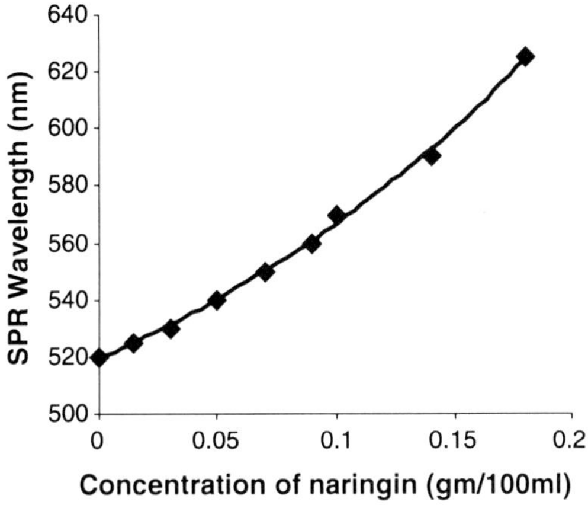

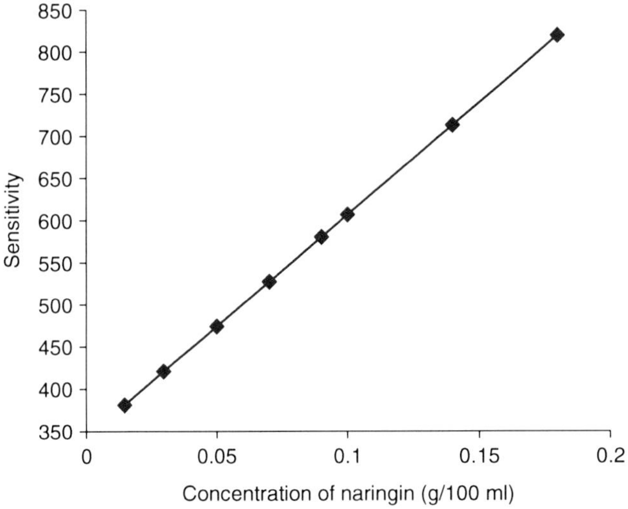

Fig. 4.10 Variation of sensitivity with the concentration of naringin [21]. Reprinted with permission Elsevier

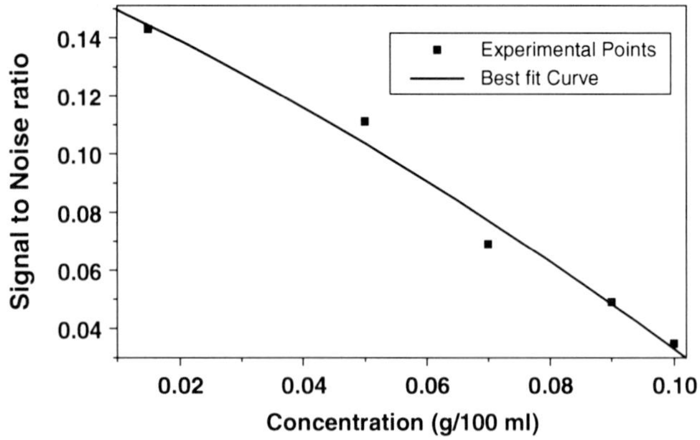

Fig. 4.11 Variation of SNR with the concentration of naringin [21]. Reprinted with permission Elsevier

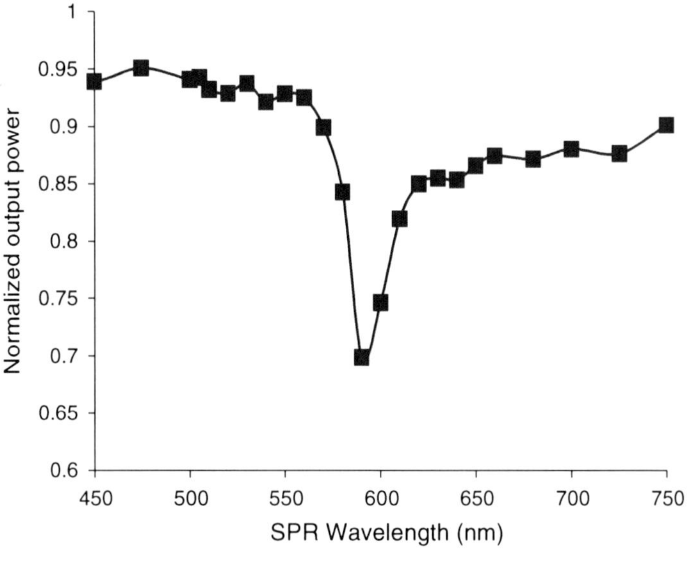

Fig. 4.12 SPR spectrum for 2.5 mM concentration of substrate and 1.0 μM concentration of pesticide [31]. Reprinted with permission Elsevier

Optical fibers with SPR technique can also be used for the detection of organophosphate pesticide, chlorphyrifos. The probe is similar to that for naringin except that the film contains acetylcholinesterase enzyme. However, the principle of detection of pesticide is different from that of naringin. It is based on the principle of competitive binding of pesticide (acting as an inhibitor) for the substrate (acetlythiocholine iodide) to the enzyme. Figure 4.12 shows the SPR spectra for pesticide concentration of 1.0 μM and a substrate concentration of 2.5 mM [31]. For this concentration of pesticide, the resonance wavelength is obtained around 590 nm. Figure 4.13 shows the variation of resonance wavelength with the concentration of

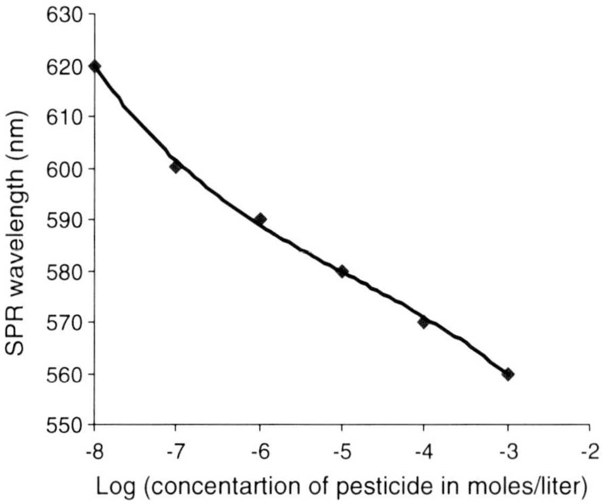

Fig. 4.13 Calibration curve of the pesticide sensor for 2.5 mM concentration of substrate [31]. Reprinted with permission Elsevier

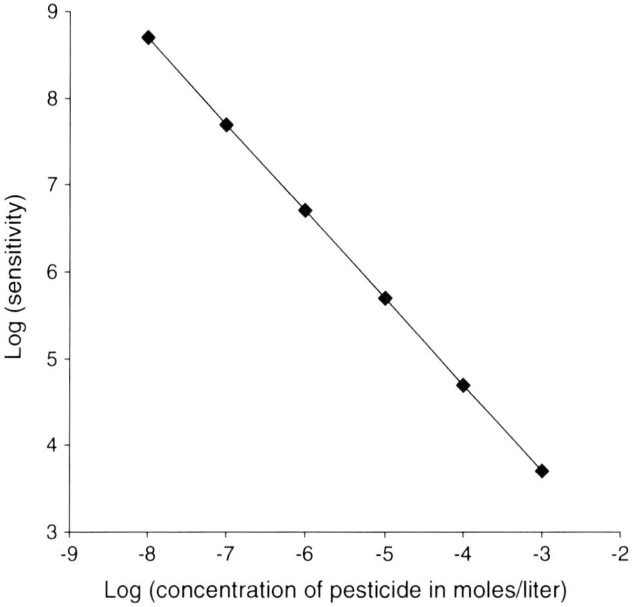

Fig. 4.14 Variation of sensitivity with the concentration of pesticide for 2.5 mM concentration of substrate [31]. Reprinted with permission Elsevier

pesticide [31]. The variation is opposite to that obtained for the detection of naringin. This implies that the decrease in the refractive index of the film occurs due to the increase in the concentration of pesticide. Figure 4.14 shows the variation of sensitivity of the probe with the concentration of pesticide. As the concentration of pesticide increases, the sensitivity of the sensor decreases. The variation of sensitivity

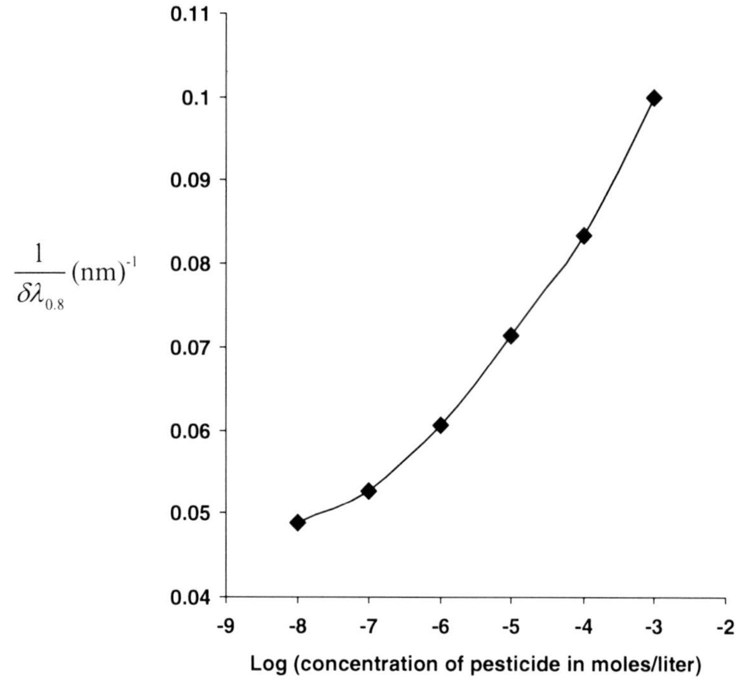

Fig. 4.15 Variation of inverse of the spectral width of the SPR curve corresponding to 80% transmission with the concentration of pesticide for 2.5 mM concentration of substrate [31]. Reprinted with permission Elsevier

with the concentration of pesticide is nonlinear. The variation of the inverse of the width of the SPR curve corresponding to 80% transmission with the concentration of pesticide is shown in Fig. 4.15. The plot implies that as the concentration of pesticide increases, the width of the SPR curve decreases and hence the detection accuracy increases. This implies that with the increase in the concentration of pesticide and hence the decrease in the refractive index of the sensing layer, the width of the SPR curve decreases. Naringin and pesticide are the two chemicals that have been chosen here to demonstrate their detection using optical fiber and SPR technique. There are many other SPR-based fiber optic sensors that have been reported in the literature. These include the measurements of salinity in water [22, 32], refractive indices of alcohols [17], BSA [24], vapor and liquid analyses [18], biomolecular interaction [33], temperature [34, 35], and many more have been reported.

5 Theoretical Modeling

Fiber optic SPR sensor comprises a high refractive index fiber core, a metal layer, and a low refractive index dielectric sensing medium (Fig. 4.6). In the case of a fiber optic SPR sensor with bimetallic layers, the single metal layer is replaced by two

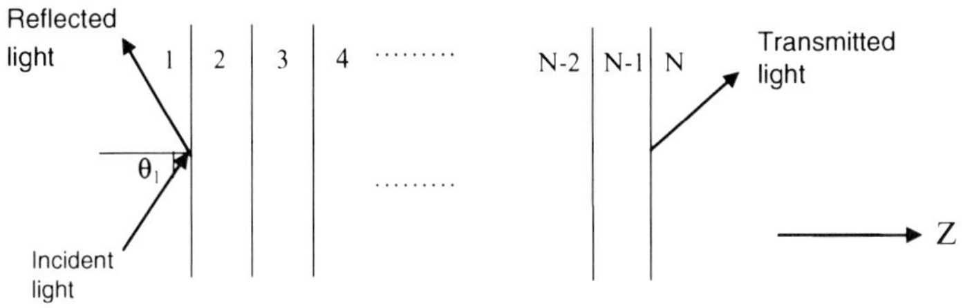

Fig. 4.16 N-layers model for the determination of the reflected light intensity [20]. Reprinted with permission Elsevier

adjacent layers of two different metals. The optical fiber generally used in SPR sensor is a step-index multimode plastic-clad silica fiber. Since the fiber optic SPR sensors are based on wavelength interrogation, the wavelength dependences of the dielectric constants of both fiber core and the metal are required for simulation. The wavelength dependence of the refractive index (n_1) of the fiber with silica core is given by the following Sellmeier dispersion relation

$$n_1(\lambda) = \sqrt{1 + \frac{a_1\lambda^2}{\lambda^2 - b_1^2} + \frac{a_2\lambda^2}{\lambda^2 - b_2^2} + \frac{a_3\lambda^2}{\lambda^2 - b_3^2}} \tag{4.7}$$

where a_1, a_2, a_3, b_1, b_2, and b_3 are the Sellmeier coefficients and λ is the wavelength [16]. For the dispersion relation of metal layer, the Drude model is used according to which the dielectric constant of a metal is given by

$$\varepsilon_{\mathrm{m}}(\lambda) = \varepsilon_{\mathrm{mr}} + \mathrm{i}\varepsilon_{\mathrm{mi}} = 1 - \frac{\lambda^2\lambda_{\mathrm{c}}}{\lambda_{\mathrm{p}}^2(\lambda_{\mathrm{c}} + \mathrm{i}\lambda)} \tag{4.8}$$

where λ_{p} and λ_{c} are the plasma and collision wavelengths, respectively.

In fiber optic SPR sensor unpolarized light is launched into the fiber. Since the geometry of the fiber is cylindrical both transverse electric (TE) and transverse magnetic (TM) polarized light with respect to core–metal interface will propagate in the fiber. As is known, surface plasmons are excited by TM polarized light and hence it will only be affected by the refractive index of the sensing medium. The transmitted power at the output end of the fiber will then be the sum of the unaffected TE polarized light and the surface plasmon affected TM polarized. The minimum normalized transmitted power, therefore, will be more than 0.5. The presence of TE polarized light does not affect the position of dip in SPR spectrum or the resonance wavelength. In the absence of TE polarized light, the normalized minimum transmitted power can be close to zero. Thus, for simulation, planar waveguide approach with TM polarized light can be used. To know the spectrum of the transmitted power, the amplitude reflection coefficient at the interface is required. This is determined using N-layer model shown in Fig. 4.16 and matrix method [20]. For the probe shown in Fig. 4.6, number of layers (*N*) is equal to 3 (fiber core, metallic layer and sensing medium). In the model, layers are assumed to be stacked along the

Z-axis which is normal to the fiber axis or core–metal interface. The arbitrary medium layer in the model is defined by thickness d_k, dielectric constant ε_k, permeability μ_k, and refractive index n_k. The ray guided in the fiber is incident on the core–metal interface at an angle θ_1. The tangential fields at the first boundary $Z = Z_1 = 0$ are related to those at the final boundary $Z = Z_{N-1}$ by

$$\begin{bmatrix} U_1 \\ V_1 \end{bmatrix} = M \begin{bmatrix} U_{N-1} \\ V_{N-1} \end{bmatrix}. \tag{4.9}$$

In (4.9), U_1 and V_1 are the tangential components of electric and magnetic fields at the boundary of the first and the second layer, respectively; U_{N-1} and V_{N-1} are the corresponding fields at the boundary of the $(N-1)$ and Nth layer and M, known as characteristic matrix of the combined structure, is given by

$$M = \prod_{k=2}^{N-1} M_k = \begin{bmatrix} M_{11} & M_{12} \\ M_{21} & M_{22} \end{bmatrix} \tag{4.10}$$

with

$$M_k = \begin{bmatrix} \cos\beta_k & (-\mathrm{i}\sin\beta_k)/q_k \\ -\mathrm{i}q_k \sin\beta_k & \cos\beta_k \end{bmatrix}, \tag{4.11}$$

where

$$q_k = \left(\frac{\mu_k}{\varepsilon_k}\right)^{1/2} \cos\theta_k = \frac{\left(\varepsilon_k - n_1^2 \sin^2\theta_1\right)^{1/2}}{\varepsilon_k} \tag{4.12}$$

$$\beta_k = \frac{2\pi}{\lambda} n_k \cos\theta_k \left(z_k - z_{k-1}\right) = \frac{2\pi d_k}{\lambda}\left(\varepsilon_k - n_1^2 \sin^2\theta_1\right)^{1/2} \tag{4.13}$$

and θ_k is the angle of the ray normal to the kth interface. The amplitude reflection coefficient (r_p) for p-polarized incident light is given by

$$r_\mathrm{p} = \frac{\left(M_{11} + M_{12}q_N\right)q_1 - \left(M_{21} + M_{22}q_N\right)}{\left(M_{11} + M_{12}q_N\right)q_1 + \left(M_{21} + M_{22}q_N\right)}. \tag{4.14}$$

Finally, the reflectance or the intensity reflection coefficient (R_p) for p-polarized light is

$$R_\mathrm{p} = \left|r_\mathrm{p}\right|^2. \tag{4.15}$$

To determine the effective transmitted power, the reflectance (R_p) for a single reflection is raised to the power of the number of reflections the specific propagating ray undergoes with the sensor interface. If the light is launched from a collimated source after focusing onto the end-face of the fiber at the axial point with the help of a microscope objective (lens), then the angular power distribution of all-guided rays launched depends on the angle (θ) of the ray and is given by

$$\mathrm{d}P \propto \frac{n_1^2 \sin\theta \cos\theta}{\left(1 - n_1^2 \cos^2\theta\right)^2} \mathrm{d}\theta. \quad (4.16)$$

For p-polarized light and all guided rays, the generalized expression for the normalized transmitted power, P_{trans}, in a fiber optic SPR sensor is

$$P_{\mathrm{trans}} = \frac{\int_{\theta_{\mathrm{cr}}}^{\pi/2} R_{\mathrm{p}}^{N_{\mathrm{ref}}(\theta)} \frac{n_1^2 \sin\theta \cos\theta}{\left(1 - n_1^2 \cos^2\theta\right)^2} \mathrm{d}\theta}{\int_{\theta_{\mathrm{cr}}}^{\pi/2} \frac{n_1^2 \sin\theta \cos\theta}{\left(1 - n_1^2 \cos^2\theta\right)^2} \mathrm{d}\theta}, \quad (4.17)$$

where

$$N_{\mathrm{ref}}(\theta) = \frac{L}{D \tan\theta}, \quad (4.18)$$

$$\theta_{\mathrm{cr}} = \sin^{-1}\left(\frac{n_{\mathrm{cl}}}{n_1}\right). \quad (4.19)$$

N_{ref} represents the total number of reflections performed by a ray making angle θ with the normal to the core–metal layer interface in the sensing region while L and D represent the length of the sensing region and the fiber core diameter respectively; θ_{cr} is the critical angle of the fiber whereas n_{cl} is the refractive index of the cladding.

6 Choice of Metals

Sensitivity, detection accuracy, reproducibility, and operating range of a sensor are the important parameters to compare it with other sensors. The best sensor is the one that has high sensitivity, detection accuracy, and operating range in addition to giving reproducible results. For a given fiber, these parameters depend on the metal used to coat the core of the fiber. Metals have complex dielectric constant. The real part of the dielectric constant ($\varepsilon_{\mathrm{mr}}$) accounts for the reflection while the imaginary part ($\varepsilon_{\mathrm{mi}}$) accounts for the absorption. The SPR spectrum is very sensitive to the values of the real and imaginary parts of the dielectric constant. Sharpness of the dip increases if the ratio ($\varepsilon_{\mathrm{mr}}/\varepsilon_{\mathrm{mi}}$) increases. In Table 4.2, the dielectric constant and the ratio of the real to imaginary parts for various metals have been tabulated [36]. It may be noted that silver has highest value of the ratio followed by copper, gold, and

Table 4.2 Values of dielectric constants for different metals

Metal	Wavelength (nm)	Dielectric constant ($\varepsilon_r + i\varepsilon_i$)	$\varepsilon_r/\varepsilon_i$
Gold (Au)	632.8	−10.92 + 1.49i	7.33
Copper (Cu)	632.8	−14.67 + 0.72i	20.4
Silver (Ag)	632.8	−18.22 + 0.48i	38.0
Aluminum (Al)	650.0	−42.00 + 16.40i	2.56

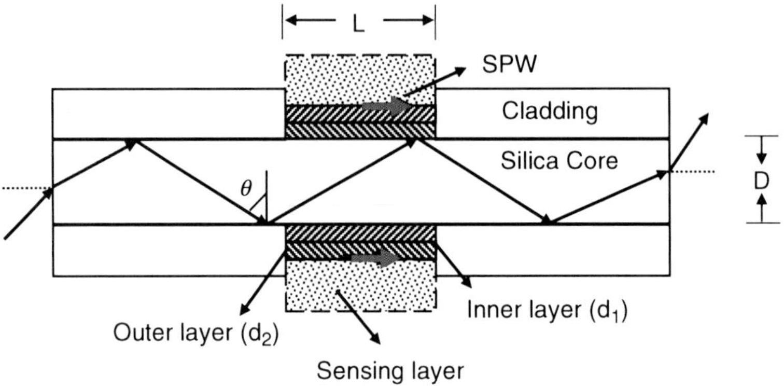

Fig. 4.17 Fiber optic SPR probe with bimetallic layers [19]. Reprinted with permission Elsevier

aluminum. This means that the sharpest peak is produced by silver which is indeed the case. Gold produces a broader SPR spectrum than silver. The disadvantage of silver and copper is that their films are not stable due to oxidation. Oxidation of silver and copper occurs as soon as these are exposed to air and especially to water, which makes it difficult to give a reproducible result and hence the sensor remains unreliable for practical applications. Thus the protection of these metals is required for stable use. These metals are protected by highly stable metals such as gold. Figure 4.17 shows a fiber optic SPR sensor utilizing a bimetallic coating on the core of the fiber. The thickness of the inner layer is d_1 while that of outer layer is d_2. The unstable metals such as silver and copper form the inner layer while the stable metals such as gold and aluminum are used for the outer layer. To analyze such a structure, simulation is carried out using the model given in Sect. 5 with $N=4$ and all guided rays launching in the fiber. Figure 4.18 shows the variation of sensitivity with the ratio of inner layer thickness to the total bimetallic thickness [$d=d_1/(d_1+d_2)$] for different bimetallic combinations [37]. The figure predicts that out of four metals the sensor with gold layer only is the most sensitive whereas the sensor with aluminum layer only is the least. As far as the sensitivities of different combinations are concerned, the Ag–Au combination is the best while Cu–Al is the worst. As expected next to Ag–Au is the Cu–Au combination which provides good sensitivity. Figure 4.19 shows the variation of SNR with the ratio of inner layer thickness to the total bimetallic thickness (d) for different bimetallic combinations [37].

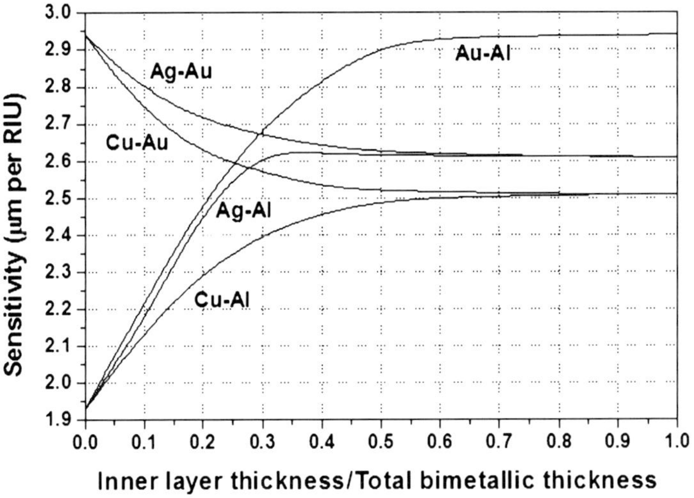

Fig. 4.18 Variation of sensitivity with the ratio of inner layer thickness to total bimetallic thickness for five combinations [37]. The probe parameters are: NA = 0.24, D = 600 μm, L = 15 mm, d_{total} = 50 nm. Reprinted with permission AIP

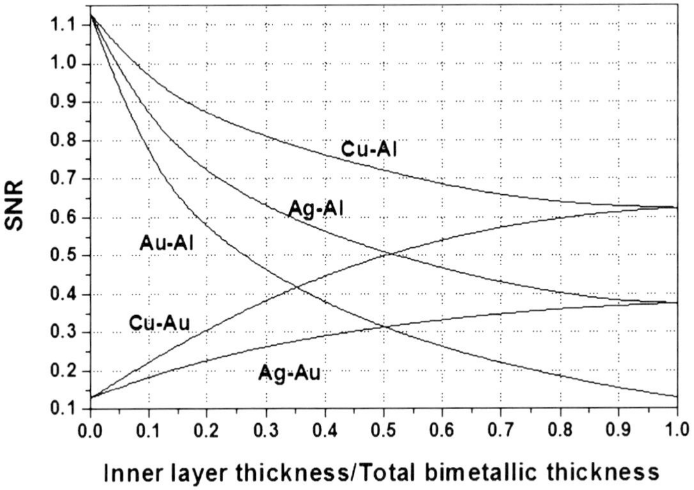

Fig. 4.19 Variation of signal-to-noise ratio with the ratio of inner layer thickness to total bimetallic thickness for five combinations [37]. The values of the probe parameters are the same as in Fig. 4.18. Reprinted with permission AIP

The variation of SNR predicts that Cu–Al is better among all the bimetallic combinations. This study implies that there is no single combination of metals that provides high values of both SNR and sensitivity simultaneously and hence one should choose the combination according to the requirement.

7 Addition of Dopants

In an SPR sensor, the resonance condition (4.4) depends upon the refractive index of the material of the prism/fiber core. The refractive index and the wavelength dependence can be changed by adding dopants in the material. Thus the sensitivity of the sensor can be enhanced or tuned by adding dopants. In the case of SPR-based fiber optic sensors, optical fiber with pure silica core is used. The addition of dopants in pure silica changes the values of the Sellmeier coefficients in (4.7). The dopants which have been tried are germanium oxide (GeO_2), boron oxide (B_2O_3), and phosphorous pentoxide (P_2O_5). The simulated results for sensitivity based on above theory and silver layer are depicted in Fig. 4.20 for different dopant concentrations [38]. The simulation predicts higher sensitivity for B_2O_3 in comparison to pure silica. For a given refractive index of the sensing medium, the sensitivity for other dopants is lower than that of pure silica. The effect of dopants on the sensitivity of the fiber optic SPR sensor remains the same irrespective of whether a single metal layer or a bimetallic configuration is used. Figure 4.21 shows bar diagram for SNR with silver fraction in bimetallic configuration for five cases [38]. It has been observed that SNR remains nearly the same for a given fraction of silver in all the five cases.

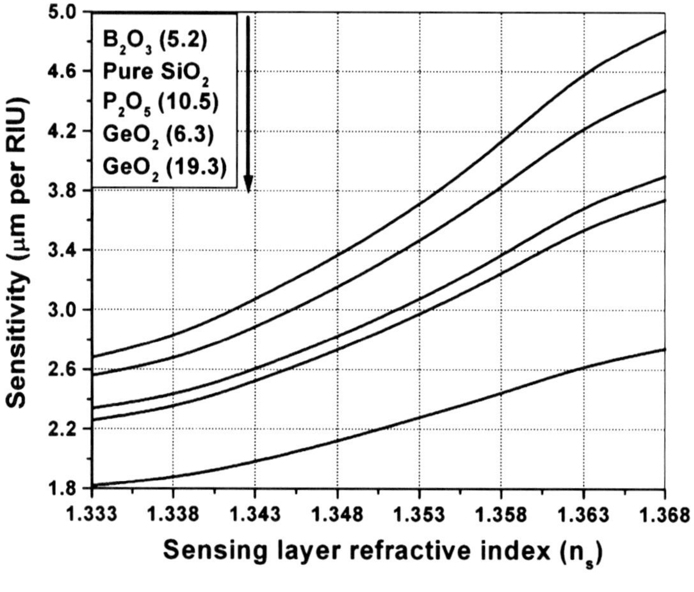

Fig. 4.20 Variation of sensitivity with sensing layer refractive index for different dopants. *Numbers* in the *brackets* are the molar concentrations of dopants in mole percent [38]. Reprinted with permission Elsevier

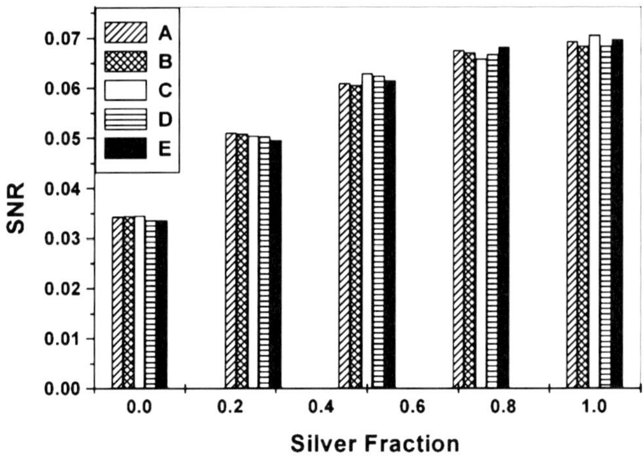

Fig. 4.21 *Bar diagram* representation of SNR with silver fraction in bimetallic coating for (**a**) Silica doped with GeO_2 (19.3); (**b**) Silica doped with GeO_2 (6.3); (**c**) Silica doped with P_2O_5 (10.5); (**d**) Pure silica with no doping; and (**e**) Silica doped with B_2O_3 (5.2). *Numbers* in *brackets* are the molar concentrations of dopant in mole percent [38]. Reprinted with permission Elsevier

8 Role of Angle of Incidence

To understand the role of angle of incidence on the sensitivity of the sensor in wavelength interrogation method, the propagation constant (K_{SP}) of surface plasmon wave has been plotted as a function of wavelength for $n_s = 1.333$ and 1.343 in Fig. 4.22a with metallic layer of gold. In the same figure, the wavelength dependence of the propagation constant of the evanescent wave (K_{ev}) has been plotted for 80° angle of incidence of the light beam incident from silica to metal. To plot these curves, the wavelength dependence of the refractive index/dielectric constant of silica and gold are determined using (4.7) and (4.8), respectively. It may be seen from the figure that, for angle of incidence equal to 80°, K_{SP} and K_{ev} curves cross each other at 0.6276 μm for $n_s = 1.333$ and at 0.6652 μm for $n_s = 1.343$. This gives a shift ($\delta\lambda_{res}$) of 37.6 nm in resonance wavelength for $\delta n_s = 0.010$. In Fig. 4.22b, same curves have been plotted for 75° angle of incidence. For this value of angle of incidence, K_{SP} and K_{ev} curves cross each other at 0.7278 μm for $n_s = 1.333$ and at 0.7928 μm for $n_s = 1.343$. This gives a shift ($\delta\lambda_{res}$) of 65.0 nm in resonance wavelength for $\delta n_s = 0.010$. Comparison of shift in resonance wavelength in the two cases suggests that the decrease in the angle of incidence increases the sensitivity of the sensor.

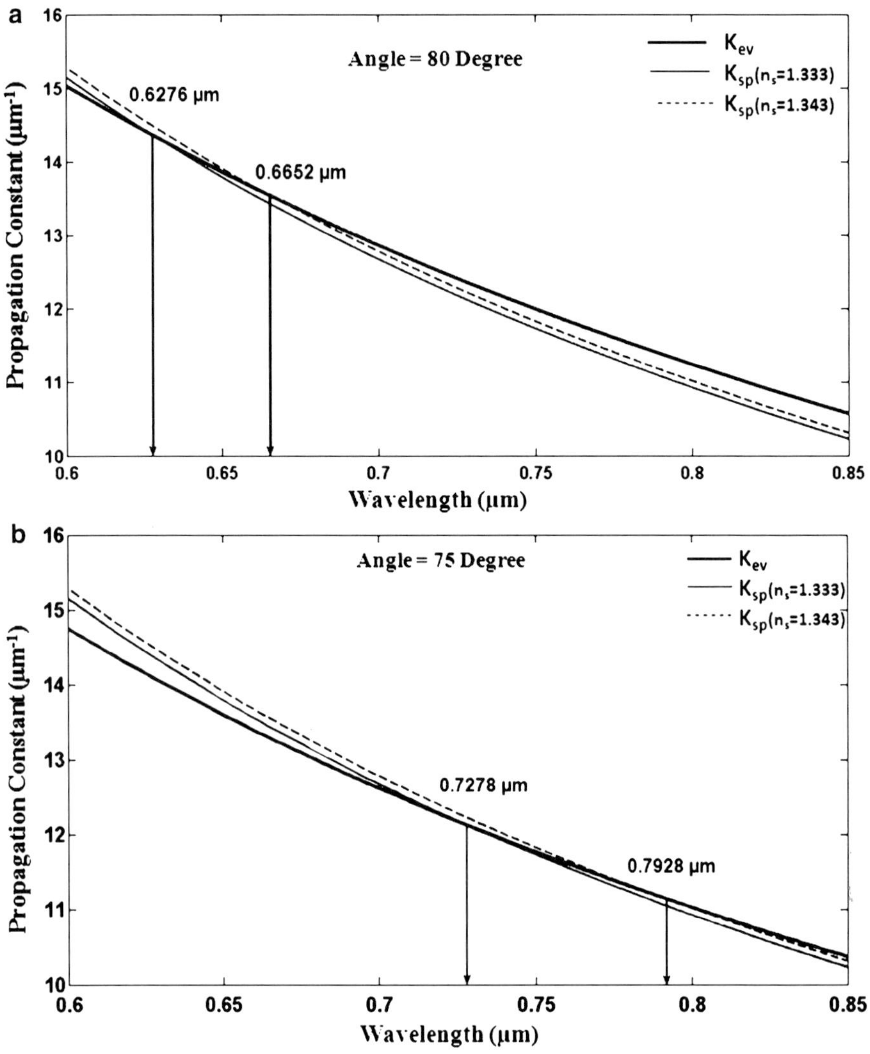

Fig. 4.22 Variation of propagation constant for SPW and evanescent wave of incident beam with wavelength for (**a**) 85° and (**b**) 75° angle of incidence

9 Change of Geometry

As seen above, the sensitivity of a fiber optic SPR sensor can be enhanced by decreasing the angle of incidence of the ray at the core–metal layer interface. This can be achieved by changing the shape or the geometry of the fiber optic SPR probe. Tapering the fiber optic SPR probe is one of the changes in the geometry of the

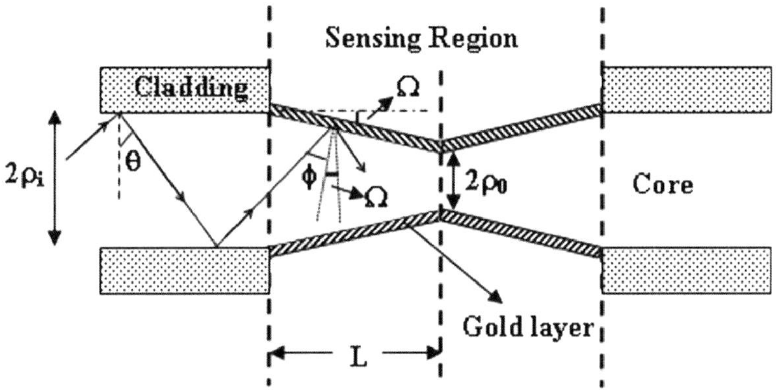

Fig. 4.23 A typical SPR-based fiber optic tapered probe [40]. Reprinted with permission Elsevier

probe that has been tried [18, 39]. A typical tapered fiber optic SPR probe is shown in Fig. 4.23 [40]. In the sensing region, the variation of core diameter has been taken as linear. It can be seen from the figure that as the ray propagates in the fiber through the tapered region, the angle of incidence of the ray with the normal to the core–metal layer interface decreases. However, the decrease in the angle of incidence increases the number of reflections per unit length which is disadvantageous because increase in the number of reflections increases the width of the SPR spectrum. In the case of linear taper, the radius of the core varies according to the following relation

$$\rho(z) = \rho_{\mathrm{i}} - \frac{z}{L}(\rho_{\mathrm{i}} - \rho_{\mathrm{o}}), \tag{4.20}$$

where ρ_{i} and ρ_{o} are the core radii for input and output end of the taper respectively; z is the distance measured from the input face of the taper and L is the length of the taper. The angle of incidence at a distance z is given as

$$\phi(z) = \cos^{-1}\left[\frac{\rho_{\mathrm{i}}\cos\theta}{\rho(z)}\right] - \Omega \tag{4.21}$$

where θ is angle of incidence of the ray before entering the taper and Ω is the taper angle. Due to the variation of angle of incidence inside the probe, the equation of transmitted power (4.17) becomes

$$P_{\mathrm{trans}} = \frac{\int_0^L \mathrm{d}z \int_{\phi_1(z)}^{\phi_2(z)} R_{\mathrm{P}}^{N_{\mathrm{ref}}(\theta,z)} \frac{n_1^2 \sin\theta\cos\theta}{(1 - n_1^2\cos^2\theta)^2}\mathrm{d}\theta}{\int_0^L \mathrm{d}z \int_{\phi_1(z)}^{\phi_2(z)} \frac{n_1^2 \sin\theta\cos\theta}{(1 - n_1^2\cos^2\theta)^2}\mathrm{d}\theta} \tag{4.22}$$

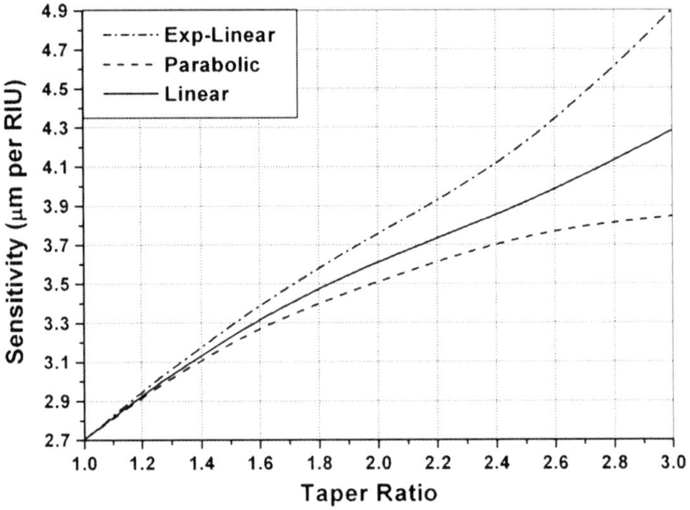

Fig. 4.24 Variation of sensitivity with taper ratio for three different taper profiles [40]. Reprinted with permission Elsevier

where

$$N_{\mathrm{ref}}(\theta, z) = \frac{L}{2\rho(z)\tan(\theta + \Omega)} \quad (4.23)$$

is the number of reflections; $\phi_1(z)$ and $\phi_2(z)$ are the angles corresponding to $\theta = \sin^{-1}(n_{\mathrm{cl}} / n_1)$ and $\pi / 2$ respectively. Figure 4.24 shows the variation of sensitivity of the tapered fiber optic SPR probe with taper ratio (ρ_i/ρ_o). In the same figure, the sensitivity has also been plotted for two other profiles, namely, parabolic and exponential–linear [40] which are mathematically written as

$$\rho_e(z) = (\rho_i - \rho_o)\left[e^{\left(-\frac{z}{L}\right)} - \frac{z}{L}\left(e^{-1}\right)\right] + \rho_o \quad \text{(Exponential - linear)}$$

$$\rho_P(z) = \left[\rho_i^2 - \frac{z}{L}\left(\rho_i^2 - \rho_o^2\right)\right]^{1/2} \quad \text{(Parabolic)}$$

An increase in the sensitivity with the increase in the taper ratio has been obtained. For a given taper ratio, the exponential–linear taper profile provides the maximum sensitivity. Thus the decrease in the angle of incidence of the guided rays with the normal to the core–cladding interface is an important parameter for the enhancement of the sensitivity.

The sensitivity can further be enhanced if an SPR probe of uniform core (with metallic coating) sandwiched between two unclad tapered fiber regions as shown in Fig. 4.25 is used [41]. In this configuration, taper region 1 is used to bring down the

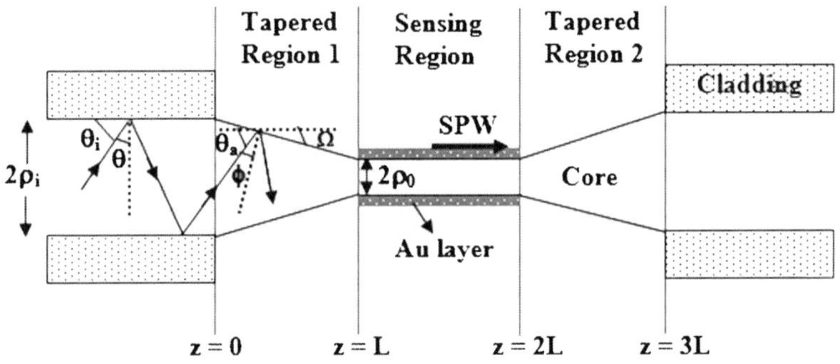

Fig. 4.25 SPR-based fiber optic probe of uniform core radius sandwiched between two tapered fiber regions [41]. Reprinted with permission IEEE

angles of the guided rays in the fiber close to the critical angle of the unclad tapered region while the taper region 2 is used to reconvert the angles of these rays to their initial values so that all the guided rays can propagate up to the output end of the fiber. In the sensing region, rays propagate with angles smaller than that in the probe shown in Fig. 4.23. For this probe, the number of reflections and the transmitted power are given by the following equations

$$N_{\text{ref}}(\theta) = L / \left[2\rho_{\text{o}} \tan\left(\theta + \Omega\right) \right]. \tag{4.24}$$

$$P_{\text{trans}} = \frac{\int_{\phi_1}^{\phi_2} R_{\text{p}}^{N_{\text{ref}}(\theta)} \dfrac{n_1^2 \sin\theta \cos\theta}{\left(1 - n_1^2 \cos^2\theta\right)^2} \mathrm{d}\theta}{\int_{\phi_1}^{\phi_2} \dfrac{n_1^2 \sin\theta \cos\theta}{\left(1 - n_1^2 \cos^2\theta\right)^2} \mathrm{d}\theta} \tag{4.25}$$

where

$$\phi_1 = \cos^{-1}\left[\rho_{\text{i}} \cos\theta_1 / \rho_{\text{o}}\right] - \Omega$$

$$\phi_2 = \cos^{-1}\left[\rho_{\text{i}} \cos\theta_2 / \rho_{\text{o}}\right] - \Omega$$

$\theta_1 = \sin^{-1}\left(n_{\text{cl}} / n_1\right)$ and $\theta_2 = \pi / 2$. Figure 4.26 shows the variation of sensitivity with taper ratio for the probe shown in Fig. 4.25 [41]. Similar to the above case, the sensitivity increases with an increase in the taper ratio. However, a significant enhancement in sensitivity is obtained in this case.

The angle of incidence of the ray with the normal to the core-cladding interface can also be decreased by bending the probe. An SPR-based fiber optic sensor with uniform semimetal coated U-shaped probe as shown in Fig. 4.27 has been analyzed using a bidimensional model [42]. To make the analysis simpler and enhance the sensitivity, all the guided rays of the p-polarized light launched in the fiber and

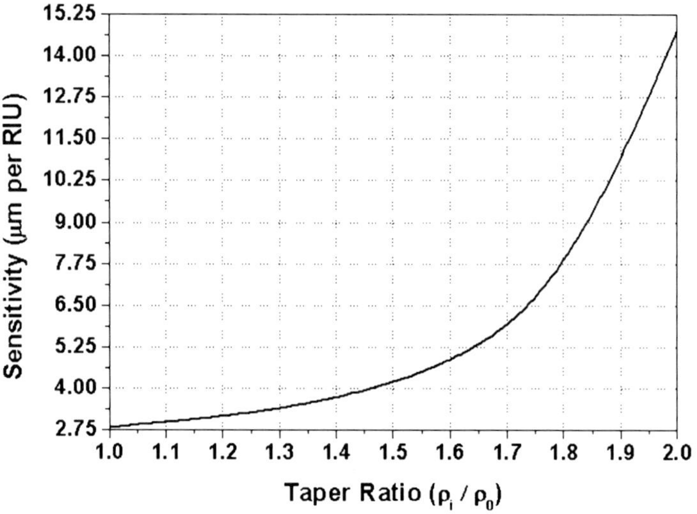

Fig. 4.26 Variation of sensitivity of the SPR probe shown in Fig. 4.25 with taper ratio [41]. Reprinted with permission IEEE

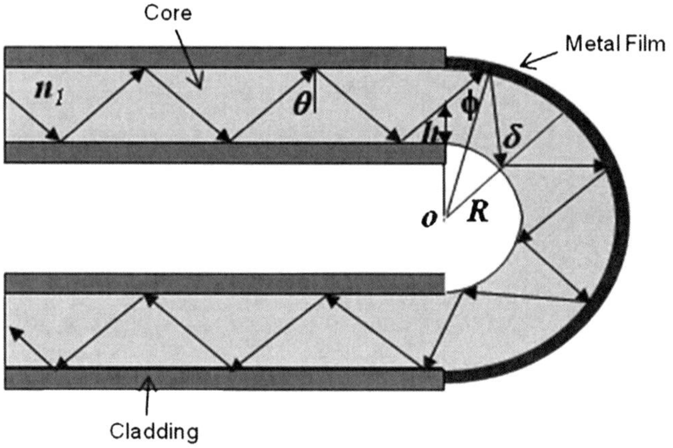

Fig. 4.27 A typical U-shaped SPR probe [42]. Reprinted with permission IOP

their electric vectors have been assumed to lie in the plane of bending of the probe. The number of reflections and the transmitted power are given by the following equations

$$N_{\mathrm{ref}} = \frac{L}{8\rho}\left[\cot\theta + \cot\left(\left(\frac{R+2\rho}{R}\right)\theta\right)\right], \tag{4.26}$$

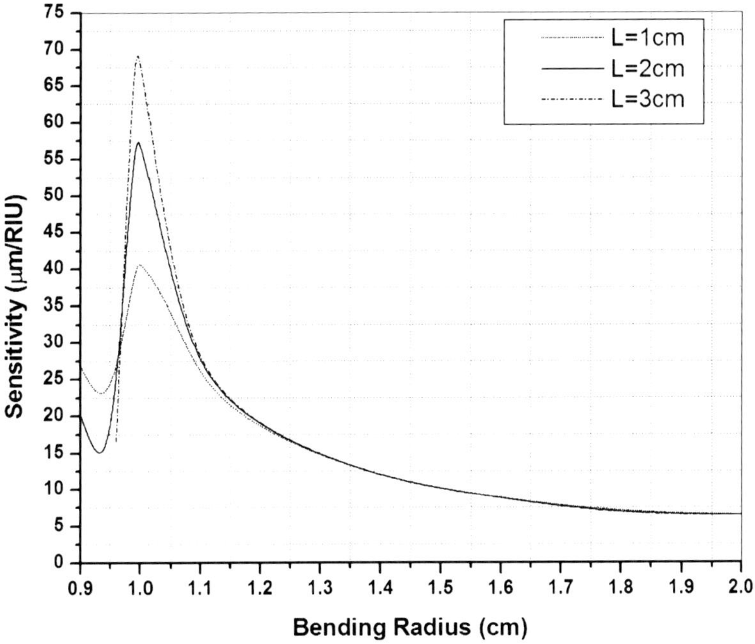

Fig. 4.28 Variation of sensitivity with bending radius for three different values of the sensing length [42]. Reprinted with permission IOP

$$P_{\text{trans}} = \frac{\int_0^{2\rho} \mathrm{d}h \int_{\phi_1(h)}^{\phi_2(h)} R_{\mathrm{P}}^{N_{\text{ref}}} \frac{n_1^2 \sin\theta\cos\theta}{\left(1 - n_1^2 \cos^2\theta\right)^2} \mathrm{d}\theta}{\int_0^{2\rho} \mathrm{d}h \int_{\phi_1(h)}^{\phi_2(h)} \frac{n_1^2 \sin\theta\cos\theta}{\left(1 - n_1^2 \cos^2\theta\right)^2} \mathrm{d}\theta}, \tag{4.27}$$

where

$$\phi(h) = \sin^{-1}\left[\frac{(R+h)}{(R+2\rho)}\sin\theta\right]. \tag{4.28}$$

h is the distance from the core-cladding interface (of the inner surface of U-region) to the point at which ray strikes the input face of the bending region; R is the bending radius of the inner circular region; $\phi_1(h)$ and $\phi_2(h)$ are the angles obtained from (4.28) corresponding to $\theta = \sin^{-1}(n_{\text{cl}} / n_1)$ and $\pi / 2$ respectively. Figure 4.28 shows the variation of sensitivity with the bending radius for three different values of the sensing length of the probe [42]. The simulated results show an increase in the sensitivity with the decrease in the bending radius. However, the increase in sensitivity is up to a certain value of the bending radius below that it starts decreasing sharply. This suggests that there exists an optimum value of the bending radius at which the sensitivity of the SPR sensor based on U-shaped probe acquires a maximum value. The trend of variation is same for all the three values of the sensing length. This is due to the independence of the angle of incidence on the sensing length. For a given bending radius, the sensitivity increases as the sensing length increases. The enhancement in sensitivity is several times more than that reported for an SPR-based fiber optic tapered probe.

All the studies discussed above use multimode optical fibers. Single-mode optical fibers have also been used in SPR sensing. Single-mode optical fiber-based SPR sensors are more sensitive and more accurate than multimode optical fiber SPR sensors. However, their fabrication is much more complex and sophisticated in comparison to those that use multimode optical fibers. Side-polished single-mode optical fiber with a thin metal over layer has been used in SPR sensing [43]. In this design, the guided mode propagating in the fiber excites the surface plasmon wave at the metal-sensing medium interface. The resonance occurs when two modes are closely phase matched. The side polished multimode optical fiber [44] and D-type single mode optical fibers have also been used for SPR sensing [45, 46]. The other designs for SPR-based fiber optic sensor includes SPR probe at one end of the fiber with the reflecting end face [32, 47] and a fiber tip [48, 49]. The photonic band-gap fibers have also been used for sensing utilizing SPR technique [50, 51].

10 Light Launching

To simplify calculations, theoretical modeling considers the propagation of only meridional rays in the fiber. However, there exist skew rays in the fiber depending on the light launching conditions. These rays do not intersect fiber axis, but follow helical path inside the fiber. To specify the trajectory of a skew ray, a second angle known as skewness angle (α) is defined. It is the angle between the normal to the core-cladding interface (PN) and the projection of the ray path on the cross-section of the fiber core as shown in Fig. 4.29. The midpoints of the ray between successive reflections touch a cylindrical surface of radius r_s, called as the inner caustic radius. It is given by

$$r_s = \rho \sin \alpha. \tag{4.29}$$

For meridional rays, α and r_s are equal to zero. To see the effect of skew rays, light launching in the fiber can be considered as shown in Fig. 4.30. Here the collimated beam makes an angle θ_s with the axis of the fiber and the lens that focuses the beam at point Q on the fiber end face. The point Q is at a distance of r_s from the fiber axis. The distance r_s can also be written as

$$r_s = f \tan \theta_s \tag{4.30}$$

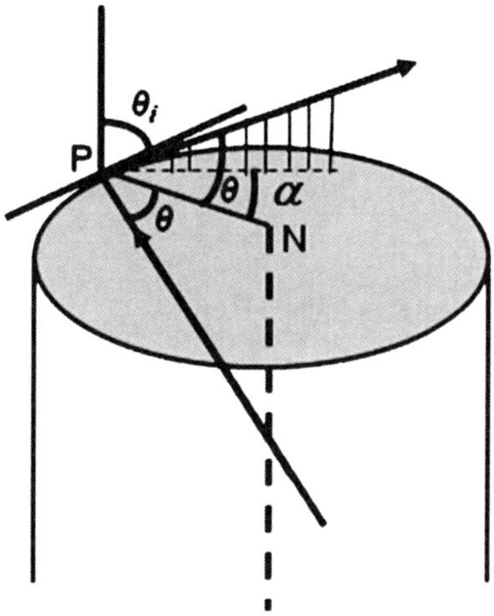

Fig. 4.29 Representation of skew ray and skewness angle in an optical fiber [52]. Reprinted with permission OSA

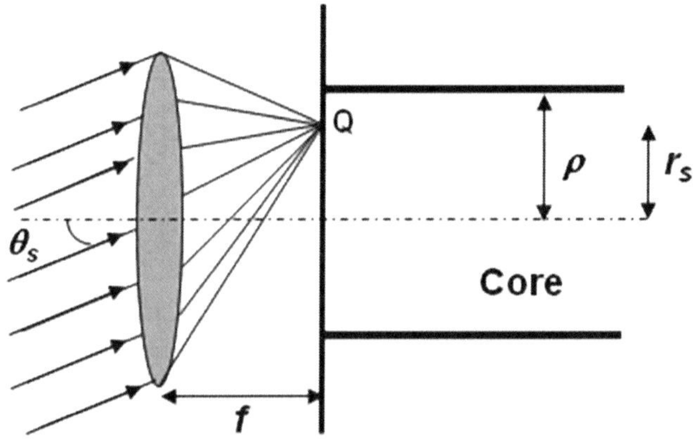

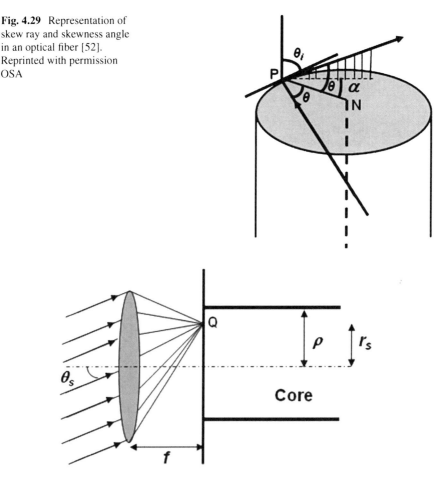

Fig. 4.30 Optical arrangement for excitation of meridional and skew rays in an optical fiber

where f is the focal length of the lens. For $r_s = \rho$, angle θ_s will have maximum value and in that case maximum number of skew rays will be excited. The number of reflections and the transmitted power are given by the following equations

$$N_{\mathrm{ref}}(\theta,\alpha) = \frac{L}{D\cos\alpha\tan\theta}, \tag{4.31}$$

$$P_{\mathrm{trans}} = \frac{\int_0^{\alpha_{\max}}\int_{\theta_{\mathrm{cr}}}^{\pi/2} R_{\mathrm{p}}^{N_{\mathrm{ref}}(\theta,\alpha)} \frac{n_1^2\sin\theta\cos\theta}{\left(1-n_1^2\cos^2\theta\right)^2}\cos\theta_s\,\mathrm{d}\theta\,\mathrm{d}\alpha}{\int_0^{\alpha_{\max}}\int_{\theta_{\mathrm{cr}}}^{\pi/2} \frac{n_1^2\sin\theta\cos\theta}{\left(1-n_1^2\cos^2\theta\right)^2}\cos\theta_s\,\mathrm{d}\theta\,\mathrm{d}\alpha}. \tag{4.32}$$

Figure 4.31 shows the variation of sensitivity with skewness parameter [52]. The sensitivity decreases as the value of skewness parameter increases irrespective of

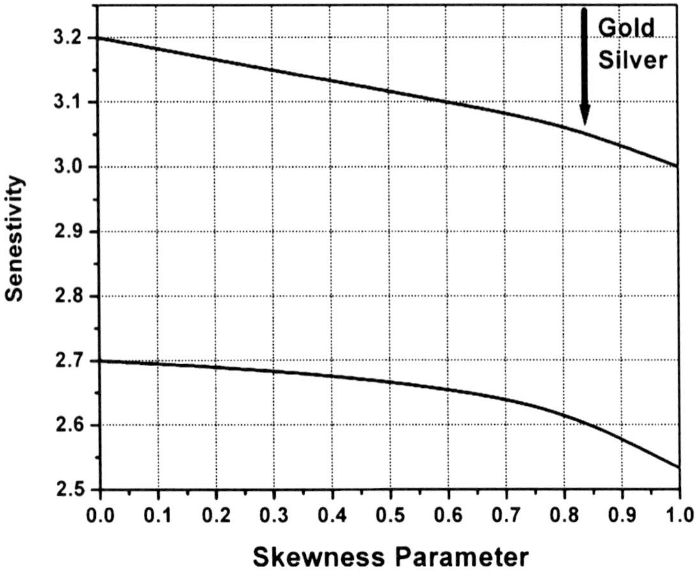

Fig. 4.31 Variation of sensitivity of an SPR-based fiber optic sensor with skewness parameter for two different metals [52]. Reprinted with permission OSA

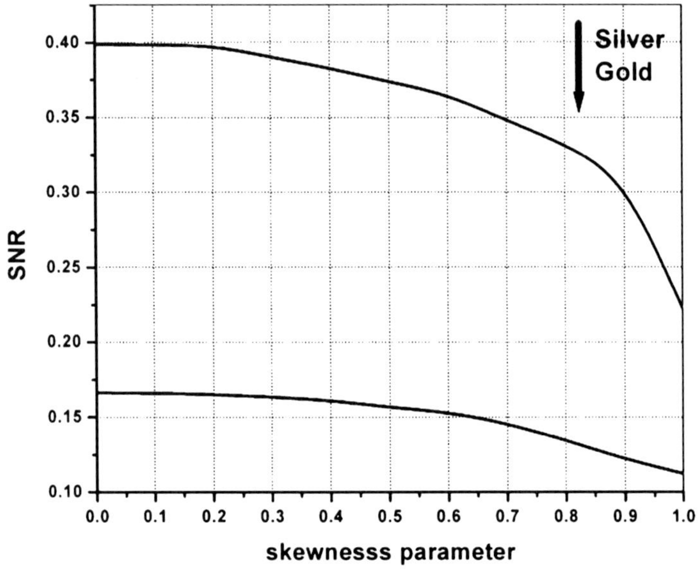

Fig. 4.32 Variation of SNR of an SPR-based fiber optic sensor with skewness parameter for two different metals [52]. Reprinted with permission OSA

the metal used for coating. As expected, the sensitivity is better in the case of gold than silver. The decrease in sensitivity for highest value of skewness parameter is about 16%. Figure 4.32 shows the variation of SNR with skewness parameter [52]. Similar to sensitivity, SNR decreases as the value of skewness parameter increases irrespective of the metal used. It is better in the case of silver than gold as expected. However, the effect of skew rays is more on SNR than on the sensitivity. In the case of gold film, the decrement is about 40% while in the case of silver it is around 30%.

11 Wavelength Tuning

In the case of metallic coatings the SPR wavelength lies in the visible region of the spectrum. Therefore these cannot be used for sensing in infrared region which is required for many applications in medicine, environment, and security. The spectral region of sensing can be brought to near infrared region if the probe is modified, for example, by using multilayer coating [53], by writing a grating on fiber core [54–58] or by exciting long and short range surface plasmons by using dielectric on both sides of the metal film [59]. If conducting metal oxide such as indium tin oxide (ITO) is used in place of metal then the wavelength tuning of the surface plasmons resonance can be done for infrared region. The theoretical analysis carried out for SPR-based fiber optic sensor utilizing ITO film predicts an increase in the sensitivity by 60% along with operating region in mid infrared region [60]. The wavelength range can further be shifted to infrared side if the geometry of the probe is changed, for example, by tapering.

12 Summary

In this chapter, we have described the principle of SPR technique that has been used for sensing. In the beginning, the technique was used for prism-based SPR sensors but later when the advantages of optical fiber were realized it was applied to optical fiber-based sensors. To enhance the sensitivity of the fiber optic SPR sensor, various designs of the probe have been described. The probe designs are tapered probes with different taper profiles, U-shaped fiber probe, and replacement of metallic film by metal oxide. These developments are likely to drive future trends in the research and development of optical fiber sensors.

Acknowledgment The present work is partially supported by the Department of Science and Technology (India).

References

1. Liedberg, B., Nylander, C. and Sundstrom, I. (1983). Surface Plasmon Resonance for Gas Detection and Biosensing, Sensors and Actuators, 4, 299–304.
2. van Gent, J., Lambeck, P.V., Kreuwel, H.J.M., Gerritsma, G.L., Sudholter, E.J.R., Reinhoudt, D.N. and Popma, T.J.A. (1990). Optimization of a Chemooptical Surface Plasmon Resonance based Sensor", Applied Optics, 29 (19), 2843–2849.
3. Stenberg, E., Persson, B., Roos, H. and Urbaniczky, C. (1991). Quantitative Determination of Surface Concentration of Protein with Surface Plasmon Resonance using Radiolabeled Proteins, Journal of Colloid and Interface Science, 143 (2), 513–526.
4. Dougherty, G. (1993). A Compact Optoelectronic Instrument with a Disposable Sensor based on Surface Plasmon Resonance, Measurement Science and Technology, 4, 697–699.
5. Ekgasit, S., Tangcharoenbumrungsuk, A., Yu, F., Baba, A. and Knoll, W. (2005). Resonance Shifts in SPR Curves of Nonabsorbing, Weakly Absorbing, and Strongly Absorbing Dielectrics", Sensors and Actuators B, 105, 532–541.
6. Chyou, J.J., Chu, C.S., Chien, F.C., Lin, C.Y., Yeh, T.L., Hsu, R.C. and Chen, S.J. (2006). Precise Determination of the Dielectric Constant and Thickness of a Nanolayer by Use of Surface Plasmon Resonance Sensing and Multiexperiment Linear Data Analysis, Applied Optics, 45 (23), 6038–6044.
7. Chiang, H.P., Chen, C.W., Wu, J.J., Li, H.L., Lin, T.Y., Sanchez, E.J. and Leung, P.T. (2007). Effects of Temperature on the Surface Plasmon Resonance at a Metal-Semiconductor Interface, Thin Solid Films, 515, 6953–6961.
8. Person, J.L., Colas, F., Compere, C., Lehaitre, M., Anne, M.L., Boussard-Pledel, C., Bureau, B., Adam, J.L., Deputier, S. and Guilloux-Viry, M. (2007). Surface Plasmon Resonance in Chalcogenide Glass based Optical System, Sensors and Actuators B, 130, 771–776.
9. Feng, W., Shenye, L., Xiaoshi, P., Zhuangqi, C. and Yongkun, D. (2008). Reflective Type Configuration for Monitoring the Photobleaching Procedure based on Surface Plasmon Resonance, Journal of Optics A: Pure and Applied Optics, 10, 095102.
10. Jorgenson, R.C. and Yee, S.S. (1993). A Fiber-Optic Chemical Sensor based on Surface Plasmon Resonance, Sensors and Actuators B, 12, 213–220.
11. Harris, R.D. and Wilkinson, J.S. (1995). Waveguide Surface Plasmon Resonance Sensors, Sensors and Actuators B, 29, 261–267.
12. Lin, W.B., Jaffrezic-Renault, N., Gagnaire, A. and Gagnaire, H. (2000). The Effects of Polarization of the Incident Light-Modeling and Analysis of a SPR Multimode Optical Fiber Sensor, Sensors and Actuators A, 84, 198–204.
13. Slavík, R., Homola, J., Ctyroký, J. and Brynda, E. (2001). Novel Spectral Fiber Optic Sensor based on Surface Plasmon Resonance, Sensors and Actuators B, 74, 106–111.
14. Piliarik, M., Homola, J., Maníková, Z. and Ctyroký, J. (2003). Surface Plasmon Resonance Sensor based on a Single-Mode Polarization-Maintaining Optical Fiber, Sensors and Actuators B, 90, 236–242.
15. Gentleman, D.J., Obando, L.A., Masson, J.F., Holloway, J.R. and Booksh, K. (2004). Calibration of Fiber Optic based Surface Plasmon Resonance Sensors in Aqueous Systems", Analytica Chimica Acta, 515, 291–302.
16. Sharma, A.K. and Gupta, B.D. (2004). Absorption-based Fiber Optic Surface Plasmon Resonance Sensor: A Theoretical evaluation, Sensors and Actuators B, 100, 423–431.
17. Mitsushio, M., Higashi, S. and Higo, M. (2004). Construction and Evaluation of a Gold-Deposited Optical Fiber Sensor System for Measurements of Refractive Indices of Alcohols, Sensors and Actuators A, 111, 252–259.
18. Kim, Y., Peng, W., Banerji, S. and Booksh, K.S. (2005). Tapered Fiber Optic Surface Plasmon Resonance Sensor for Analyses of Vapor and Liquid Phases, Optics Letters, 30 (17), 2218–2220.

19. Sharma, A.K. and Gupta, B.D. (2005). On the Sensitivity and Signal to Noise Ratio of a Step-Index Fiber Optic Surface Plasmon Resonance Sensor with Bimetallic Layers, Optics Communications, 245, 159–169.
20. Gupta, B.D. and Sharma, A.K. (2005). Sensitivity Evaluation of a Multi-Layered Surface Plasmon Resonance-based Fiber Optic Sensor: A Theoretical Study, Sensors and Actuators B, 107, 40–46.
21. Rajan, Chand, S. and Gupta, B.D. (2006). Fabrication and Characterization of a Surface Plasmon Resonance based Fiber-Optic Sensor for Bittering Component—Naringin, Sensors and Actuators B, 115, 344–348.
22. Diaz-Herrera, N., Esteban, O., Navarrete, M.C., Haitre, M.L. and Gonzalez-Cano, A. (2006). In Situ Salinity Measurements in Seawater with a Fibre-Optic Probe, Measurement Science and Technology, 17, 2227–2232.
23. Sharma, A.K., Jha, R. and Gupta, B.D. (2007). Fiber-Optic Sensors based on Surface Plasmon Resonance: A Comprehensive Review, IEEE Sensors Journal, 7 (8), 1118–1129.
24. Yu-Cheng, L., Yu-Chia, T., Woo-Hu, T., Tsui-Shan, H., Ko-Shao, C. and Shu-Chuan, L. (2008). The Enhancement Method of Optical Fiber Biosensor based on Surface Plasmon Resonance with Cold Plasma Modification, Sensors and Actuators B, 133, 370–373.
25. Kanso, M., Cuenot, S. and Louarn, G. (2008). Sensitivity of Optical Fiber Sensor based on Surface Plasmon Resonance: Modeling and Experiments, Plasmonics, 3(2–3), 49–57.
26. Navarrete, M.C., Diaz-Herrera, N., Gonzalez-Cano, A. and Esteban, O. (2010) A Polarization Independent SPR Fiber Sensor, Plasmonics, 5, 7–12.
27. Yan, J., Lu, Y., Wang, P., Gu, C., Zheng, R., Chen, Y., Ming, H. and Zhan, Q. (2009). Improving the Sensitivity of Fiber-Optic SPR Sensor via Radially Polarized Beam Excitation, Chinese Optics Letters, 7(10), 909–911.
28. Yanase, Y., Araki, A., Suzuki, H., Tsutsui, T., Kimura, T., Okamoto, K., Nakatani, T., Hiragun, T. and Hide, M. (2010) Development of an Optical Fiber SPR Sensor for Living Cell Activation, Biosensors and Bioelectronics, 25, 1244–1247.
29. Kretchmann, E. and Reather, H. (1968). Radiative Decay of Non-Radiative Surface Plasmons Excited by Light, Naturforsch, 23, 2135–2136.
30. Zynio, S.A., Samoylov, A.V., Surovtseva, E.R., Mirsky, V.M. and Shirsov, Y.M. (2002). Bimetallic Layers Increase Sensitivity of Affinity Sensors based on Surface Plasmon Resonance, Sensors, 2, 62–70.
31. Rajan, Chand, S. and Gupta, B.D. (2007). Surface Plasmon Resonance based Fiber-Optic Sensor for the Detection of Pesticide, Sensors and Actuators B, 123, 661–666.
32. Gentleman, D.J. and Booksh, K.S. (2006). Determining Salinity using a Multimode Fiber Optic Surface Plasmon Resonance Dip-Probe, Talanta, 68, 504–515.
33. Matsushita, T., Nishikawa, T., Yamashita, H., Kishimoto, J. and Okuno, Y. (2008). Development of New Single-Mode Waveguide Surface Plasmon Resonance Sensor using a Polymer Imprint Process for High-Throughput Fabrication and Improved Design Flexibility, Sensors and Actuators B, 129, 881–887.
34. Ozdemir, S.K. and Sayan, G.T. (2003). Temperature Effects on Surface Plasmon Resonance: Design Considerations for an Optical Temperature Sensor, Journal of Lightwave Technology, 21 (3), 805–814.
35. Sharma, A.K. and Gupta, B.D. (2006). Theoretical Model of a Fiber Optic Remote Sensor based on Surface Plasmon Resonance for Temperature Detection, Optical Fiber Technology, 12, 87–100.
36. Mitsushio, M., Miyashita, K. and Higo, M. (2006). Sensor Properties and Surface Characterization of the Metal-Deposited SPR Optical Fiber Sensors with Au, Ag, Cu and Al, Sensors and Actuators A, 125, 296–303.
37. Sharma, A.K. and Gupta, B.D. (2005). On the Performance of Different Bimetallic Combinations in Surface Plasmon Resonance based Fiber Optic Sensors, Journal of Applied Physics, 101, 093111.

38. Sharma, A.K., Rajan and Gupta, B.D. (2007). Influence of Dopants on the Performance of a Fiber Optic Surface Plasmon Resonance Sensor, Optics Communications, 274, 320–326.
39. Grunwald, B. and Holst, G. (2004). Fibre Optic Refractive Index Microsensor based on White-Light SPR Excitation, Sensors and Actuators A, 113, 174–180.
40. Verma, R.K., Sharma, A.K. and Gupta, B.D. (2008). Surface Plasmon Resonance based Tapered Fiber Optic Sensor with Different Taper Profiles, Optics Communications, 281, 1486–1491.
41. Verma, R.K., Sharma, A.K. and Gupta, B.D. (2007). Modeling of Tapered Fiber-Optic Surface Plasmon Resonance Sensor with Enhanced Sensitivity, IEEE Photonics Technology Letters, 19 (22), 1786–1788.
42. Verma, R.K. and Gupta, B.D. (2008). Theoretical Modeling of a Bi-dimensional U-Shaped Surface Plasmon Resonance based Fibre Optic Sensor for Sensitivity Enhancement, Journal of Physics D: Applied Physics, 41, 095106.
43. Homola, J. and Slavik, R. (1996). Fibre-Optic Sensor based on Surface Plasmon Resonance, Electronics Letters, 32(5), 480–482.
44. Lin, H.Y., Tsai, W.H., Tsao, Y.C. and Sheu, B.C. (2007). Side-Polished Multimode Fiber Biosensor based on Surface Plasmon Resonance with Halogen Light", Applied Optics, 46(5), 800–806.
45. Chiu, M.H., Shih, C.H. and Chi, M.H. (2007). Optimum Sensitivity of Single-Mode D-Type Optical Fiber Sensor in the Intensity Measurement, Sensors and Actuators B, 123, 1120–1124.
46. Chiu, M.H. and Shih, C.H. (2008). Searching for Optimal Sensitivity of Single-Mode D-Type Optical Fiber Sensor in the Phase Measurement, Sensors and Actuators B, 131, 596–601.
47. Suzuki, H., Sugimoto, M., Matsui, Y. and Kondoh, J. (2008). Effects of Gold Film Thickness on Spectrum Profile and Sensitivity of a Multimode-Optical-Fiber SPR Sensor, Sensors and Actuators B, 132, 26–33.
48. Kurihara, K., Ohkawa, H., Iwasaki, Y., Niwa, O., Tobita, T. and Suzuki, K. (2004). Fiber-Optic Conical Microsensors for Surface Plasmon Resonance using Chemically Etched Single-Mode Fiber, Analytica Chimica Acta, 523, 165–170.
49. Abrahamyan, T. and Nerkararyan, Kh. (2007). Surface Plasmon Resonance on Vicinity of Gold-Coated Fiber Tip, Physics Letters A, 364, 494–496.
50. Hassani, A. and Skorobogatiy, M. (2007). Design Criteria for Microstructured-Optical-Fiber-based Surface-Plasmon-Resonance Sensors, Journal of Optical Society of America B, 24 (6), 1423–1429.
51. Gauvreau, B., Hassani, A., Fehri, M.F., Kabashin, A. and Skorobogatiy, M. (2007). Photonic Bandgap Fiber-based Surface Plasmon Resonance Sensors, Optics Express, 15(18), 11413–11426.
52. Dwivedi, Y.S., Sharma, A.K. and Gupta, B.D. (2006). Influence of Skew Rays on the Sensitivity and Signal-to-Noise Ratio of a Fiber Optic Surface-Plasmon-Resonance Sensor: A Theoretical Study, Applied Optics, 46 (21), 4563–4569.
53. Allsop, T., Neal, R., Mou, C., Brown, P., Saied, S., Rehman, S., Kalli, K., Webb, D.J., Sullivan, J., Mapps, D. and Bennion, I. (2009). Exploitation of Multilayer Coatings for Infrared Surface Plasmon Resonance Fiber Sensors, Applied Optics, 48(2), 276–286.
54. Nemova, G. and Kashyap, R. (2006). Fiber Bragg Grating Assisted Surface Plasmon Polariton Sensor, Optics Letters 31 (14), 2118–2120.
55. Shevchenko, Y.Y. and Albert, J. (2007). Plasmon Resonances in Gold Coated Tilted Fiber Bragg Gratings. Optics Letters 32 (3), 211–213.
56. Tripathi, S.M., Kumar, A., Marin, E. and Meunier, J.P. (2008). Side Polished Optical Fiber Grating based Refractive Index Sensors Utilizing the Pure Surface Plasmon, Journal of Lightwave Technology, 26 (13), 1980–1985.
57. Ding, J., Shao, L., Su, H. and Ruan, S. (2008). A Highly Sensitive Refractive Index Sensor based on the Long Period Grating Pair with a Fiber Taper in Between, International Conference on Advanced Infocomm Technology'08, China.

58. Kashyap, R. and Nemova, G. (2009). Surface Plasmon Resonance based Fiber and Planer Waveguide Sensors, Journal of Sensors, 2009, 645162.
59. Buckley, R. and Berini, P. (2007). Figure of Merit for 2D Surface Plasmon Waveguides and Applications to Metal Strips, Optics Express, 15(19), 12174–12182.
60. Verma, R.K. and Gupta, B.D. (2010). Surface Plasmon Resonance based Fiber Optic Sensor for Infrared Region using Conducting Metal Oxide Film, Journal of Optical Society of America A, 27, 846–851.

Chapter 5
Fabrication and Application of Plasmonic Silver Nanosheet

Kaoru Tamada, Xinheng Li, Priastute Wulandari, Takeshi Nagahiro, Kanae Michioka, Mana Toma, Koji Toma, Daiki Obara, Takeshi Nakada, Tomohiro Hayashi, Yasuhiro Ikezoe, Masahiko Hara, Satoshi Katano, Yoichi Uehara, Yasuo Kimura, Michio Niwano, Ryugo Tero, and Koichi Okamoto

K. Tamada (✉)
Research Institute of Electrical Communication, Tohoku University, 2-1-1 Katahira, Aoba-ku, Sendai 980-8577, Japan

Department of Electronic Chemistry, Tokyo Institute of Technology, 4259 Nagatsuta, Midori-ku, Yokohama 226-8502, Japan

Institute of Materials Chemistry and Engineering, Kyushu University, 6-10-1 Hakozaki, Higashi-ku, Fukuoka 812-8581, Japan
e-mail: tamada@ma.ifoc.kyushu-u.ac.jp

X. Li • P. Wulandari • K. Michioka • M. Toma • K. Toma • T. Hayashi • T. Nagahiro
Department of Electronic Chemistry, Tokyo Institute of Technology, 4259 Nagatsuta, Midori-ku, Yokohama 226-8502, Japan

D. Obara • T. Nakada • S. Katano • Y. Uehara • Y. Kimura • M. Niwano
Research Institute of Electrical Communication, Tohoku University, 2-1-1 Katahira, Aoba-ku, Sendai 980-8577, Japan

Y. Ikezoe • M. Hara
Department of Electronic Chemistry, Tokyo Institute of Technology, 4259 Nagatsuta, Midori-ku, Yokohama 226-8502, Japan

Flucto-Order Functions Research Team, RIKEN, 2-1 Hirosawa, Wako 351-0198, Japan

R. Tero
Institute of Molecular Science, Okazaki 444-8585, Japan

Toyohashi University of Technology, Toyohashi 444-8585, Japan

K. Okamoto
Department of Electronic Science and Engineering, Kyoto University, Katsura Campus, Kyoto 615-8510, Japan

Institute of Materials Chemistry and Engineering, Kyushu University, 6-10-1 Hakozaki, Higashi-ku, Fukuoka 812-8581, Japan

C.D. Geddes (ed.), *Reviews in Plasmonics 2010*, Reviews in Plasmonics,
DOI 10.1007/978-1-4614-0884-0_5,

1 Introduction

There is no doubt that metal nanoparticles are one of key materials for future bottom-up nanotechnology due to their unique chemical, electronic, and optical properties [1–3]. Among a huge number of studies proposed in these years, studies related to localized surface plasmon resonance (LSPR) with metal particles or arrays excite tremendous importance to various fields of science and technology, especially for highly efficient optoelectronics and biosensing device applications [4–6] (Fig. 5.1). LSPR excited by absorption of selective wavelength of light generates locally intensified electromagnetic (EM) fields on the nanoparticles, which can enhance fluorescence photoemission [7] or second-harmonic generations [8] or Raman signals [9]. Recent report revealed that even electric light (EL) emission [10] or photocurrent generations [11] are improved under plasmon excitation conditions.

One of the remarkable characteristics of LSPR is the interparticle coupling. When a metal nanoparticle is placed in close proximity to one another, the coupling of LSPR between particles occurs [12]. This interparticle plasmon coupling results in a tremendous red-shift of the LSPR band and further enhancement of the electromagnetic field [13]. The relation between the interparticle distance and the peak shift has been investigated precisely for one-dimensionally (1D) ordered arrays on solid substrate, such as paired disks [14–16]. Here, the shape and size of arrays and the gap distances are controlled on demand by the fabrication with electron beam

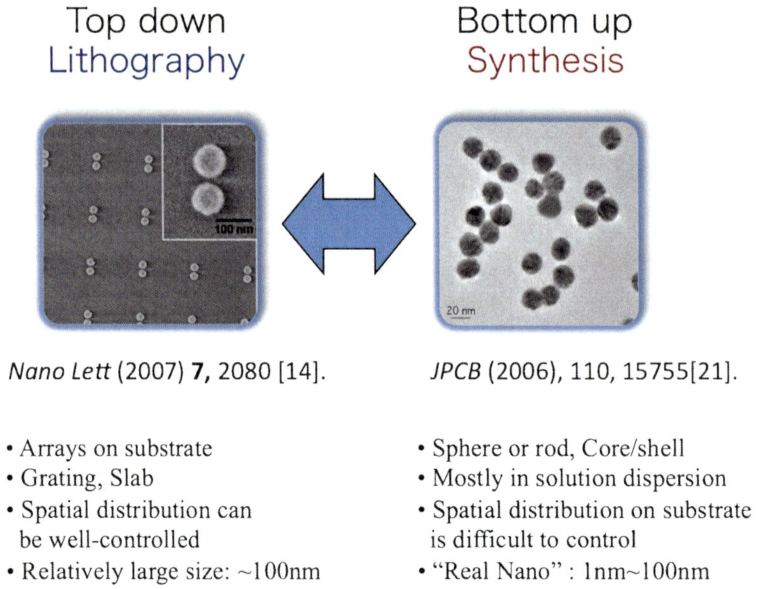
Top down
Lithography
Bottom up
Synthesis
100 nm
20 nm
Nano Lett (2007) **7,** 2080 [14].
JPCB (2006), 110, 15755[21].
• Arrays on substrate
• Grating, Slab
• Spatial distribution can be well-controlled
• Relatively large size: ~100nm
• Sphere or rod, Core/shell
• Mostly in solution dispersion
• Spatial distribution on substrate is difficult to control
• "Real Nano" : 1nm~100nm

Fig. 5.1 Nanostructured materials for excitation of localized surface plasmon resonances (LSPR)

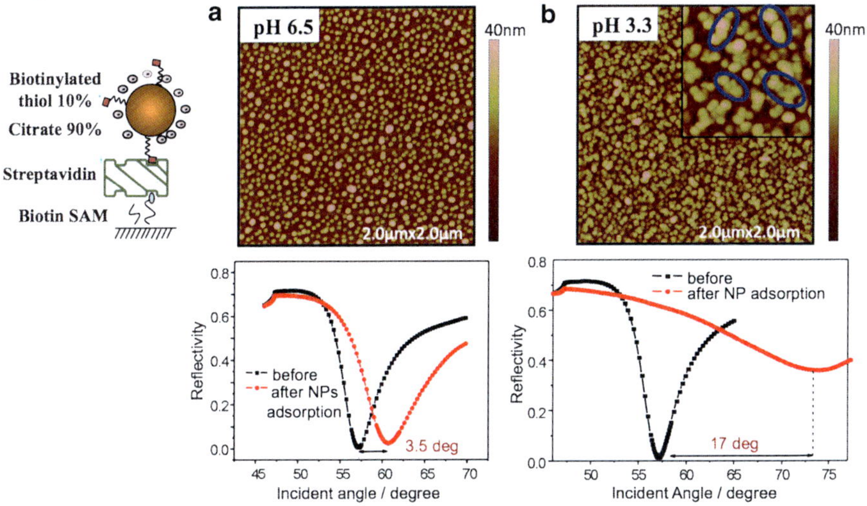

Fig. 5.2 Control of the interparticle distance of gold nanoparticles on substrate by pH. Tapping mode AFM images and SPR angular curves of biotin-capped gold particles immobilized on gold surface at (**a**) pH 6.5 and (**b**) pH 3.3 [22]

lithography. On the other hand, most of the studies for interparticle coupling between chemically synthesized nanoparticles have been conducted in dispersion because of a difficulty to immobilize the particles as regular arrays with constant gap distances [13, 17, 18].

In our previous study, we utilized DNA-capped gold nanoparticles for the enhancement of surface plasmon resonance signals on DNA microarray [19, 20]. Target DNA capped gold nanoparticles are hybridized with probe DNA on gold surface, which results in 10 times large reflectivity increase compared with that of original target DNA. This enhancement effect can be well described by the Maxwell–Garnett theory for the effective dielectric constants when the number of immobilized particles is still low and the particles are isolated on the surface [21]. On the other hand, when the number of the particles increases and the particles form the aggregates, the deviation from the theoretical prediction becomes obvious. In the following study, we controlled systematically the interparticle distance of immobilized gold nanoparticles by solution pH via electrostatic repulsion force (Fig. 5.2) [22]. This approach worked well even for the case of immobilization by avidin–biotin reaction; we could reproducibly fabricate the surface both with isolated particles and with aggregates at the saturated adsorption. On aggregates, the experimental SPR angle shift was much larger than the predicted value, and a significant increase of the minimum intensity ("dip-up") was found. This phenomenon can be reasonably interpreted by enhanced effective dielectric constants (especially imaginary part) by dipole interaction between neighboring particles, i.e., by local plasmon coupling. However, for the case of spontaneous adsorption, the formation of regular arrays with constant gap distance was still not achieved [23].

Recently, we found that two dimensional (2D) crystalline sheet could be fabricated at air–water interface by simply spreading homogeneously sized silver nanoparticles (AgNPs) capped by myristates (AgMy) [24]. Here, the interparticle distance of AgNPs is exactly determined by the length of capping organic molecules. This technique produces enormously wide (more than several 10 cm) homogeneous films with single-particle thickness, which can be transferred on hydrophobic substrates reproducibly by Langmuir–Schaefer method. This Ag nanosheet has an extremely sharp localized plasmon resonance band (as sharp as a dispersion in a good solvent) at $\lambda_{max} = 470$ nm. The excitation of LSPR depends on metal core size, core-to-core distance and surrounding dielectric materials. Thus, our Ag nanosheet, in which all these parameters are well-controlled, can be excellent candidate to study fundamental plasmonic science with nanocolloidal particles.

Silver is known to have a superior optical property than gold. For example, silver has much smaller imaginary part of the dielectric function for visible light compared with gold, which results in quite sharp plasmon responses. The plasmonic enhancement factor in spectroscopic measurements is also known to be higher for silver compared with gold [25]. Only the weakness of silver is low resistance to oxidation, which causes unstable optical response occasionally. However, strongly adsorbed ligands on metal, such as carboxylate and thiol derivatives promise long-term stability of the particles preventing them being oxidized [26–28]. We also refer some chemistry on metal nanoparticle, such as the ligand exchange and the molecular coordination of adsorbate molecules focusing on the difference between gold and silver in this review.

2 Synthesis and Characteristics of Silver Nanoparticles

AgMy are synthesized by thermal reduction of silver acetate precursor in the melt of myristic acid (the melting point: 80°C) as a reaction medium as well as a capping agent (Fig. 5.3) [29]. A mixture of silver acetate 40 mg (2.4 mmol) and myristic acid 330 mg (14.4 mmol) was heated to the boiling point of myristic acid (250°C) in an oil bath without solvents. The melt was stirred by a magnetic stirrer under a N_2 purge. During the reaction, the color of the melt gradually changes from colorless to brown to purple, indicating the formation of highly concentrated silver nanoparticles.

Figure 5.4 is the result of dynamic light scattering (DLS) measurement to test the particle size during the reaction [29]. After reaching the target temperature, a small amount of reaction wax was sampled every minute. As shown in Fig. 5.4, the distribution of the particle size was quite dependent on the reaction time. The nanometer-sized particles started forming by 6 min (Fig. 5.4c) and the reaction was completed by 7 min (Fig. 5.4d). When the reaction time became 8 min and more (Figs. 5.4e

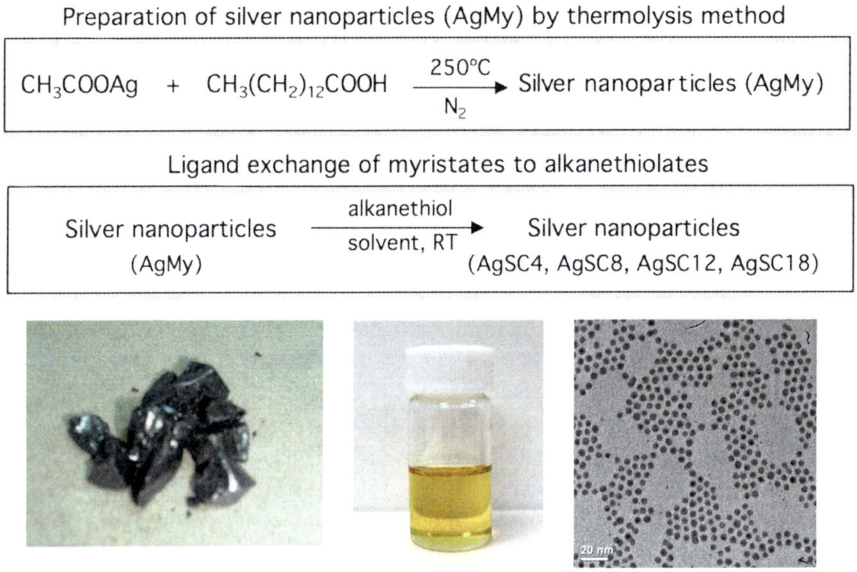

Fig. 5.3 Gram-scale synthesis of uniformly sized AgMy and their ligand exchange. Dried sample (*left*), redispersion in toluene (*middle*) and TEM image of AgMy, core size: 4.8 ± 0.1 nm (*right*) [29]

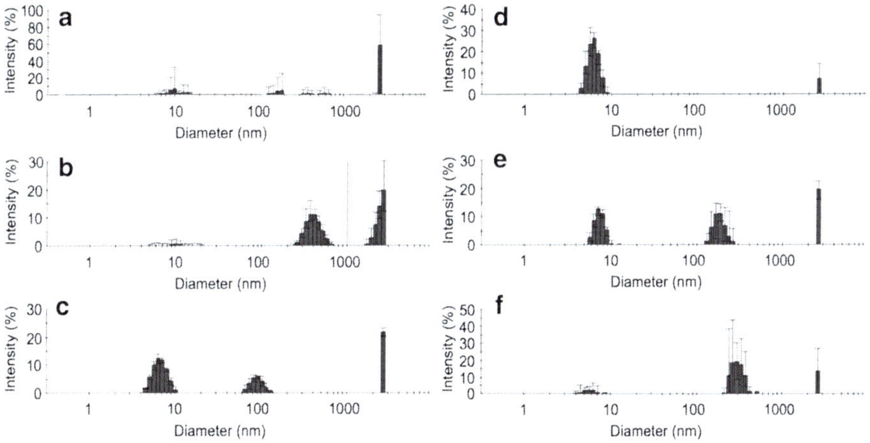

Fig. 5.4 Particle size analysis of AgMy monitored by DLS. Reaction time at 250°C; (**a**) 2 min, (**b**) 4 min, (**c**) 6 min, (**d**) 7 min, (**e**) 8 min, (**f**) 10 min [29]

and 5.4f), the particles started aggregating. The optimum reaction time (7 min) was not exactly the same for all runs, but the reaction time profile was found in common. The yield of AgMy was also dependent greatly on the reaction time. The yield at the optimum reaction time (Fig. 5.4d) was 90% or more, i.e., almost all silver atoms participated in the nanoparticle formation. On the other hand, too short or long. Reaction time resulted in a low yield, such as 25–35%.

The formation of homogeneously sized AgMy was confirmed by UV–vis spectroscopy by the sharp plasmon band in toluene as well as TEM images (Figs. 5.3 and 5.8). Uniformities in core size and capping layer thickness are clearly shown in the images. The average diameter of the Ag core is estimated to be 4.8 ± 0.1 nm. The quality of our particles was quite identical or better compared with another myristate-capped silver nanoparticles in the previous report synthesized from silver myristate precursors [30]. We have also succeeded to obtain the same quality of AgMy in scale-up synthesis, in which silver acetate 11.6 g and myristic acid 94.2 g were mixed by mechanical stirring in an N_2 purged flask. Thus, gram-scale homogeneous silver nanoparticles can be produced by this simple method.

Before use, AgMy was repeatedly (more than 3 times) purified by the reprecipitation with ethanol in toluene (or hexane) to remove the excess myristic acids. The particles as the precipitates are separated from a top clear solution by low-speed centrifugation (15°C, 6,000 rpm, 10 min). Then, the particle aggregates are removed from the redispersed solution in toluene (or hexane) by high-speed centrifugation (15°C, 20,000 rpm, 10 min). Both toluene and hexane are good solvents for AgMy, however, toluene produces the low yield but more homogeneous particles, while hexane produces the high yield but less homogeneous particles.

AgMy exhibits a unique property concerning a ligand exchange. The particle is synthesized at 250°C, i.e., the myristate cappings can stay on silver nanoparticles at such high temperature. However, they can be exchanged to thiol derivatives even at room temperature owing to their chemical potentials. This characteristic of the particle enables us to control the thickness and surface functionality of capping layers on silver cores on demand. As the typical procedure, AgMy solution (25 mg/50 ml) was mixed with an alkanethiol solution ($CH_3(CH_2)_nSH$, $n=3$, 7, 11, 17, 2 mM, 50 ml) and stirred at room temperature for 3 h, then purified by the same procedure as described above. The complete ligand exchange reaction was confirmed by FTIR measurements.

Here, we note an interesting result concerning the dispersibility of Ag nanoparticles in organic solvents after ligand exchange [29]. Figure 5.5 shows the list of good solvents for Ag nanoparticles with alkanethiolate cappings. It is found that the polarity of the particle changes to the polar side when the chain length of alkanethiolate capping becomes shorter than C8, where nonpolar solvent (e.g., hexane) is not good solvent anymore but instead they are well dispersed in hexane/ethanol mixed solvent. This phenomenon can be nicely interpreted by LSPR (Fig. 5.5 bottom). The LSPR is known to extend only less than the core radius from the metal particle surface [31]. The thickness of 1-octanethiol (C8-SH) self-assembled

Nanoparticles	Solvent	Capping molecules
AgS-C4	Hexane / EtOH = 1/1	$CH_3(CH_2)_3SH$
AgS-C6	Hexane / EtOH = 1/1	$CH_3(CH_2)_5SH$
AgS-C8	Hexane	$CH_3(CH_2)_7SH$
AgS-C10	Hexane	$CH_3(CH_2)_9SH$
AgS-C16	Hexane	$CH_3(CH_2)_{15}SH$

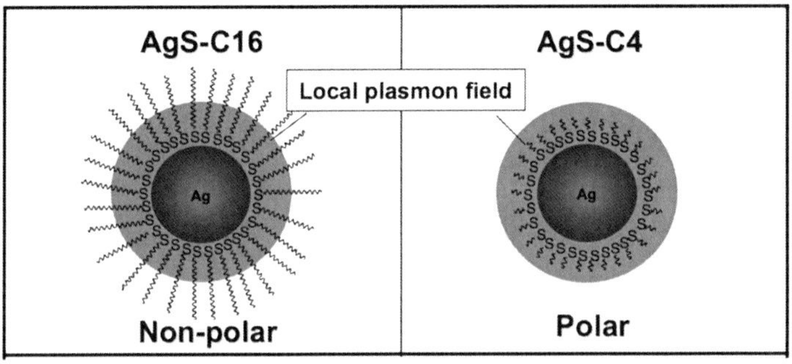

Fig. 5.5 List of good solvent for Ag nanoparticles capped by alkanethiols and schematic drawing of correlation between thickness of capping layers and polarity of the particles [29]

monolayer is ca. 1.1 nm in previous report [32]. At the carbon number of less than 8, the plasmon field penetrating out to the solvent phase through the alkanethiolate capping influences the polarity and dispersibility of the particles.

Recently, we found the ligand exchange on AgMy does work only with thiol derivatives but not with disulfides [33]. This fact surprises us since the reaction of thiol and disulfide on silver NPs is expected to be identical to that on gold [26]. We found the answer of this mystery in our FTIR data of carboxylate-capped gold and silver nanoparticles, which revealed completely different coordination of carboxylate bound on gold and silver nanoparticles (Fig. 5.6) [34]. The shift of carboxylate stretching frequency confirmed that carboxylate bound on gold nanoparticles forms unidentate coordination by the bonding intermediate between ionic and covalent, while that on silver nanoparticles shows the bridging coordination by ionic bonding. The different ionic radius and electronegativity of metals may be attributed to the different coordination of carboxylates. From the fact that carboxylates on AgNP can be exchanged only by thiols, the adsorption energy of carboxylates on AgNP is assigned between that of thiol and disulfide (thiol > carboxylate > disulfide) unlike that on AuNP (thiol, disulfide > carboxylate).

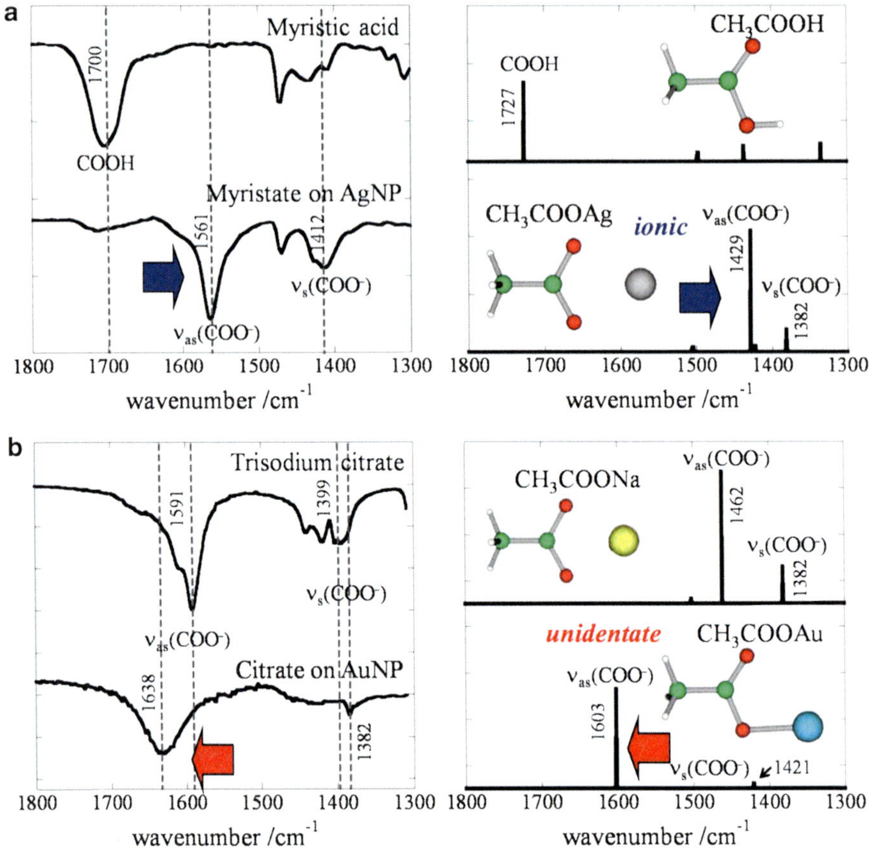

Fig. 5.6 Shift of carboxylate IR spectra by adsorption on metallic NP, and MO calculation of corresponding metal acetate; (**a**) myristatic acid and myristate on AgNP and (**b**) trisodium citrate and citrate on AuNP [34]

3 2D Crystalline Sheet (Plasmonic Nanosheet)

AgMy can spread on an air–water interface by forming a stable monolayer owing to their amphiphilic property. The amphiphilicity of AgMy is mainly attributed to the property of myristate cappings, however, even alkanethiolate-capped nanoparticles prepared by the ligand exchange also exhibit an amphiphilicity depending on the alkyl chain length. As mentioned in Chap. 4, the penetration of LSPR from the silver cores creates a certain level of hydrophilicity to the hydrophobic out-shells, which realizes successful spreading on water.

The following is the typical procedure to fabricate 2D crystalline sheet: In order to fabricate the nanosheet composed of AgMy, a 600–700 ml of AgMy solution (0.25–0.3 mg/ml) was spread on water in LB trough at room temperature (T = 24°C). The nanosheet was transferred by Langmuir–Schaefer method on hydrophobic

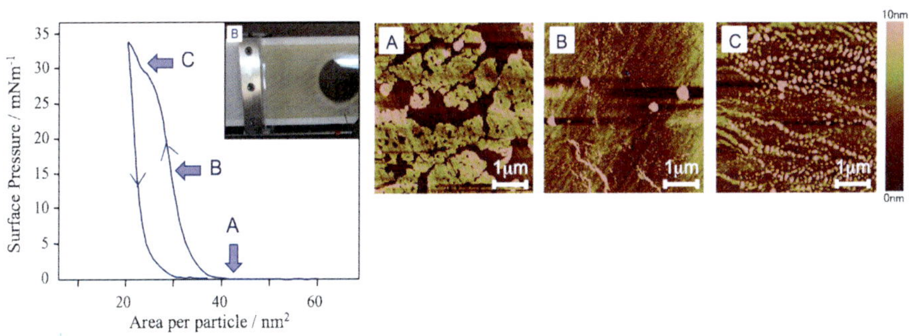

Fig. 5.7 Π–A isotherm of AgMy layer at air–water interface (T=24°C) and AFM images of transferred sheets on hydrophobic quartz by Langmuir–Schaefer method at various surface pressure (A–C) [24]

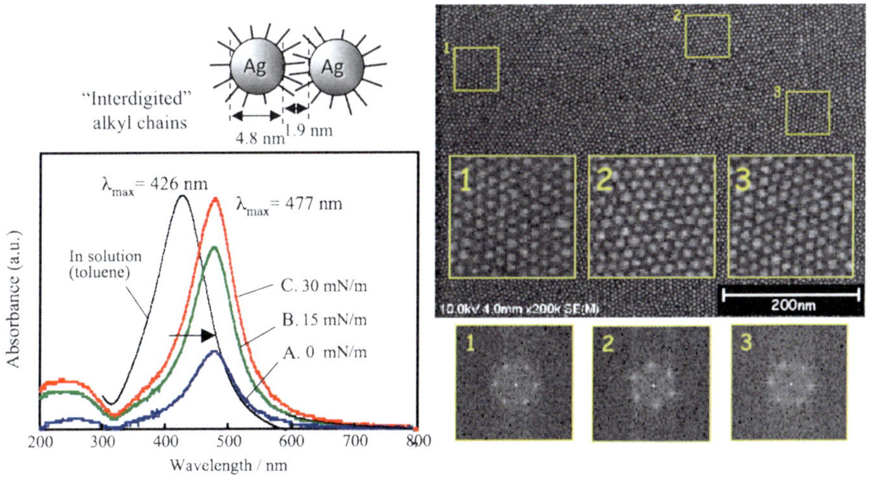

Fig. 5.8 Plasmon absorption bands of AgMy sheet transferred on quartz at various surface pressures (A–C) (*left*) and SEM and FFT images of AgMy sheet transferred at B (15 mN/m) (*right*) [24]

quartz substrates modified with hexamethyldisilazane (HMDS). Figure 5.7 is a typical Π–A isotherm of AgMy film. The inserted arrows (A, B, C) on the Π–A curve are the positions, where the AgMy layers were transferred on the substrates. From the AFM image, solid-like domains are confirmed even at A (Π=0 mN/m). By compression, the domains are fused into a large continuous sheet as shown in the image B (Π=15 mN/m). When the Π–A curve reached at a collapse point, line-like multilayer defects appeared on the surface, which was parallel to the compression bar (perpendicular to the compression direction). The inserted picture B is a top view of the AgMy sheet on the trough at 15 mN/m. A vivid yellow color originating from plasmon absorption band was visible even in monolayer.

Figure 5.8 (left) is the UV–vis transmission spectra of AgMy sheets transferred on quartz at A–C in Fig. 5.7. Compared with the spectrum in toluene, the spectra of

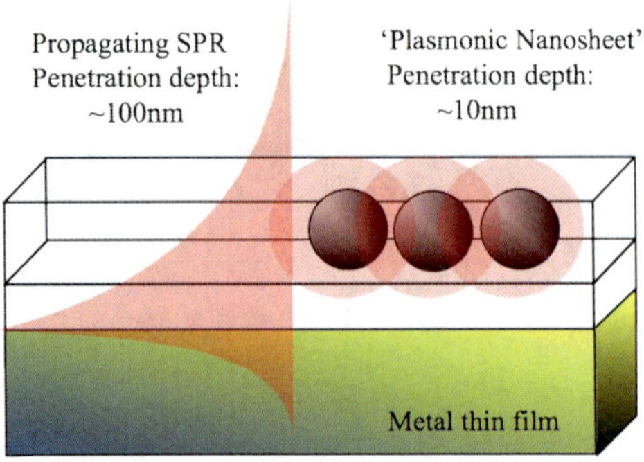

Fig. 5.9 Schematic drawing of "plasmonic Ag nanosheet"

AgMy sheets were largely red-shifted to the longer wavelength, ca. 50 nm, while the peak width was rather reduced (sharpened) by the sheet formation. The peak position of plasmon absorption band did not change by surface pressures for deposition. Figure 5.8 (right) is an SEM and FFT images of the AgMy sheet at 15 mN/m (at B in Fig. 5.7). Here, the densely packed structure composed of two dimensional crystalline domains (SEM images) and the core-to-core distance of 6.7 ± 0.1 nm (FFT images) are shown. The distance between edge to edge of the particles is found to be ca. 1.9 nm. Since the molecular length of myristate is estimated to be 1.86 nm, long alkyl chains of myristate cappings are assumed to form "interdigited structure" in the sheet by hydrophobic interactions on water.

The most significant characteristic of our AgMy sheet is the sharpened plasmon band. This extraordinary result suggests a homogeneous coupling of LSPR in the sheet. The SEM image in Fig. 5.8 indicates that the film is not composed of a single crystalline but multicrystallines with numbers of domain boundaries. It comes as a surprise to learn that these defects do not induce a broadening of the plasmon absorption band. We assume that LSPR is "*delocalized*" in the sheet (homogeneously coupled in plane, "*plasmonic nanosheet*"). This 2D sheet has a distinguished property with both advantage of propagating SPR and LSPR. Propagating SPR excited on flat metal thin film can trap bulk light at metal/dielectric interface within ~100 nm region (Fig. 5.9 left), although special optical systems, such as a prism coupler or a grating coupler are required. On the other hand, LSPR can condense photon energy at the particle interface in 10–100 times narrow region (nanometer-thickness) without coupler systems, although the light being trapped by individual particles is also spatially limited to the nano-area. Our plasmonic nanosheet can trap bulk light homogeneously in nano-thickness and excite quite strong electromagnetic field on entire surface without any expensive, complicated instrumentations (Fig. 5.9 right).

The LSPR band of Ag nanosheet was tuned by the interparticle distance of AgNPs via the length of capping organic molecules as shown in Fig. 5.10. The red shift of LSPR band followed exponential law against the interparticle distance in a similar manner to previous reports of metal nanodisk pairs [14]. Figure 5.11 is a comparison of our results of 2D sheet with other publications of 1D system, which

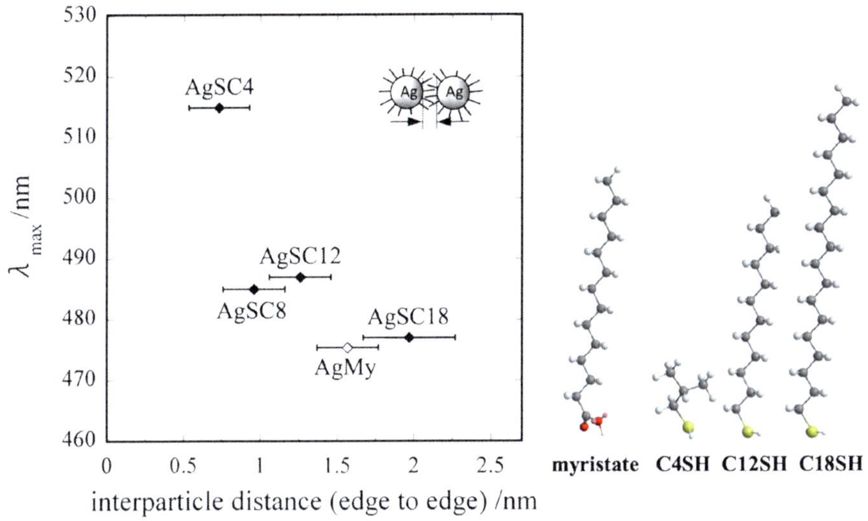

Fig. 5.10 LSPR peak position (λ_{max}) vs. interparticle distance in Ag nanosheets [35]

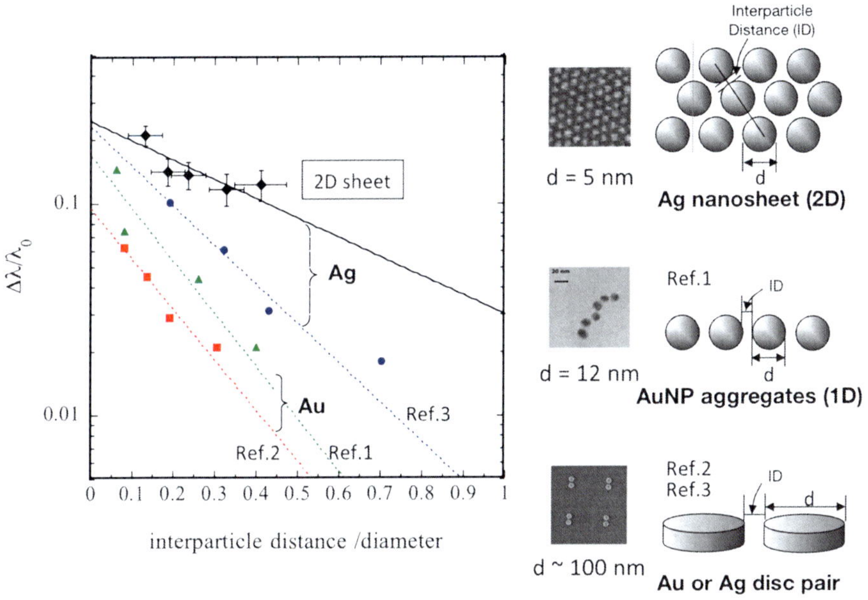

Fig. 5.11 Fractional LSPR peak shift vs. the ratio of interparticle distance to particle diameter and exponential fit [35]. (Ref. 1) [13]. (Ref. 2) [14, 16]. (Ref. 3) [15]

presents several important aspects of LSPR in these systems. Firstly, independently of particle (or disk) size, all the plots followed a liner regression in the exponential plots. Second, the *y*-intercept of the progression line of silver particles and disks appeared slightly larger than that of gold systems as expected from the stronger

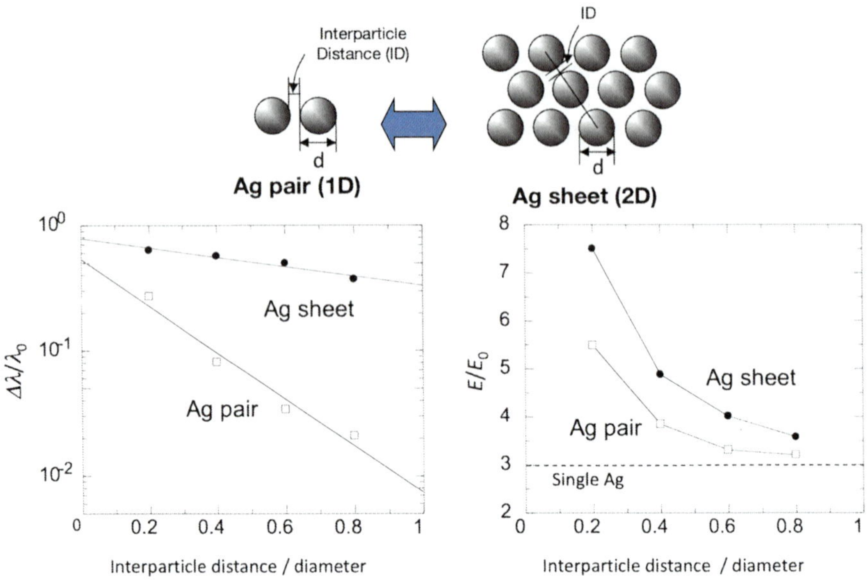

Fig. 5.12 Prediction of LSPR peak position and maximum electromagnetic (EM) field for AgNP pair (1D) and Ag nanosheet (2D) by FDTD method [35]

LSPR field. However, most remarkable finding is that the slope of the 2D sheet was much smaller than those of 1D systems. This result indicates that the AgNPs in 2D sheet interact each other at the distance more than the diameter.

The simulation by finite difference time domain (FDTD) method guaranteed the reliability of our experimental result (Fig. 5.12); here, the structural and optical parameters of AgMy are utilized for both 1D and 2D systems [35]. The difference in slope of LSPR peak shift vs. interparticle distance was rather clear in the simulation, where the interaction of the particles more than 6 times longer distance compared with the particle pairs (e.g., 30 nm) is predicted (Fig. 5.12 left). The simulation also confirmed additional amplification of local electromagnetic field in 2D sheet compared with that on paired particles, although the level of amplification is not remarkably high (only twice, Fig. 5.12 right). The simulation data indicate that LSPR in 2D sheet is less sensitive to the small interparticle separation gap compared with 1D systems. That is to say, for colorimetric biosensor applications (the detection by color change of NPs), the utilization of paired particles is more efficient rather than 2D networked systems. On the other hand, for the detection via LSPR-enhanced optical signals of metallic NPs (e.g., plasmon field-enhanced fluorescence spectroscopy or surface-enhanced Raman scattering), 2D sheet can provide more stable and homogeneous (and strong) enhancement field on entire surface, which is definitely of great advantage for such applications. The sharpened LSPR band in our nanosheet, regardless of the small defect in crystalline domains, is one of the characteristics of the delocalized LSPR.

4 Applications

Based on the characteristics of 2D Ag nanosheet described above, we propose two directions of potential applications; one is a replacement of Kretchmann type of optical configurations (metal thin film and prism) to Ag nanosheet (without prism) for biosensor application [36]. The EM field under excitation of propagating SPR on flat metal thin film is 100–300 nm in visible light region, which can be used to monitor biomolecular interaction at the interface through the reflectivity change by the resonance angle shift or by the enhanced fluorescence photoemission [37]. Figure 5.13 is a result of one-shot experiments to confirm the field enhanced fluorescence photoemission on our Ag nanosheet. The EM field excited on Ag sheet is strongly localized in a few nm region at the interface and produces strong photoemission from lipid monolayer. Here, the resonance wavelength of the sheet can be tuned by the interparticle distances according to the excitation wavelength of the dyes. Another application is for optoelectronic devices, such as solar cells [38–40]. For the case of electronic devices, the control of interfacial electronic structure and energy level alignment will be another important challenge [41], however, at least a concern of random aggregation of NPs which causes inhomogeneous opto-electric response will be solved by the use of regularly arrayed metallic NPs.

In our latest study of ultrahigh vacuum scanning tunneling microscopy (UHV-STM), a spike-like STM-light emission (STM-LE) from AgNPs was detected reproducibly on long-range ordered AgNP arrays but not on the randomly located AgNP (Fig. 5.14) [42]. The coupling between the vibrational state of the molecules in the gap and corrective excitation of LSPR by tunneling electrons on 2D array seems to be responsible for this photoemission.

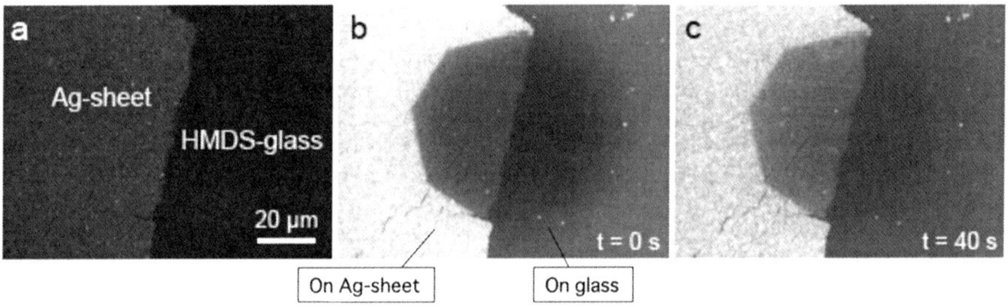

Fig. 5.13 Field enhanced fluorescence images on Ag nanosheet composed of AgNP 2D crystalline (Ex/Em: 445 nm/507 nm). DOPC monolayers with NBD-DOPE dye (0.5 wt%) was spread on hydrophobic glass substrate (HMDS modified) with half-covered Ag nanosheet by a vesicle fusion technique (**a**). Stronger photoemission was observed on Ag nanosheet compared with on HMDS-glass. The (**b**) and (**c**) are the images after photobreaching to confirm continuous DOPC monolayer formation crossing the boundary

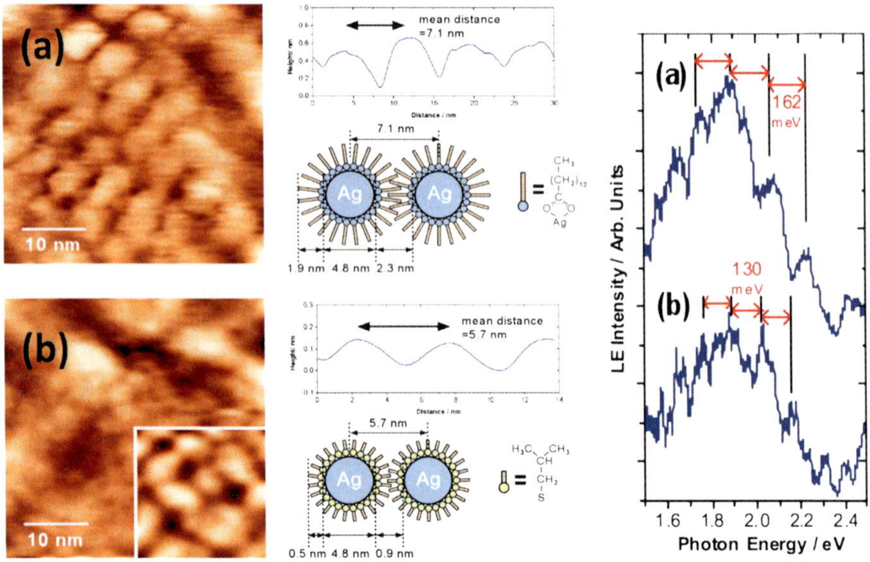

Fig. 5.14 STM images and STM-LE spectra of AgMy (**a**) and AgSC4 (**b**) 2D nanosheet transferred on C12SH SAM/gold. The STM images were obtained with the sample bias of 1.36 V and tunneling current of 0.16 nA, while the LE spectra were obtained with the sample bias of 2.3 V, the tunneling current of 2.0 nA and the exposure time of 100 s [42]

At the end of this chapter, we introduce one application of Ag nanosheet as a marker for lateral remote photocatalytic reaction of TiO_2 on substrates [43]. Ag nanosheet composed of AgMy (organic/metal hybrid material) is sensitive to photocatalytic reaction. They change the color from yellow to gray when the decomposition of myristate cappings and the fusion of silver cores take place. Ag nanosheet was deposited on top of a TiO_2 modified substrate to monitor the photo-degradation process under UV irradiation. Here, the data of TiO_2 nanotube anatase crystals synthesized by anodic oxidation was systematically compared with that of commercial TiO_2 powders. UV spectra and SEM/AFM images confirmed that only 10–20% of TiO_2 nanotubes on surface decomposes the entire surface of AgNPs nanosheet (Figs. 5.15 and 5.16). The SEM images revealed the reactions even in regions a few microns distant from TiO_2 nanotubes. This lateral remote reaction was only found for TiO_2 nanotubes but never for the commercial TiO_2 powder. This difference in remote photocatalytic activity between the TiO_2 nanotubes and powders is probably correlated to the different life time of radicals produced on each surface, although a further investigation is necessary. In any case, we found that homogeneous Ag nanosheet with sharp plasmon absorption band is quite useful to monitor the nanoscale of surface reactions.

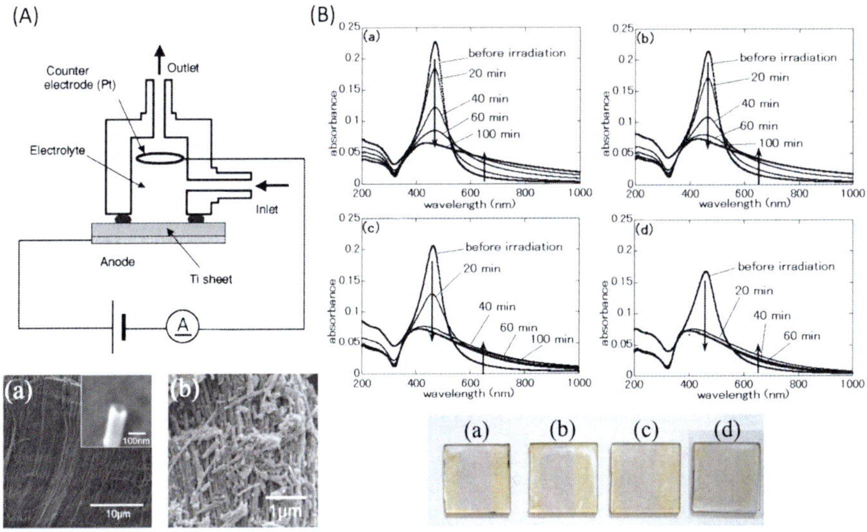

Fig. 5.15 (**A**) Anodic reaction cell and SEM images of TiO_2 nanotubes; before (**a**) and after (**b**) annealing. (**B**) Transmission spectrum change of Ag nanosheets deposited on TiO_2 nanotube-modified quartz substrates by UV irradiation. Adsorption time allowed for TiO_2 nanotubes; (**a**) 1 min, (**b**) 3 min, (**c**) 6 min, and (**d**) 10 min. A photograph of the sample substrates was taken after UV irradiation, (**a**)–(**c**) 100 min and (**d**) 60 min, in which only the center part of the surface (1 cm width) was exposed to UV light [43]

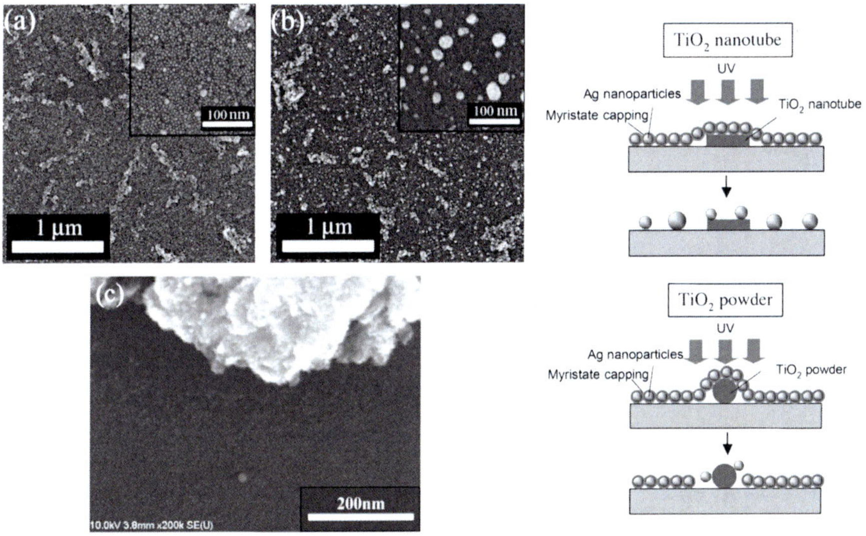

Fig. 5.16 SEM images of Ag nanosheet on TiO_2 nanotube-modified Si(100) wafer before (**a**) and after (**b**) 100 min UV irradiation, and that on TiO_2 powder-modified Si(100) wafer after 100 min UV irradiation (**c**). The *right* is the schematic drawing of lateral remote photocatalytic reactions of TiO_2 nanotube and powder [43]

5 Summary

In this review, we summarized the recent progress of Ag nanosheet related topics. Recent years, the evolution of chemical synthesis enables us to have a great variety of nanostructured materials, e.g., nanoparticles, nanorods, nanowires, etc. The advantage of these synthesized materials is not only industrial productivity, but also their unique property as *real-nano* materials. They are definitely smaller than the features fabricated by top-down lithography. However, a difficulty to control the spatial position on solid substrates has been a principle disadvantage of such bottom-up nanostructured materials up to now. When the object is nanoscale, *self-assembly* or *self-organization* will be a key to construct long-range ordered regular arrays [44–47]. Recently, many of new ideas of advanced plasmon devices have been developed by the use of gratings or confined metal gap structures fabricated by top-down nanotechnology [48–52]. We believe that self-assembled nanomaterials with tunable plasmon resonances will also stimulate various future device concepts [53].

Acknowledgments This work was supported by the Grant-in-Aid for Scientific Research (B), JSPS (21310067), Grant to promote basic research by research personnel in private-sector business, JST, and Nation-wide Cooperative Research Projects, RIEC, Tohoku University.

References

1. Schmid G, Talapin DV, Shevchenko EV (2004) Nanoparticles: From theory to applications. Weinheim, Germany: 251–298.
2. Bohren CF, Huffman DR (1998) Absorption and scattering of light by small particles. Vol. 1, John Wiley & Sons, Inc., New York.
3. Toshima, N, Yonezawa T (1998) Bimetallic nanoparticles - novel materials for chemical and physical applications. New J. of Chem. 22 (11): 1179–1201.
4. Haynes CL, Van Duyne RP (2001) Nanosphere lithography: A versatile nanofabrication tool for studies of size-dependent nanoparticle optics. J. Phys. Chem. B 105(24): 5599–5611. (b) Haes AJ, Zou B, Schatz GC, Van Duyne RP (2004) A nanoscale optical biosensor: The long range distance dependence of the localized surface plasmon resonance of novel metal nanoparticles. J. Phys. Chem. B 108: 109–116.
5. Kawata S (2001) Near-field optics and surface polaritons. Springer, Berlin.
6. Mirkin CA, Letsinger RL, Mucic RC, Storhoff JJ (1996) A DNA-based method for rationally assembling nanoparticles into macroscopic materials. Nature 382 (6592): 607–609. (b)Taton TA, Mirkin CA, Letsinger RL (2000) Secanometric DNA array detection with nanoparticle probes. Science, 289(8): 1757–1760.
7. Lakowicz JR, Geddes CD, Gryczynski I, Malicka J, Gryczynski Z, Aslan K, Lukomska J, Matveeva E, Zhang JA, Badugu R, Huang J (2004) Advances in surface enhanced fluorescence. J. Fluorescence, 14(4): 425–441. (b)Aslan K, Geddes CD (2008) A review of an ultrafast and sensitive bioassay platform technology: Microwave-accelated metal-enhanced fluorescence. Plasmonics, 3: 89–101. (c)Zhang J, Fu Y, Chowdhury MH, Lakowicz JR (2007) Metal-enhanced single-molecule fluorescence on silver particle monomer and dimer: Coupling effect between metal particles. Nano Lett. 7(7): 2101–2107.

8. Tsuboi K, Fukuba S, Naraoka R, Fujita K, Kajikawa K (2006) Second-harmonic spectroscopy of surface immobilized gold nanospheres above a gold surface supported by self-assembled monolayers. J. Phys. Chem. B, 125(17): 174703. (b)Abe S, Kajikawa K (2006) Linear and nonlinear optical properties of gold nanospheres immobilized on a metallic surface. Phys. Rev. B, 74(3), 035416.
9. Nie S, Emroy SR (1997) Probing single molecules and single nanoparticles by surface-enhanced Raman scattering. Science 275: 1102–1106.
10. Okamoto K, Niki I, Shvartser A, Narukawa Y, Mukai T, Scherer A (2004) Surface-plasmon-enhanced light emitters based on InGaN quantum wells. Nature Materials, 3(9): 601–605.
11. Xie XN, Xie Y, Gao X, Sow CH, Wee ATS (2009) Metallic nanoparticle network for photocurrent generation and photodetection. Adv. Mater. 21: 3016–3021.
12. Zheng YB, Juluri BK, Mao X, Walker TR, Huang TJ (2008) Systematic investigation of localized surface plasmon resonance of long-range ordered Au nanodisk arrays. J. Appl. Phys. 103: 014308.
13. Sendroiu IE, Mertens SFL, Schiffrin D (2006) Plasmon interactions between gold nanoparticles in aqueous solution with controlled spatial separation. Phys. Chem. Chem. Phys. 8: 1430–1436.
14. Jain PK, Huang W, El-Sayed, MA (2007) On the universal scaling behavior of the distance decay of plasmon coupling in metal nanoparticle pairs: a plasmon ruler equation. Nano Lett. 7(7): 2080–2088.
15. Gunnarsson L, Rindzevicius T, Prikulis J, Kasemo B, Kall M, Zou S, Schatz GC (2005) Confined plasmons in nanofabricated single silver particle pairs: Experimental observations of strong interparticle interactions. J. Phys. Chem. B 109: 1079–1087.
16. Jain PK, El-Sayed MA (2008) Surface plasmon coupling and its universal size scaling in metal nanostructures of complex geometry: elongated particle pairs and nanosphere trimers. J. Phys. Chem. C, 112: 4954–4960. (b)Huang W, Qian W, Jain PK, El-Sayed MA (2007) The effect of plasmon field on the coherent lattice phonon oscillation in electron-beam fabricated gold nanoparticle pairs. Nano Lett., 7(19): 3227–3234.
17. Reinhard BM, Siu M, Agarwal H, Alivisatos P, Liphardt J (2005) Calibration of dynamic molecular rulers based on plasmon coupling between gold nanoparticles. Nano Lett. 5: 2246–2252.
18. Sendroiu IE, Schiffrin DJ, Abad JM (2008) Nanoparticle organization by a Co(II) coordination chemistry directed recognition reaction. J. Phys. Chem. C, 12: 10100–10107.
19. Nakamura F, Ito M, Manna A, Tamada K, Hara M, Knoll W (2006) Observation of hybridization on a DNA array by surface plasmon resonance imaging using Au nanoparticles. Jpn. J. Appl. Phys. 45(2A): 1026–1029. (b)Tamada K, Nakamura F, Ito M, Li X, Baba A (2007) SPR-based DNA Detection with Metal Nanoparticles. Plasmonics, 2(4): 185–191.
20. Ito M, Nakamura F, Baba A, Tamada K, Ushijima H, Lau KHA , Manna A, Knoll W (2007) Enhancement of surface plasmon resonance signals by gold nanoparticles on high-density DNA microarrays. J. Phys. Chem. C 111(31): 11653–11662.
21. Li X, Tamada K, Baba A, Knoll W, Hara M (2006) Estimation of Dielectric Function of Biotin-Capped Gold Nanoparticles via Signal Enhancement on Surface Plasmon Resonance. J. Phys. Chem. B 110(32): 15755–15762.
22. Li X, Tamada K, Baba A, Hara M (2009) pH controlled two dimensional gold nanoparticle aggregates for systematic study of local surface plasmon coupling. J. Nanoscience and Nanotechnology, 9(1), 408–416.
23. Okamoto T, Yamaguchi I, Kobayashi T (2000) Local plasmon sensor with gold colloid monolayers deposited upon glass substrates. Optics Lett. 25(6): 372–374.
24. Tamada K, Michioka K, Li X, Ikezoe Y, Saito M, Otsuka K, (2009) Bioapplication of plasmonic nanosheet. Proc.of SPIE, 7213: 72130E.
25. Knoll W (1998) Interfaces and thin films as seen by bound electromagnetic waves. Annu. Rev. Phys. Chem. 49: 569–638.
26. Porter, L.A, Ji D, Westcott SL, Graupe M, Czernuszewicz RS, Halas NJ, Lee TR (1998) Gold and silver nanoparticles functionalized by the adsorption of dialkyl disulfides. Langmuir 14(26):7378–7386.

27. Nagasawa H, Maruyama M, Komatsu T, Isoda S, Kobayashi T (2002) Physical characteristics of stabilized silver nanoparticles formed using a new thermal-decomposition method. Phys. Status Solid A, 191: 67–76.
28. Yamamoto M, Kashiwagi Y, Nakamoto M (2006) Size-controlled synthesis of monodispersed silver nanoparticles capped by long-chain alkyl carboxylates from silver carboxylate and tertiary amine. Langmuir, 22: 8581–8586.
29. Keum CD, Ishii N, Michioka K, Wulandari P, Tamada K, Furusawa M, Fukushima H (2008) A gram scale synthesis of monodispersed silver nanoparticles capped by carboxylates and their ligand exchange. J. Nonlinear Opt. Phys. & Mater. 17(2): 131–142.
30. Nagasawa H, Maruyama M, Komatsu T, Isoda S, Kobayashi T (2002) Physical characteristics of stabilized silver nanoparticles formed using a new thermal-decomposition method. Phys. Status Solidi A 191: 67–76.
31. Kelly KL, Coronado E, Zhao LL, Schatz G (2003) The optical properties of metal nanoparticles: the influence of size, shape, and dielectric environment. J. Phys. Chem. B, 107: 668–677.
32. Porter MD, Bright TB, Allara DL, Chidsey CED (1987) Spontaneously organized molecular assemblies. 4. Structural characterization of n-alkyl thiol monolayers on gold by optical ellipsometry, infrared spectroscopy, and electrochemistry. J. Am. Chem. Soc. 109: 3559–3568.
33. Obara D, Nakada T, Hayashi T, Akiyama H, Tamada K, to be submitted.
34. Wulandari P, Nagahiro T, Michioka K, Tamada K, Ishibashi K, Kimura Y, Niwano M (2008) Coordination of Carboxylate on Metal Nanoparticles Characterized by Fourier Transform Infrared Spectroscopy. Chem. Lett. 37(8): 888–889.
35. Toma M, Toma K, Michioka K, Ikezoe Y, Obara D, Okamoto K, Tamada K (2011) Collective plasmon modes excited on a silver nanoparticle 2D crystalline sheet, Phys. Chem. 13:7459–7466.
36. Liedberg B, Lundstrom L (1993) Principles of biosensing with an extended coupling matrix and surface plasmon resonance. Sensor and Actuators B, 11, 63–72. (b)Homola J (2003) Present and future of surface plasmon resonance biosensors. Anal. Bioanal. Chem, 377: 528–539.
37. Liebermann T, Knoll W (2000) Surface-plasmon field-enhanced fluorescence spectroscopy. Colloids and surfaces A, 171: 115–130. (b)Tawa K, Yao DF, Knoll W(2005) Matching base-pair number dependence of the kinetics of DNA-DNA hybridization studied by surface plasmon fluorescence spectroscopy. Biosensors and Bioelectronics, 21: 322–329. (c)Robelek R, Niu LF, Schmid EL, Knoll W (2004) Multiplexed hybridization detection of quantum dot-conjugated DNA sequences using surface plasmon enhanced fluorescence microscopy and spectrometry. Anal. Chem. 76: 6160–6165.
38. Rand BP, Peumans P, Forrest SR (2004) Long-range absorption enhancement in organic tandem thin-film solar cells containing silver nanoclusters. J. Appl. Phys. 96(12): 7519–7526.
39. Pillai S, Catchpole KR, Trupke T, Green MA (2007) Surface plasmon enhanced silicon solar cells. J. Appl. Phys. 101 (9): 093105.
40. Wen C, Ishikawa K, Kishima M, Yamada K (2000) Effects of silver particles on the photovoltaic properties of dye-sensitized TiO2 thin films. Solar Energy Materials & Solar Cells 61: 339–351.
41. Ishii H, Sugiyama K, Ito E, Seki K (1999) Energy level alignment and interfacial electronic structures at organic metal and organic organic interfaces. Adv. Mater. 11: 605–6025.
42. Katano S, Toma M, Toma K, Tamada K, Uehara Y (2010) Nanoscale coupling of photons to vibrational excitation of Ag nanoparticle 2D array studied by scanning tunneling microscope light emission spectroscopy, Phys. Chem. 12:14749–14753.
43. Nagahiro T, Ishibashi K, Kimura Y, Niwano M, Hayashi T, Ikezoe Y, Hara M, Tatsuma T, Tamada K (2010) Ag nanoparticle sheet as a marker of lateral remote photocatalytic reactions. Nanoscale 2: 107–113.
44. Taleb A, Petit C, Pileni MP (1998) Optical properties of self-assembled 2D and 3D superlattice of silver nanoparticles. J. Phys. Chem. B, 102, 2214–2220.

45. Ikezoe Y, Kumashiro Y, Tamada K, Matsui T, Yamashita I, Shiba K, Hara M, Growth of giant two-dimensional crystal of protein molecules from a three-phase contact line. Langmuir, 24(22):12836–12841.
46. Feng CL, Zhong XH, Steinhart M, Caminade AM, Majoral JP (2008) Functional quanum-dot/dendrimer nanotubes for sensitive detection of DNA hybridization. Small, 4(5): 566–571.
47. Acharya H, Sung J, Sohn BH, Kim DH, Tamada K, Park C (2010) Tunable surface plasmon band pf position selective Ag and Au nanoparticles in thin block copolymer micelle films. Chem. Mater 21: 4248–4255.
48. Lamprecht B, Schider G, Lechner RT, Ditlbacher H, Krenn JR, Leitner A, Aussenegg FR (2000) Metal nanoparticle gratings: influence of dipolar particle interaction on the plasmon resonance. Phys. Rev. Lett. 84(20): 4721–4724.
49. Chiu NFC, Lin CW, Lee JH, Kuan CH, Wu KC, Lee CK (2007) Enhanced luminescence of organic/metal nanostructure for grating coupler active long-range surface plasmonic device. Appl. Phys. Lett. 91: 083114.
50. Papanikolaou N (2007) Optical properties of metallic nanoparticle arrays on a thin metallic film, Phys. Rev. B, 75: 235426.
51. Cui XQ, Tawa K, Hori H, Nishi J (2010) Tailored Plasmonic Gratings for Enhanced Fluorescence Detection and Microscopic Imaging. Adv. Func. Mater. 20 (4): 546–553.
52. Sivaramakrishnan S, Chia PJ, Yeo YC, Chua LL, Ho PKH (2007) Controlled insulator-to-metal transformation in printable polymer composites with nanometal clusters. Nature Materials 6: 149–155.
53. Eurenius L, Hagglund C, Olsson E, Kasemo B, Chakarov D Grating formation by metal-nanoparticle-mediated coupling of light into waveguided modes. Nature Photonics 2: 360–364.

Chapter 6
Nanomaterial-Based Long-Range Optical Ruler for Monitoring Biomolecular Activities

Paresh Chandra Ray, Anant Kumar Singh, Dulal Senapati, Sadia Afrin Khan, Wentong Lu, Lule Beqa, Zhen Fan, Samuel S.R. Dasary, and Tahir Arbneshi

1 Introduction

Optical ruler-based distance measurements are essential tool in everyday activities of biological and biophysical laboratories for tracking biomolecular conformational changes and in cell biology [1–3]. Förster resonance energy transfer (FRET) [4, 5] between fluorophores and fluorescence-quenching molecular compounds are well-characterized distance-dependent optical ruler, which are known to be capable of reporting biological processes on the nanometer length scale [1–7]. Single-molecule FRET measurements have been used to probe the conformational changes of RNA or protein molecules and track RNA–protein interactions [1, 6, 7]. Although FRET optical ruler is used routinely at the single-molecule detection limit, the length scale is limited by the nature of the dipole–dipole mechanism, which is on the order of maximum 10 nm [6]. Due to distance limitation, normal FRET optical ruler will not be suitable for the investigation of the structural dynamics of large biological systems, where the donor and acceptor distances are much more than 10 nm. Recently, several groups including ours have reported that, the limitations of FRET optical ruler can be overcome with a long-range spectroscopic ruler based [8–26] on the induced dipole coupling of noble metal nanoparticles and fluorescent dyes. Since last few years several groups are developing gold nanomaterial-based long-range optical ruler for monitoring biological processes [20–56]. This chapter highlights recent advances on the development of nanomaterials surface energy transfer (NSET) long-range optical ruler for probing RNA folding transition states, screening pathogens DNA/RNA and monitoring DNA/RNA cleavage.

P.C. Ray (✉) • A.K. Singh • D. Senapati • S.A. Khan • W. Lu • L. Beqa • Z. Fan • S.S.R. Dasary • T. Arbneshi
Department of Chemistry, Jackson State University, Jackson, MS, USA
e-mail: paresh.c.ray@jsums.edu

C.D. Geddes (ed.), *Reviews in Plasmonics 2010*, Reviews in Plasmonics,
DOI 10.1007/978-1-4614-0884-0_6,

2 Short Range Optical Ruler

FRET [4], in which excited fluorophores transfer energy to neighboring chromophores, are well characterized in photochemistry and have been used in a wide range of applications in analytical biochemistry. FRET results from dipole–dipole interactions and is thus strongly dependent on the center-to-center separation distance. Spectral overlap between donor emission and acceptor absorption is one of the requirements for normal FRET process. If the distance R between the donor and acceptor is not too small, the transfer occurs due to the coulombic interaction. For a large enough R, the matrix element for the transfer reduces to an interaction between the transition dipoles of the two dyes and has an R^{-3} dependence, and hence the rate behaves like R^{-6}. As a result, optical ruler FRET is used for measuring distances in the 3–8 nm range [1–7].

About 42 years ago [5], FRET was introduced by Stryer and Haugland as an optical ruler. Since then it has been used to address a wide range of biological questions, including detecting molecular interactions and conformational changes. As a result, FRET optical ruler has been used in many assays [1–7] involving enzymes, antibodies, and nucleotides. FRET technology is very convenient and also applied routinely at the single-molecule detection limit, the length scale for detection using Förster method is limited by the nature of the dipole–dipole mechanism and the quantum efficiency of energy transfer can be written as

$$\Phi_{\mathrm{EnT}} = \frac{1}{1+\left(\frac{R}{R_0}\right)^n}. \tag{6.1}$$

where R is the distance between donor and acceptor, R_0 is the distance between donor and acceptor at which the energy transfer efficiency is 50%. In the case of Förster or dipole–dipole energy transfer, $n=6$ and $R_0 \sim$ 3–8 nm for typical fluorophores [1–7]. Cases where probes with a diameter of 30–40 nm can be tolerated, the limitations of FRET can be overcome with a dynamic molecular ruler based on the distance-dependent plasmon coupling of two noble metal nanoparticles or NSET [8–26].

3 Long-Range Optical Ruler Using Nanomaterial

Recently, several groups including ours [8–27] have reported that nanomaterial-based plasmon ruler (as shown in Scheme 6.1) NSET (as shown in Scheme 6.1) ruler are capable of measuring distances more than twice of FRET distance. The metal has a continuum of electron-hole pair excitations and as a result, the density of these excitations is large, being proportional to the excitation energy. Experimentally, several groups, including ours [8, 9, 19, 20, 23, 26], found an R^{-4}

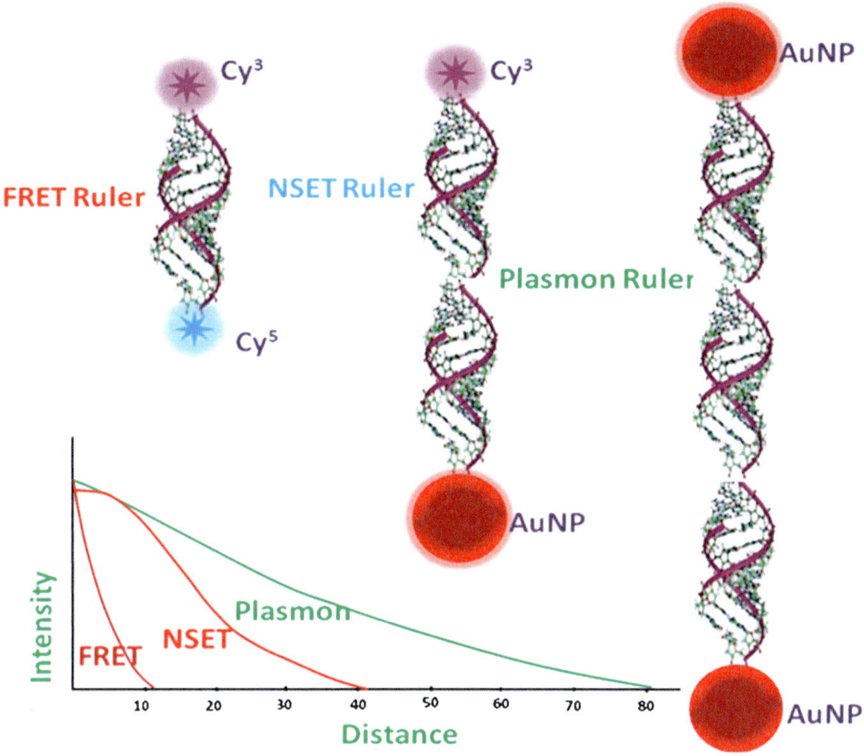

Scheme 6.1 Schematic representation of FRET, NSET, and plasmon ruler and their distance-dependent properties

dependence of the rate of energy transfer from the oscillating dipole to the continuum of electron-hole pair excitations in the metal.

4 Plasmon Ruler

Nanoparticle plasmon refers to the collective oscillation of the free electrons within a metal nanoparticle. Usually, when two nanoparticles are brought into proximity, around 2.5 times the particle diameter, their plasmons couple in a distance-dependent manner, as reported recently [15–19, 27]. The distance dependence plasmon coupling has been reported for different nanostructures [15–19, 27]. Reinhard et al. have shown that [17] pairs of noble metal nanoparticles-based plasmon rulers offer exceptional photostability and brightness, and their result indicates that it can be used to measure distances via the distance dependence of their plasmon coupling till about 100 nm (as shown in Fig. 6.1).

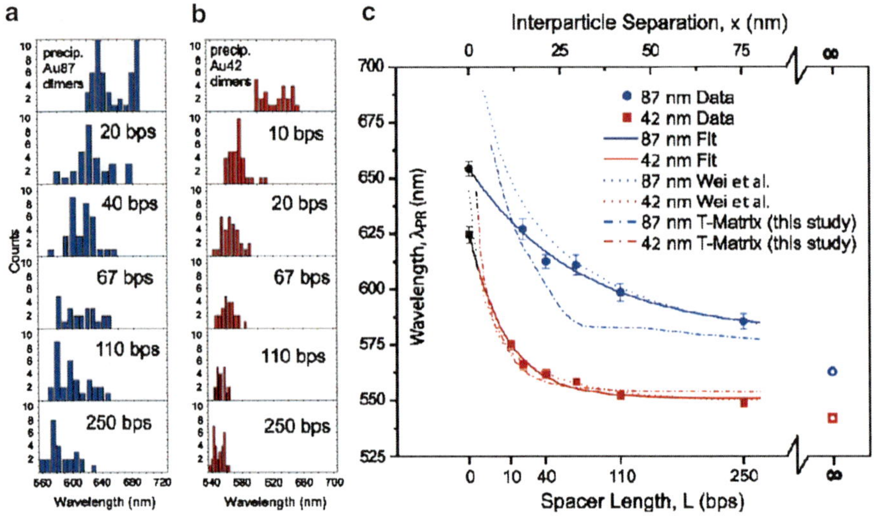

Fig. 6.1 Plasmon resonance vs. interparticle separation. Distributions of measured plasmon resonance wavelengths for selected dsDNA spacer lengths for (**a**) 87-nm Au plasmon rulers and (**b**) 42-nm Au plasmon rulers. (**c**) Plot of the average plasmon resonance as a function of spacer length, L, (*bottom axis*) and approximated interparticle distance x (*top axis*) for 42- (*red squares*) and 87-nm (*blue circles*) plasmon rulers (reprinted from [17], with permission)

El-Sayed et al. [18, 27] have shown that a simple model based on dipole–dipole interactions can explain the distance decay of plasmon coupling in metal nanostructures, using the dependence of the single-particle polarizability on the cubic power of the particle dimension and the decay of the particle plasmon near-field as the cubic power of the inverse distance. The long-range plasmon coupling behavior has a weak dependence on the particle separation and varies $1/R^3$ within dipole approximation. So the plasmon ruler provides a much longer interaction range as compared to FRET, which has $1/R^6$ dependence.

5 NSET Ruler

When an acceptor nanoparticle and a donor organic dye are brought into proximity, then there will be dipole-surface type energy transfer from dye molecular dipole to nanometal surface which can be termed as NSET [8–14, 20–26]. NSET is a through space mechanism like FRET and it arises from the damping of the fluorophore's oscillating dipole by the gold metal's free electrons. Like FRET [1–7], the interaction for NSET [8–14, 20–26] is dipole–dipole in nature. On the other hand, NSET is geometrically different because an acceptor nanoparticle has a surface and an isotropic distribution of dipole vectors to accept energy from the donor [8–14, 20–26]. This arrangement increases the probability of energy transfer and accounts

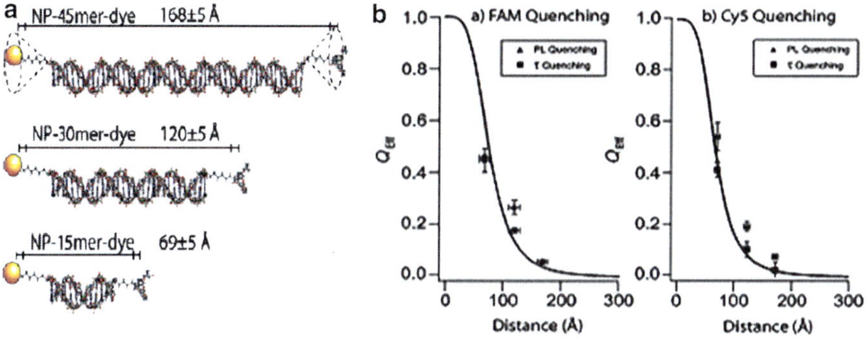

Fig. 6.2 (**A**) Scheme of DNA binding to a 1.5 nm Au/NP. By varying the length of the DNA strand, the terminal dye fluorophore is separated from the Au/NP by discrete distances (168, 120, and 69 Å). (**B**) Quenching data for FAM (**a**) and Cy5 (**b**) based upon photoluminescence (2) and lifetimes (9) overlaid on top of a theoretical curve generated (reprinted from [23], with permission)

for the enhanced efficiency of NSET over FRET. In the case of NSET, a small dipole in the excited fluorophore induces a large dipole in the particle, leading to an enhancement in the energy-transfer efficiencies.

Jennings et al. [20, 23, 25, 26] reported quenching efficiency of the fluorophore at discrete distances using three different lengths of dsDNA-dye, two dyes of different energies and 1.5 nm gold NPs. Their steady state and picosecond time-resolved spectroscopy measurements points out that the NSET follows $1/R^4$ distance dependence (as shown in Fig. 6.2). Their report clearly demonstrated [23] the first successful application of an NSET, more than doubles the traditional Förster range.

To understand the limitation of our long-range NSET ruler in terms of length of RNA and size of gold nanoparticle (Au/NP), we have used *HCV genome RNA* of different lengths [8]. Scheme 6.2 shows a schematic diagram of the NSET ruler and their operating principles for RNAs of different lengths. After hybridization, by varying the RNA lengths, the separation distance between gold nanoparticle and Cy3 dye can be systematically varied between 8 and 50 nm, by varying the number of base pairs [8]. The distance from the center of the molecule to the metal surface is estimated by taking into account size of the fluorescent dye, 0.32 nm for each base pair, and 1.8 nm for Au–S distance + base pair to dye distance. As a result, after hybridization the distance becomes 8.2 nm for 20 bp RNA, 14.6 nm for 40 bp RNA, 24.2 nm for 70 bp RNA and 40.2 nm for 120 bp RNA. We have assumed a linear ds-RNA strand configuration because ds-RNA is known to be rigid having a persistence length of 90 nm [8]. Figure 6.3 shows how the quenching efficiency varies with the increase in the distance between gold nanoparticle and Cy3 dye for gold nanoparticles of different particle sizes. Our result shows that the distance-dependent quenching efficiency is highly dependent on the particle size. Figure 6.3b shows how R_0 (distance at which the energy transfer efficiency is 50%) value varies with the size of gold nanoparticles.

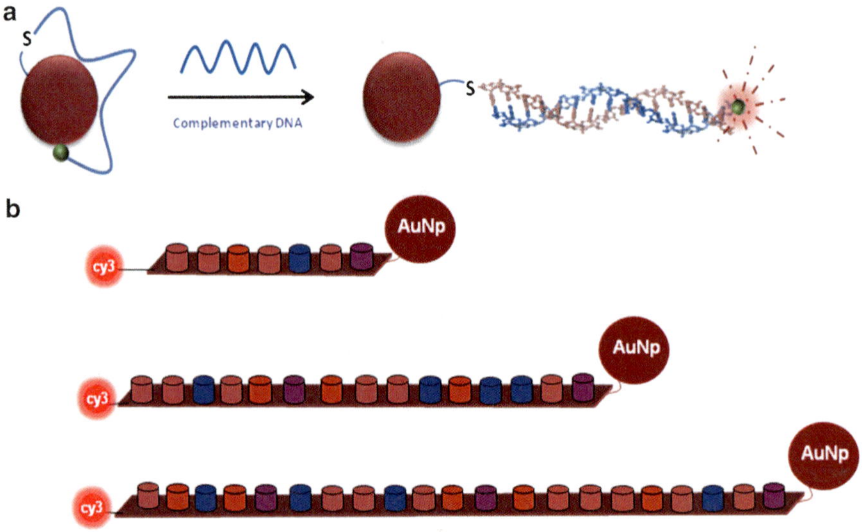

Scheme 6.2 (**a**) Schematic representation of the RNA hybridization process, when one end of the RNA is covalently coupled via thiol–gold chemistry. (**b**) Schematic illustration of 5′JOE and 3′-SH modified DNA of different lengths (reprinted from [8], with permission)

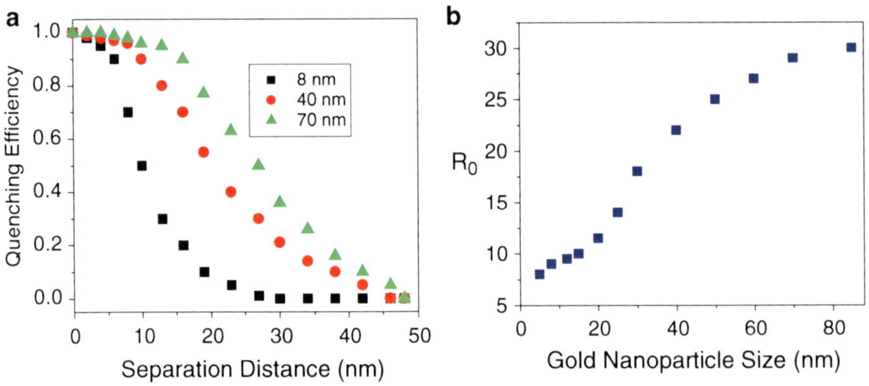

Fig. 6.3 (**a**) Variation of the quenching efficiency with distance between gold nanoparticle and Cy3 dye, (**b**) Variation of R_0 with the size of gold nanoparticle (reprinted from [8], with permission)

Our results indicate that one can tune the distance ranging all the way from 8 nm, which is very near to the accessible distance conventional FRET (6 nm), to about 40 nm by choosing GNPs of different diameters. Although, in general the interaction between nanoparticle and dye are quite complex, the situation can be due to the light induced oscillating dipole moments in each gold particle, and their instantaneous $(1/r)^3$ coupling results in a repulsive or attractive interaction, modifying the plasmon resonance of the system.

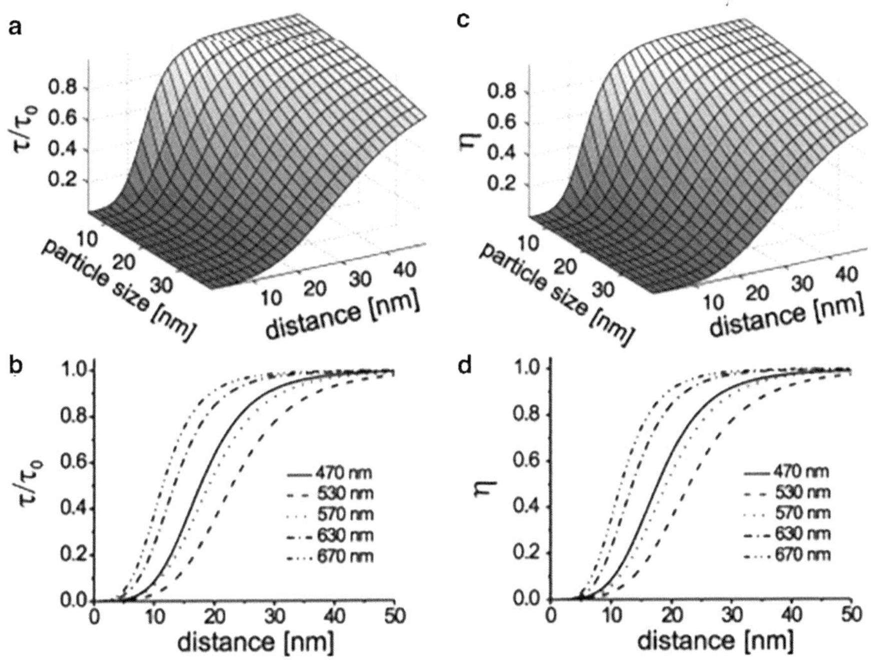

Fig. 6.4 (**a**, **c**) Fluorescence lifetime τ normalized to the unperturbed lifetime τ_0 and quantum efficiency η of an emitter at different distances from the surface of a GNP of various diameters. (**b**, **d**) Normalized lifetime and quantum efficiency of an emitter placed at different separations from the surface of a GNP of diameter 15 nm and for various emission wavelengths (reprinted from [22], with permission)

Recently, Seelig et al. [22] reported that nanoparticle-induced lifetime modification can serve as a nanoscopic ruler for the distance range well beyond 10 nm, which is the upper limit of FRET.

Their result shows that lifetime (τ_r) highly depends on the particle size as well as the distance between nanoparticle and dyes. As shown in Fig. 6.4, their data indicate that for bigger size nanoparticle (20–40 nm in diameters), τ_r is highly sensitive to small changes in the dye–particle distance even if they are separated by up to 40 nm, which explains our observation of high variation of R_0 with gold nanoparticle size. Our observations also point that when one chooses the experimental parameters for optimizing the NSET sensitivity, it is very important to take into account the effect of the gold nanoparticle size, which strongly affects the quenching efficiency and distance-dependent NSET. One has to remember that due to the effect of surface charge, surface coverage, and mutual strand interaction on the bending properties of individual RNA strand, the apparent length of the oligonucleotides can be smaller than their expected molecular length.

To understand the distance-dependent quenching process, we tried to fit our data (as shown in Fig. 6.5) with the theoretical modeling using general FRET and dipole-to-NSET used by Jeening et al. [20, 23]. Our results also point out that NSET model provides a better description of the distance dependence of the quenching

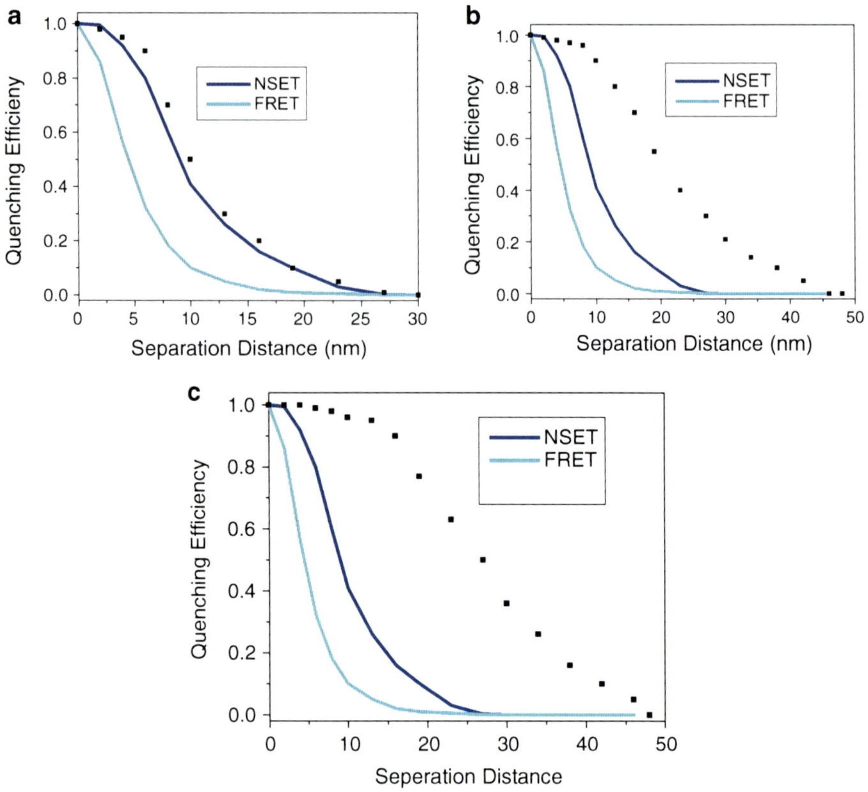

Fig. 6.5 Fitting data for variation of the quenching efficiency with distance using FRET, NSET, and experimental values for (**a**) 8 nm, (**b**) 40 nm, and (**c**) 70 nm gold nanoparticles (reprinted from [8], with permission)

efficiencies for 8 nm gold nanoparticle (as shown in Fig. 6.5a), but agreement is very poor for 40 and 70 nm gold nanoparticles (as shown in Fig. 6.5b, c). We also tried to fit our data with dipole-to-metal-particle energy transfer (DMPET) model as used by Pons et al. [28], but in this case also agreement is good with experimental data for 8 nm gold nanoparticle only. Our results also point out that DMPET and NSET models provide a better description of the distance dependence of the quenching efficiencies for 8 nm gold nanoparticle, but agreement is poor for 40 and 70 nm gold nanoparticles, where the measured values were always larger than the predicted ones.

Now, to evaluate how the distance-dependent behavior of long-range NSET ruler varies with the shape of nanomaterial, we have used ds-RNA of different lengths attached to gold nanorod ($\sigma = 3.2$) via thiol–gold chemistry in one end and the other end attached to Cy3 dye (as shown in Scheme 6.3). We have varied the distance from 3 to 50 nm by varying the number of base pairs in RNA, and we have assumed a linear ds-RNA strand configuration because ds-RNA is known to be rigid having a persistence length of 90 nm [8].

Scheme 6.3 Schematic illustration of Cy3 and gold nanorod-modified ds-RNA of different lengths, used in this manuscript

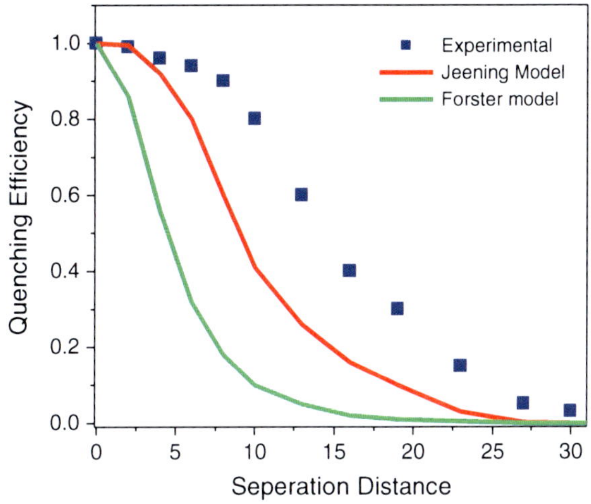

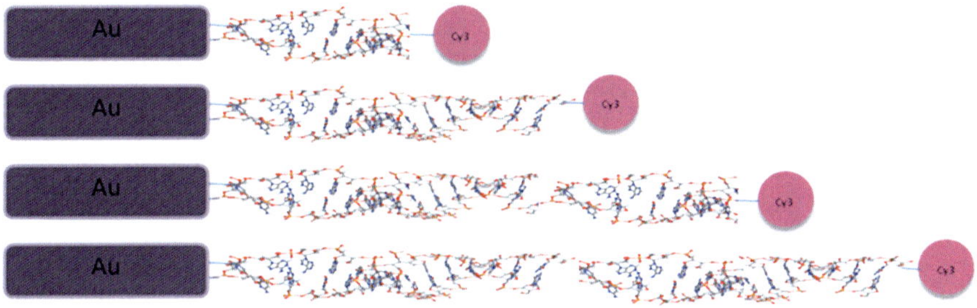

Fig. 6.6 Fitting of data for the variation of the quenching efficiency with distance using FRET, NSET theoretical model by Jeening and experimental values for gold nanorods of aspect ratio 2.0

Figure 6.6 shows how the quenching efficiency varies with the distance between gold nanorod and Cy3 dye. Our results show that 10% quenching efficiency even at a distance of 25 nm, which is more than twice as far as FRET can be measured. To understand the distance-dependent quenching process, we tried to fit our data (as shown in Fig. 6.6) with theoretical modeling using general FRET and dipole-to-NSET used by Jeening et al. [20, 23]. Our results indicate that the long distance quenching rate is better described with a slower distance-dependent quenching rate than the classical $1/R^6$ characteristic of Förster energy transfer. Our results also point out that though NSET model provides a better description of the distance dependence of the quenching efficiencies for very small gold nanoparticle, agreement is very poor for gold nanorods, where the measured values were always larger than the predicted ones. We also tried to fit our data with DMPET model as used by Pons et al. [28], but in this case, it is also not in good agreement with experimental data. Our observations also point that when one chooses the experimental parameters for optimizing the NSET sensitivity, it is very important to take into account the effect of the gold nanoparticle size and shapes, which strongly affects the quenching

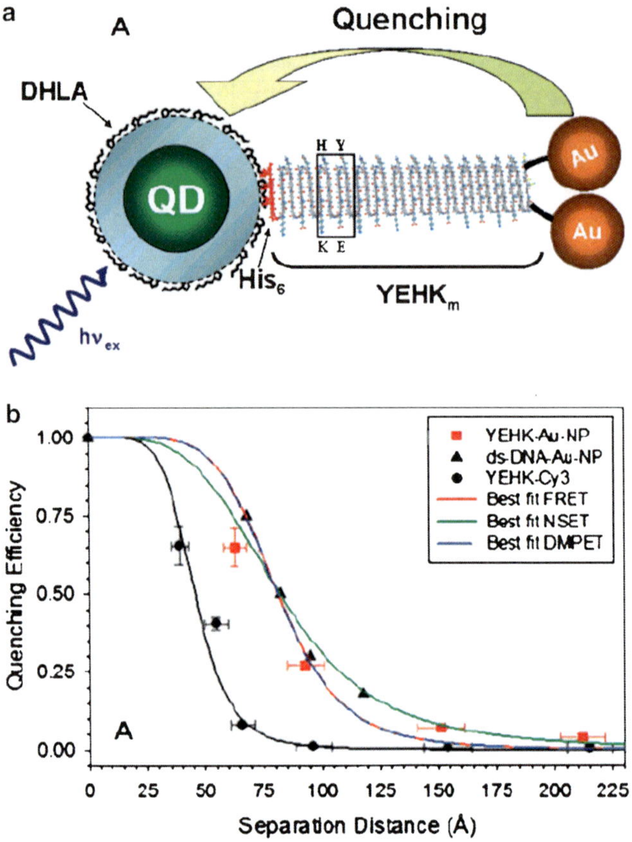

Fig. 6.7 (**a**) Schematic representation of the QD–peptide–Au/NP bioconjugates. The C-terminal His6 coordinates to the QD surface while the cysteines are used as attachment sites for 1.4 nm monomaleimide-functionalized Au/NPs. Repeat units of 5, 7, 14, or 21 were used; YEHK7 as shown with a single core YEHK boxed. The Au/NPs are separated from the cysteine thiol by a maximum of 8 Å and from each other by a maximum of 18 Å, (**b**) PL quenching efficiency vs. *R* for QD–YEHK5,7,14,21–Au-NP conjugates (*red squares*) and QD–dsDNA–Au-NP from [14] (*black triangles*) together with best fits using FRET (*red line*), DMPET (*blue line*), and NSET (*green line*). The quenching efficiencies for QD–YEHK1,3,5,7,14,21–Cy3 conjugates from [33] along with a fit using Förster FRET formalism are also shown (*black dots* and *black line*). Comparison between fits and data using *R*0 either as a fitting parameter (A), or the experimental values deduced from the spectral overlap and QY

efficiency and distance-dependent NSET. One has to remember that due to the effect of surface charge, surface coverage, and mutual strand interaction on the bending properties of individual RNA strand, the apparent length of the oligonucleotides can be smaller than their expected molecular length.

Pons et al. [28] reported Au/NP-driven QD PL quenching extends over a large distance range, much larger than what was predicted and measured for dye–dye and QD–dye pairs, as shown in Fig. 6.7. Their results indicate that it may be possible to extend the energy transfer rate far beyond the range allowed by "classic" dye-to-dye FRET pairs. Their results indicate that the QD PL quenching is mainly due to non-radiative energy dissipation by the Au/NP without any significant modifications of the QD radiative rate. They have also reported that the dipole to metal NSET model provides a better description of the distance dependence of the quenching efficiencies, even though agreement is only qualitative since the measured values were always larger than the predicted ones. Basko et al. [29] suggested that there could be efficient energy transfer from the excited states of a quantum well to a layer of organic molecules in its proximity. The transfer takes place essentially through the Förster mechanism, i.e., by coulombic interaction. They found the transfer to be highly efficient. They used the dipolar approximation for evaluating the transition matrix element and then used the Fermi golden rule to evaluate the rate. Their calculations led to an R^{-4} dependence of the rate on the distance for the case of a quasicontinuum of excitations of quantum dots. For the transfer to discrete low energy quantum dot excitations, they found an exponential dependence on the distance.

6 Portable NSET Probe

The photograph of the minimized system components [40, 47, 49] for the NSET sensor configuration to detect environmental toxin is shown in the Fig. 6.8a. For fluorescence excitation, we have used a continuous wavelength Melles-Griot green laser pointer (18 Lab 181) operating at 532 nm, as an excitation light source. The laser pointer can maintain 10–13 h with two AAA size batteries, and its maximum output power is ~5 mW. This light source has a capability that can minimize whole sensor configuration. The total size of the sensor configuration was 5″×8″×5″, including the laser pointer, optical fiber, and OOI spectrometer in the aluminum box. This probe consists of total of seven optical fibers, each having 200 μm core

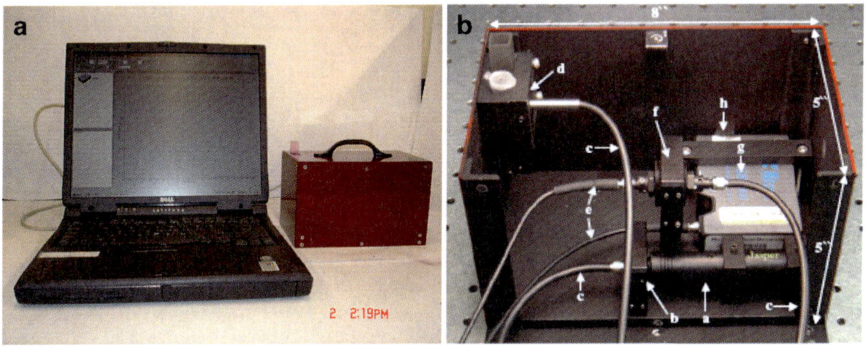

Fig. 6.8 (**a**) Photograph of the minimized NSET probe box, connected with laptop computer for data acquisition. (**b**) The photograph of the system components for the LIF sensor configuration (Reprinted from [11] and [14] with permission)

diameters with one launching fiber and six surrounding collecting fibers (as shown Fig. 6.8b). Excitation light source was first attenuated using appropriate neutral density (ND) filter and coupled to the excitation arm of the Y-shaped reflection probe through a plano-convex lens (*f*: ~4.5 mm) as shown in Fig. 6.8b. A typical laser energy at the sample was adjusted to ~1.3 mW with 0.3 ND filter.

The collected emission signal was transmitted through an online filter module to a 600 μm core diameter UV grade fused silica auxiliary fiber prior to feeding it to an Ocean Optics spectrometer. A low resolution OOI spectrometer with 600/1 mm grating (cover 200–850 nm) was used in this work. An OOI spectrometer was interfaced with a note book computer though USB port. The LIF spectrum was collected with Ocean Optics data acquisition software. The ON/OFF (Push /Pull) switch of the laser pointer is located at the other end of laser pointer, and it is fixed in such a way that one can operate it from outside of the aluminum box.

6.1 NSET Probe for DNA/RNA Hybridization Detection

Recently, we have demonstrated that [8] long-range NSET optical assay can be used to detect Hepatitis C virus (HCV) RNA. After a pioneering work by Mirkin et al. [31] and Alivisatos et al. [32], several groups [8–12, 21–25, 33] from academia and industry are working on the detection of DNA and RNA sequence using different nanotechnology-based techniques.

Scheme 6.4 shows a schematic diagram of the nanoparticle probes and their operating principles for single RNA hybridization detection process [8]. Dye tagged ss-RNA is adsorbed onto the gold nanoparticle and as a result the fluorescence from the dye is completely quenched by gold nanoparticle [8–15]. Our experimental data showed a quenching efficiency of nearly 100% when the fluorophore was statically adsorbed on the particle (static quenching). Thus, when an oligonucleotide molecule is firmly tethered to a particle, the fluorophore at the

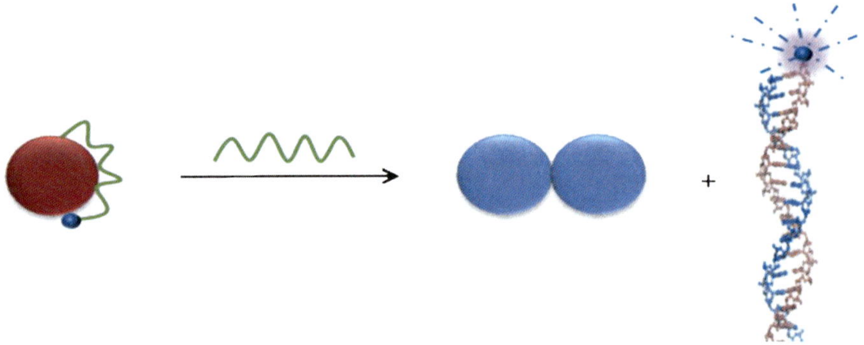

Scheme 6.4 Schematic representation of the RNA hybridization process (reprinted from [8], with permission)

distal end can look back and adsorb on the same particle. Upon target binding, due to the duplex structure, the double-strand (ds) RNA does not adsorb onto gold and the fluorescence persists (as shown in Scheme 6.4).

This structural change generates a fluorescence signal that is highly sensitive and specific to the target RNA. In this case, instead of covalently binding the probe-RNA to the gold nanoparticle, as we used for understanding the distance-dependent NSET process, the probe-RNA is being reversibly adsorbed to the gold nanoparticle. Thus, upon hybridization of the target-RNA, the resultant ds-RNA is being completely released to the solution instead of having left tethered to the gold particle. As shown in Fig. 6.9a, we observed a very distinct emission intensity change after hybridization even at less than 100 pM concentration of probe fluorophore-tagged ss-RNA [8].

When target RNA with complementary sequence is added to the probe RNA, a clear colorimetric change from red to blue–gray color is observed within few minutes (as shown in Fig. 6.10a), and we noted the plasmon band shifted from 512 to 750 nm (as shown in Fig. 6.10b). Figure 6.10c shows the TEM image after hybridization and our results indicate that gold nanoparticles undergo aggregation after hybridization, and it is due to the presence of sodium chloride. Gold nanoparticles in solution are typically stabilized by adsorbed negative citrate ions whose repulsion prevents the strong van der Waals attraction between gold particles from causing them to aggregate. As soon as the ds-RNA separated from gold nanoparticle, aggregation of gold nanoparticle happens. This is due to the screening effect of the salt, which minimizes electrostatic repulsion between the nanoparticles. Our zeta potential measurement using Zetasizer NanoZS shows that negatively charged surface (Zeta potential −63.5 mV) of the nanoparticles reduced surface charge (Zeta potential −9.8 mV) in the presence of 0.08 M NaCl. So after hybridization, due to the presence of around 0.4 M NaCl, screening effect of salt leads to more linked particles and hence larger damping of the surface plasmon absorption of Au nanoparticle surfaces. This structural change generates a fluorescence signal that is highly sensitive and specific to the target RNA. Figure 6.9a also illustrates single-mismatch detection capability. Our result indicates that our NSET probes are highly specific in discriminating against noncomplementary DNA sequences and single-base mismatches. The addition of noncomplementary nucleic acids had no effect on the fluorescence, and a single-base mismatch reduced the fluorescence intensity by 90% (in comparison with the fluorescence intensity of perfectly matched targets). So our NSET probes will be applicable to rapid detection of single-nucleotide polymorphisms (SNPs) in genomic RNA, an exciting prospect for eliminating time-consuming and expensive gel sequencing procedures that are currently the standard protocol.

To evaluate whether our NSET probe is capable of measuring target RNA concentration quantitatively, we performed NSET intensity measurement at different concentrations of target RNA. As shown in Fig. 6.9b, the NSET emission intensity is highly sensitive to the concentration of target RNA and the intensity increased linearly with concentration. Linear correlation was found between the emission intensity and concentration of target DNA over the range of 20–450 pM (as shown in Fig. 6.9c). So our NSET probe can provide a quantitative measurement of HCV

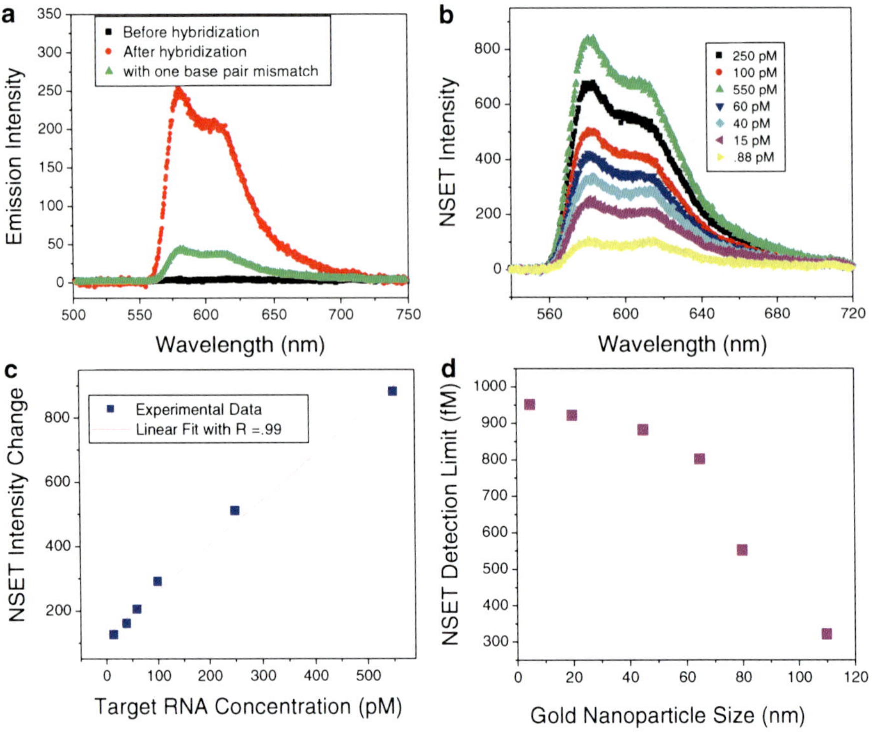

Fig. 6.9 (**a**) Plot of fluorescence intensity vs. wavelength for 40 pM 5′-Cy3 modified HCV RNA (Cy3-5′UGUACUCACCGGUUCCGCAGACCAC-3′) adsorbed onto 45 nM gold nanoparticle solution before hybridization, after hybridization with complementary RNA which has one base pair mismatch at the end, after hybridization with exact complementary RNA; (**b**) Fluorescence response upon addition different concentration of target RNA on 150 nm probe RNA; (**c**) Plot of fluorescence intensity vs. target RNA concentration in picomoles. Linear correlation exist over the range of 15–550 pM with $R = 0.994$; (**d**) Variation of NSET sensitivity with gold nanoparticle size (reprinted from [8], with permission)

RNA concentration in a sample. Figure 6.9d shows how the NSET detection limit for HCV RNA detection varies with particle size. The detection limits for NSET have chosen when NSET intensity change is 12, before and after hybridization. As we can note from the figure, NSET sensitivity highly depends on the particle diameter. As the particle diameter increases, the detection limit becomes better and better and our data indicate that NSET can detect RNA as low as 300 fM, when the particle size is 110 nm. This variation of sensitivity efficiency with particle size can be due to several factors, and these are (1) reduced surface area of smaller nanoparticles will limit the dye-tagged RNA accommodation on the gold nanoparticles; and (2) increasing overlap between nanoparticle absorption with Cy3 emission and K_{SV} values with particle size. So our experimental data suggests that the combination of Cy3-modified RNA with proper size of gold nanoparticles can potentially lead to highly sensitive optical biosensors.

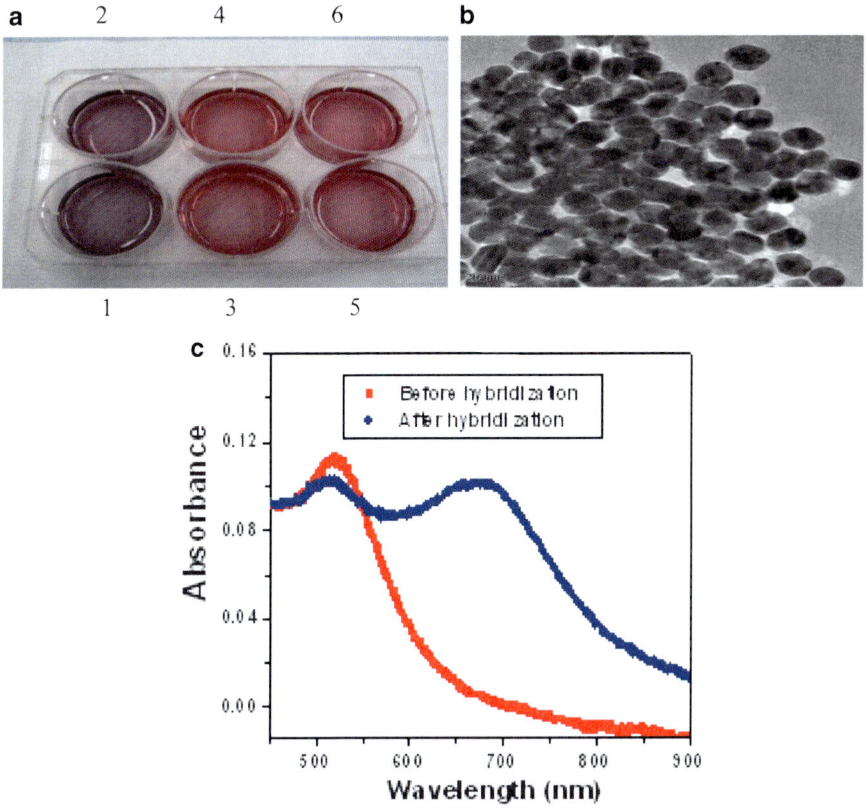

Fig. 6.10 (**a**) Photograph showing colorimetric change upon addition of (1) 40 nM complementary RNA; (2) 10 nM complementary RNA, (3) 3 nM complementary RNA; (4) 1 nM complementary RNA; (5) 40 nM complementary RNA with one base pair mismatch and (6) only gold nanoparticle. (**b**) TEM image of gold nanoparticle after hybridization. (**c**) Absorption profile of RNA-coated Au nanoparticles before and after hybridization with 10 nM concentration of probe RNA (reprinted from [30], with permission)

6.2 NSET for DNA/RNA Cleavage Detection

Recently, we have reported [13] a gold nanoparticle-based long-range NSET optical ruler assay to monitor the cleavage of DNA by nucleases. Scheme 6.5 shows a schematic diagram of the nanoparticle probes and their operating principles for DNA cleavage process when the ssDNA is attached to gold nanoparticle through S-linkage. Our assay consists of two separate steps (1) Cy3-labeled DNA-coated gold nanoparticles were hybridized with the complementary nucleic acid containing one nucleotide less. The hybridization was followed by the fluorescence quenching of the dye by gold nanoparticles; (2) Au/NP/dye DNA duplex was treated with S1 nuclease to cleave the DNA and kinetics of the cleavage process was monitored from the recovery of the dye fluorescence.

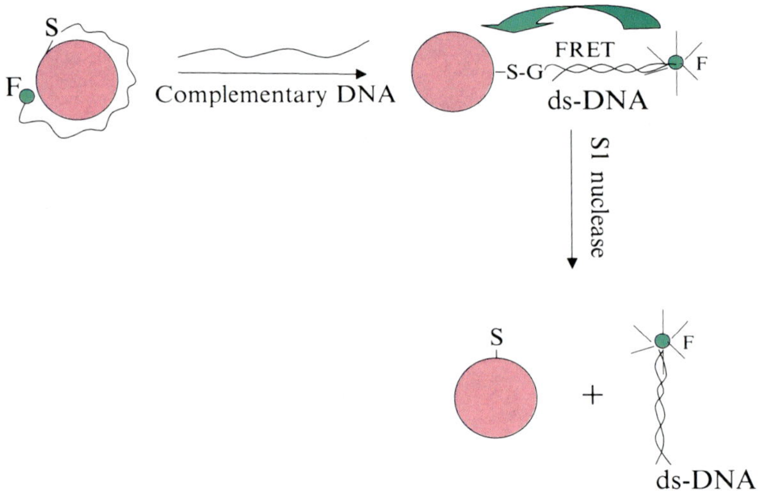

Scheme 6.5 Schematic illustration of DNA cleavage process when DNA is attached to gold nanoparticle through –SH group. The *circles* represent the gold nanoparticle and F represents the fluorophore (reprinted from [13], with permission)

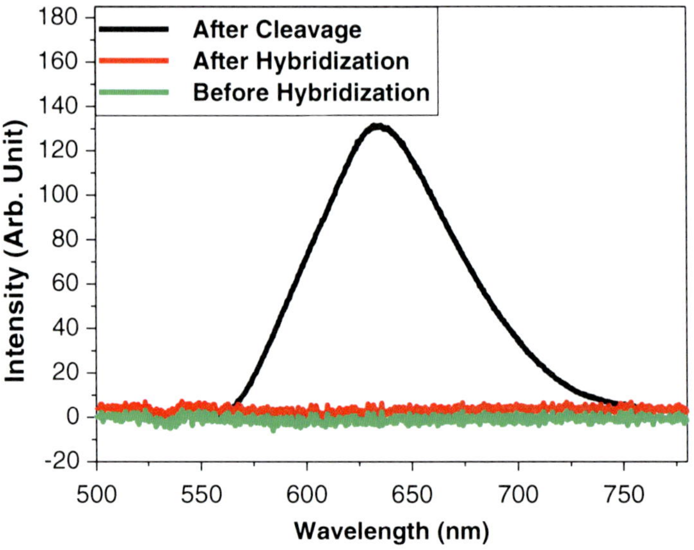

Fig. 6.11 Plot of fluorescence intensity vs. wavelength for 5′-Cy3 modified GAAAAACCCCT TTTTT-3′ oligonucleotide (reprinted from [13], with permission)

As shown in Fig. 6.11, the fluorescence from Cy3-labeled nucleic acid is totally quenched by gold nanoparticles after hybridization and the recovery of the fluorescence signal was observed only after DNA cleavage. We observed a very distinct LIF intensity change after cleavage even at 250-fM concentration of fluorophore-tagged ds-DNA. Fluorescence signal enhancement is observed by a factor of 120 after the cleavage reaction in the presence of S1 nuclease. The assay we have reported here enables an extremely high signal to background ratio (~120). As a control experiment, Cy3-labeled ss-DNA-coated gold nanoparticles were treated with S1 nuclease to cleave the ss-DNA. The addition of S1 nuclease had no effect on the fluorescence quenching; only a weak and broad background was observed. This indicates that though S1 nuclease can cleave the ssDNA, it can affect the adsorbed fluorophore position and as a result the fluorescence was quenched totally.

Alivisatos et al. [15–17] have demonstrated that the plasmon ruler is effective in measuring distances till 100 nm range, which makes it a unique tool for studying macromolecular conformational changes at the single-molecule level. The ability of the long-range optical ruler to operate over extended time scales and to be effective in measuring distances over the 1- to 100-nm range makes it a unique tool for studying biological process at the single-molecule level. As shown in Fig. 6.12, their results clearly show that DNA cleavage by the EcoRV restriction enzyme can be monitored using long-range optical ruler [16]. Their results match nicely with the prior ensemble kinetic studies and single-molecule FRET experiments, which indicates that the Au particles do not significantly perturb the system. They have also demonstrated that [16] the plasmon ruler can be used of studying fluctuations in bending of DNA, which can be used to determine the potential energy change between the straight and bent DNA states. The inherent brightness of long-range plasmon rulers can be an excellent candidate for monitoring the dynamics of biological processes and biopolymers.

Gill et al. [34] reported nucleic acid-functionalized CdSe/ZnS quantum dots (QDs) hybridized with the complementary Texas-Red-functionalized nucleic acid. The hybridization was monitored using fluorescence resonance energy transfer from the QDs to the dye units. Treatment of the QD/dye DNA duplex structure with DNase I resulted in the cleavage of the DNA and the recovery of the fluorescence properties of the CdSe/ZnS QDs. Their results indicate that luminescence properties of the QDs were only partially recovered due to the nonspecific adsorption of the dye onto the QDs. Skewis et al. [19] reported the influence of spermidine on the cleavage kinetics of RNase A at the single-molecule level using pairs of RNA-tethered 40 nm gold nanoparticles as shown in Fig. 6.13.

Their results point out that because of their high temporal resolution and the ability to follow the cleavage of individual RNA molecules, the RNA plasmon rulers provide information about relative stabilities of weakly stabilized subpopulation and their lifetimes. They have demonstrated that efficient RNA cleavage rates for all investigated concentrations (1 nM–5 mM). Time-resolved cleavage experiments (temporal resolution: ~10 ms) indicate that with increasing spermidine concentration, cleavage is delayed and discrete subpopulations with longer lifetimes emerge. The reduction of RNase A cleavage rates and appearance of subpopulations that indicate transient structural RNA stabilization confirmed that RNA-binding enzyme activity is regulated through spermidine-induced changes in the charge and structure of the RNA substrate.

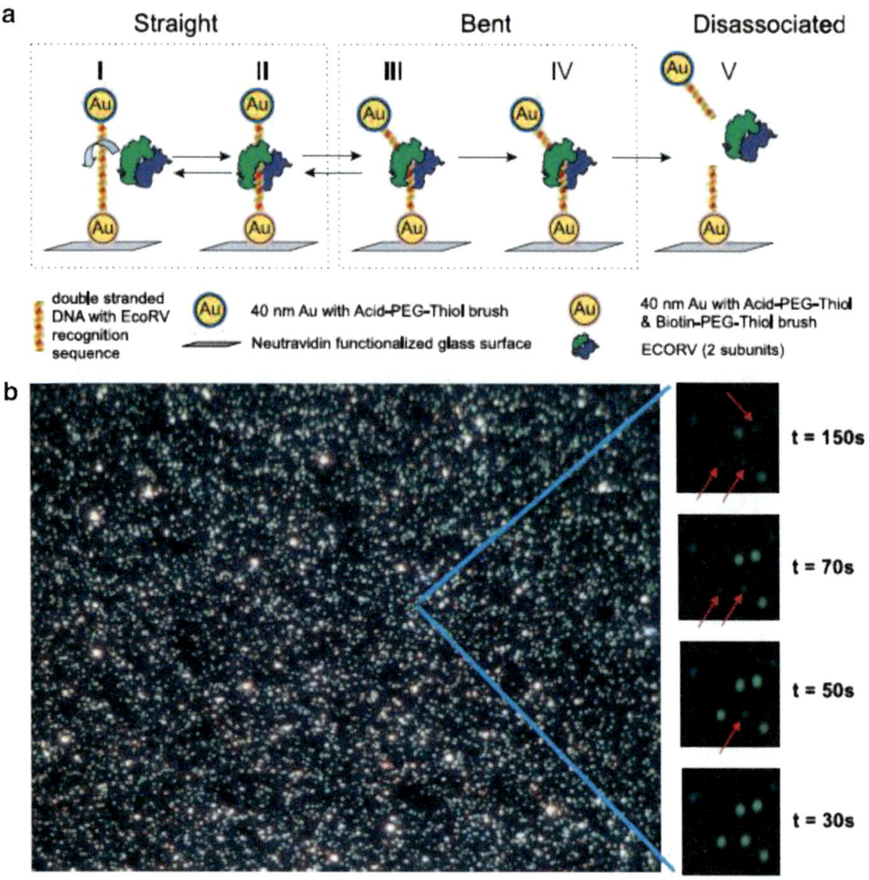

Fig. 6.12 Highly parallel single EcoRV restriction enzyme digestion assay. (**a**) The plasmon rulers are immobilized with one particle to a glass surface through biotin–neutravidin chemistry. The homodimeric EcoRV enzyme binds nonspecifically to DNA bound between the particles (I), translocates and binds to the target site (II), bends the DNA at the target site by 50° (III), cuts the DNA in a blunt-ended fashion by phosphoryl transfer (54) (IV), and subsequently releases the products (V). (**b**) A 150 × 100-μm field of view with surface immobilized plasmon rulers. Individual dimers are visible as *bright green dots*. Dimer dissociation upon EcoRVcatalyzed DNA cleavage leads to a strong change in scattering intensities. The dimers are converted into monomers as shown for selected particles (*red arrows*, *right side bar*). EcoRV is added at $t = 0$ s (reprinted from [16], with permission)

6.3 NSET for Monitoring Mg^{2+}-Dependent RNA Folding

Recently, we have demonstrated [9] that gold nanoparticle-based fluorescence resonance energy transfer can be used to track the folding of RNA (as shown in Fig. 6.14). As a model system, the conformational changes of two-helix junction RNA molecules induced by the binding of Mg^{2+} ions studied by measuring time-dependent

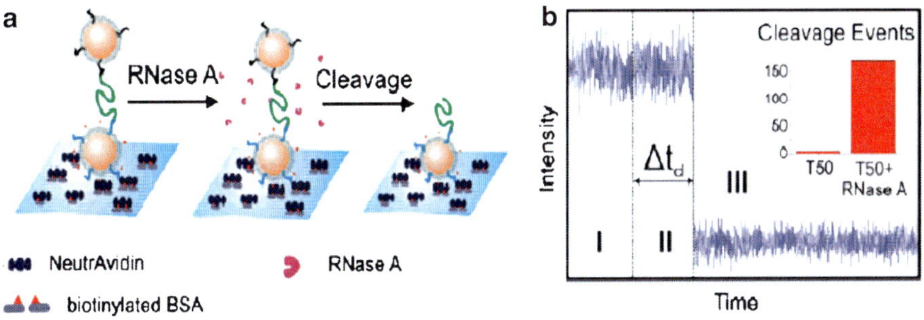

Fig. 6.13 Plasmon ruler RNase A cleavage assay. (**a**) The RNA plasmon rulers are bound to the surface of a glass flow chamber using a bovine serum albumin (BSA)–Biotin–NeutrAvidin surface chemistry. Upon addition of RNase A, the RNA tether is cleaved, and the dimer converted into a monomer. (**b**) Single RNA plasmon ruler cleavage trajectory (recorded at 96 Hz). (I) The plasmon ruler is first incubated in buffer containing spermidine at defined concentrations (0–5 mM), (II) the buffer is exchanged with a 1 nM RNase A solution, causing (III) a strong drop in intensity upon RNA cleavage. *Inset*: Number of cleavage events for flushing with/without enzyme

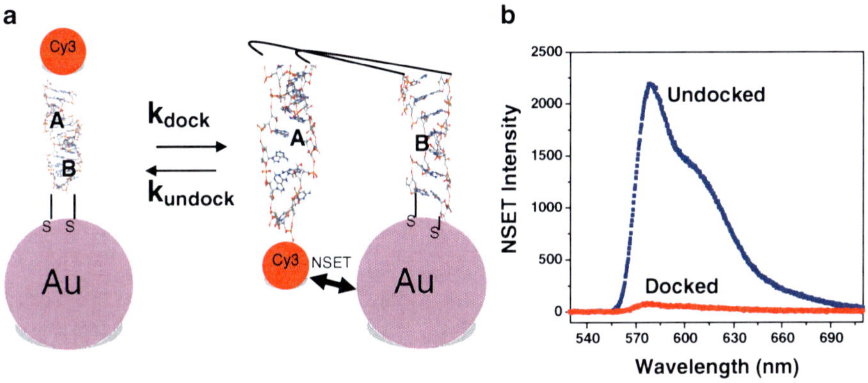

Fig. 6.14 (**a**) Schematic representation of NSET assay for RNA folding tracking, (**b**) Time-dependent NSET intensity in the presence of 15 mM Mg^{2+} at 25°C (Reprinted from [9], with permission)

fluorescence signal (as shown in Fig. 6.15). The transition from an folded to an open configuration changed the distance between gold nanoparticles and the dye molecule attached to the ends of two helices in the RNA junction and as a result, the unfolding process monitored from the change of fluorescence intensity. To probe the transition states involved during transition, unfolding of two-helix junction RNA molecules induced by the Mg^{2+} ions was studied by measuring time-dependent NSET signal (as shown in Fig. 6.15). Our experimental result shows clearly that there are four separate states involved during docking to undocking transition. RNAs are found to switch very slowly between the four states. The emission intensity changes with time for each state follow first order kinetics, for docked configu-

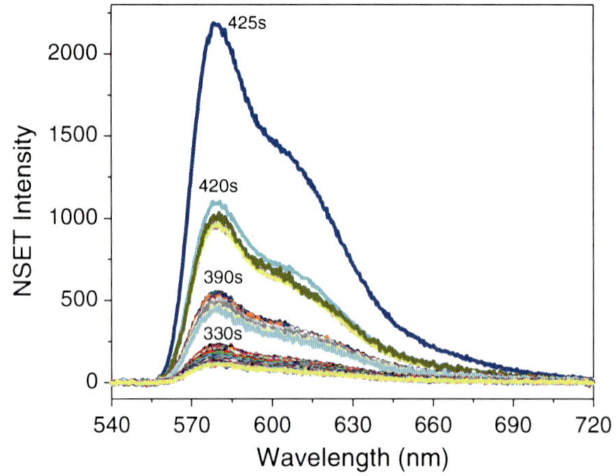

Fig. 6.15 Time-dependent (in 0.1 s) NSET intensity in the presence of 15 mM Mg^{2+} at 25°C, showing undocking process (reprinted from [9], with permission)

ration. Lowest emission intensity configuration is docked configuration (D), where the two domains are in contact and at a 70° angle as reported in the crystal structure. Since in this configuration, Cy3 dye is very near to the gold nanoparticle, the fluorescence is almost completely quenched by gold nanoparticle. Highest emission intensity configuration is an extended configuration that represents the undocked state (U), where gold nanoparticle and Cy3 dye is about 43 nm far. Other two states are transition state through which RNA folding takes place from D → U state. To understand the structure of these two transition states, we measure distance-dependent NSET. Our time-dependent experiments (as shown in Fig. 6.16) indicate that quenching efficiency is about 98% in D state, 85% in first transition state, 45% in second transition state, and 0% in U state.

From the distance-dependent curve as shown in Fig. 6.15, we estimate distance between gold nanoparticle and Cy3 is ~3 nm in D state, 5 nm in first transition state and 12 nm in second transition state (as shown in Fig. 6.16). From our distance-dependent NSET and time-dependent RNA folding data, we propose that during folding from D → U state, the first transition state can be a state where the two domains are in direct contact at a parallel 0° angle, and start allowing the insertion of hydrated metal ions (C).

The second transition state is the state, where the two domains surfaces are apart at a 0° angle (NC), allowing the insertion of high amount of hydrated metal ions (as shown in Fig. 6.16). Four states RNA folding process has been proposed before for smaller two-way junction RNA by Zhuang et al. [121]. By fitting the time-dependent fluorescence intensity change with first order reaction kinetics, we measured the rate constant for each state. Rate constant for $K_C = 0.008\ s^{-1}$, $K_{NC} = 0.18\ s^{-1}$, $K_U = 2\ s^{-1}$. Our measured K_C and K_{NC} are slower than reported values for smaller two-way junction RNA by Zhuang et al. [36]. And it is due to the fact that for general FRET, the interaction between donor (Cy3 dye) and acceptor (Cy5 dye) is only dipole–dipole coupling, where as in case of NSET, gold nanoparticle and Cy3 are bound by strong electrostatic interaction.

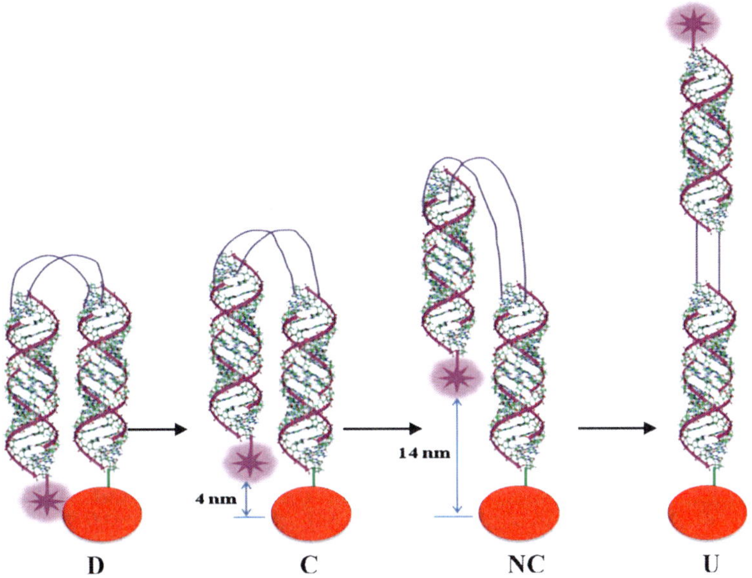

Fig. 6.16 Schematic representation of possible transition states involved in docked to undocked state

Seferos et al. [37] demonstrated that oligonucleotide-modified gold nanoparticle probes hybridized to fluorophore-labeled complements can be used as both transfection agents and cellular "nano-flares" for detecting mRNA in living cells. Their results indicate that nano-flares exhibit high signaling, low background signal and sensitive to changes in the number of RNA transcripts present in cell. Bates et al. [38] reported magnesium-ion-mediated RNA–RNA loop–receptor interactions using self-assembled nanowires. Jennings et al. [25] demonstrated that NSET can be used to measure Mg^{2+}-induced conformational changes for a hammerhead ribozyme and confirmed their measurements using FRET. As shown in Fig. 6.17, their results indicate that NSET is able to enhance the understanding of the different kinetic pathways for ribozyme. Their report clearly shows that NSET can be used for kinetic and structural measurements on the hammerhead ribozyme. They have also shown that NSET has advantageous for the measurement of very fast processes that are otherwise difficult to observe. The measured rate constant of binding hammerhead ribozyme was similar with the literature reports in same pH conditions.

Zhang et al. [35] reported a quantum-dot (QD)-based nanosensor that can be used in FRET assays of RRE IIB RNA-Rev peptide interactions. Their results indicate that QD-based nanosensor offers the distinct advantages of not inhibiting the Rev-RRE interaction, high sensitivity, improved accuracy, and simultaneous FRET-related two-parameter detection. This QD-based nanosensor provides a new approach to study the effects of inhibitors upon Rev-RRE interaction, and it may have a wide applicability in the development of new drugs against HIV-1 infection.

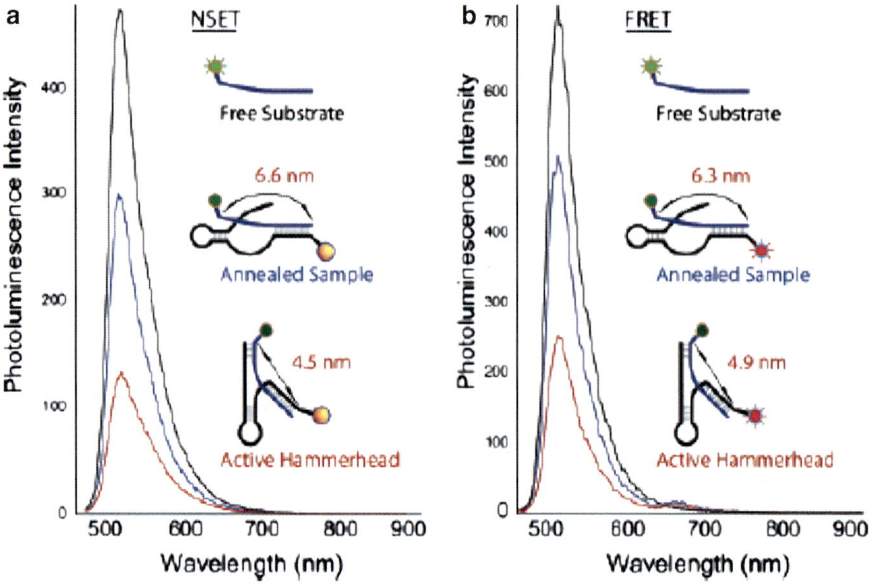

Fig. 6.17 Photoluminescence quenching correlated structural changes in the hammerhead ribozyme via energy transfer from FAM to NG (NSET, part **a**) and from FAM to AF647 (FRET, part **b**). The most intense spectrum shows the intensity of the substrate alone and then after annealing with quencher-bound ribozyme (*blue*). Finally, the solution is adjusted to 20 mM Mg^{2+} and PL is measured again (*red*) (reprinted from [25], with permission)

7 Summary, Outlook, and Future Needs

In conclusion, in this chapter, we have discussed recent efforts on the development of a nanomaterial-based long-range optical ruler for probing biological dynamics, screening pathogen DNA/RNA and monitoring DNA/RNA cleavage. Our experimental observation demonstrated gold nanoparticles-based long-range optical ruler strategies at distances more than triple distances achievable using traditional FRET methods. This long-range feature allows the development of new type biosensors and bioassays that are not possible by using general FRET process. Since noble metal nanostructures exhibit great range, versatility, and systematic tunability of their optical attributes, developing long-range NSET optical ruler using suitable material, can be readily used to other areas of biology and medicine.

To take full advantage of the long distance range of the plasmon ruler, new strategies need to be developed to improve the size and shape homogeneity of the nanoparticles. It is probably possible to improve the NSET sensitivity by several orders of magnitudes by choosing proper materials and detection systems. Continued optimization of different parameters is necessary to monitor biomolecules in complex environments. Future advances require continued innovations by chemists in close collaboration with experts in medical and toxicological fields.

The possible hazards associated with nanomaterials can be significant. Therefore, an understanding of biological response and environmental remediation is necessary. In addition, given the influence of impurity on a wide range of nanoparticle properties, a well-characterized analytical technique that can detect and quantify impurities will be important.

The possible hazards associated with nanomaterials can be significant. Therefore, an understanding of biological response and environmental remediation is necessary. To assess the safety of complex multicomponent and multifunctional nanomaterials, we need to develop validated models capable of predicting the release, transport, transformation, accumulation, and uptake of engineered nanomaterials in the environment. The availability of routine analytical methods that address these issues is a key to gaining a better understanding of the mechanisms of nanoparticle formation and reactivity. In addition, given the influence of purity on a wide range of nanoparticle properties, analytical techniques that can detect and quantify impurities will be important for pursuing greener approaches.

Acknowledgments Dr. Ray thanks NSF-PREM grant # DMR-0611539, ARO grant # W911NF-06-1-0512 and NIH-SCORE grant # S06GM 008047 for their generous funding.

References

1. Royer C A, Probing Protein Folding and Conformational Transitions with Fluorescence, *Chem. Rev.*, **2006**, *106*, 1769–1784.
2. Martí A A, Jockusch S, Stevens N, Ju J and Turro N J, Fluorescent Hybridization Probes for Sensitive and Selective DNA and RNA Detection, *Acc. Chem. Res.*, **2007**, *40* ,402–407.
3. Ho H H, Najari A and Leclerc M, Optical Detection of DNA and Proteins with Cationic Polythiophenes, *Acc. Chem. Res.*, **2008**, *41*, 168–178.
4. Forster, T., Intermolecular energy transference and fluorescence, *Ann. Physik*, 1948, 2, 55–7.
5. Stryer L and Haugland R P, Energy transfer: a spectroscopic ruler, *Proc Natl Acad Sci USA*. **1967**, 58, 719–726.
6. Michalet X, Weiss S, and Jäger M, Single-Molecule Fluorescence Studies of Protein Folding and Conformational Dynamics, *Chem. Rev.*, **2006**, *106*, 1785–1813.
7. Barbara P F, Gesquiere A J, Park S-J, and Lee J Y, Single-Molecule Spectroscopy of Conjugated Polymers, *Acc. Chem. Res.*, **2005**, *38*, 602–610.
8. Griffin, J., Singh, A. K., Senapati, D., Patsy Rhodes Mitchell, K., Robinson, B., Yu, E and Ray, P. C., Size and Distance Dependent NSET Ruler for Selective Sensing of Hepatitis C virus RNA, *Chem. Eur. J.*, **2009,** 15, 342–351.
9. Griffin, J.; Ray, P. C., Gold Nanoparticle Based NSET For Monitoring Mg^{2+} Dependent RNA Folding, *J. Phys. Chem. B.*; (Letter)**;** **2008**; 112, 11198–11201.
10. Sen T, Suparna S and Patra A, Surface energy transfer from rhodamine 6G to gold nanoparticles: A spectroscopic ruler, *Appl. Phys. Lett*, **2007**, 91, 41304.
11. Darbha, G K, LE, Glenn, E, Anderson, Y R, Preston F, Mitchell K, Ray P. C., Miniaturized NSET Sensor for Microbial Pathogens DNA and Chemical Toxins, *IEEE Sensor Journal*, **2008**, 8, 693–701.
12. Darbha G K, Ray A and Ray P C, Gold-nanoparticle-based miniaturized FRET Probe for rapid and ultra-sensitive detection of mercury in soil, water and fish, *ACS Nano*, **2007**, 3, 208–214.
13. Ray, P. C.; Fortner, A.; Darbha, G. K. Gold Nanoparticle Based FRET Assay for the Detection of DNA Cleavage, *J. Phys. Chem. B*, **2006**, *110*, 20745–20748.

14. Kim, C. K.; Kalluru, R. R.; Singh, J. P.; Fortner, A.; Griffin, J.; Darbha, G. K.; Ray, P. C. Gold Nanoparticle Based Miniaturized Laser Induced Fluorescence Probe for Specific DNA Hybridization Detection: Studies on Size Dependent Optical Properties, *Nanotechnology,* **2006**, *17*, 3085.
15. C Sönnichsen, B M Reinhard, J Liphardt & A Paul Alivisatos , A molecular ruler based on plasmon coupling of single gold and silver nanoparticles, *Nature Biotechnology* **2005**, **23**, 741–745.
16. B. M. Reinhard, S Kunchakarra, A Mastroianni, A Paul Alivisatos & J Liphardt, Use of plasmon coupling to reveal the dynamics of DNA bending and cleavage by single EcoRV restriction enzymes, *Proc. Natl Acad. Sci. USA*, 2007, **106**, 2667–2672.
17. Reinhard, B. M., Siu, M., Agarwal, H., Alivisatos, A. P., Liphardt, J, Calibration of Dynamic Molecular Rulers Based on Plasmon Coupling between Gold Nanoparticles, *Nano Letter*, **2007**, 5, 2246.
18. Jain, P. K., Huang, W., El-Sayed, M. A., On the Universal Scaling Behavior of the Distance Decay of Plasmon Coupling in Metal Nanoparticle Pairs: A Plasmon Ruler Equation, *Nano Letter*, **2007**, 7, 2080.
19. Skewis, L. R.; Reinhard, B. M., Spermidine Modulated Ribonuclease Activity Probed by RNA Plasmon Rulers, *Nano Lett.,* **2008**, 8, 214–220.
20. Nanometal Surface Energy Transfer in Optical Rulers, Breaking the FRET Barrier , C. S. Yun, A. Javier, T. Jennings, M. Fisher, S. Hira, S. Peterson, B. Hopkins, N. O. Reich, G. F. Strouse, *J. Am. Chem. Soc.* **2005**, *127*, 3115–3119.
21. D. J. Maxwell, J. R. Taylor and S. Nie, Self-Assembled Nanoparticle Probes for Recognition and Detection of Biomolecules , *J. Am. Chem. Soc.* **2002**, 124 9606.
22. Seelig, J.; Leslie, K.; Renn, A.; Kuhn, S.; Jacobsen, V.; van de Corput, M.; Wyman, C.; Sandoghdar, V., Nanoparticle-Induced Fluorescence Lifetime Modification as Nanoscopic Ruler: Demonstration at the Single Molecule Level, , *Nano Letter*, **2007**, 7, 685.
23. Jennings, T. L.; Singh, M. P.; Strouse, G. F, Fluorescent Lifetime Quenching near $d = 1.5$ nm Gold Nanoparticles: Probing NSET Validity, *J. Am. Chem. Soc.* **2006**, 128, 5462.
24. B. Dubertret, M. Calame, A. J. Libchaber, Single-mismatch detection using gold-quenched fluorescent oligonucleotides , *Nat. Biotechnol.*, **2001**, 19, 365–370.
25. T. L. Jennings, J. C. Schlatterer, M. P. Singh, N. L. Greenbaum, and G. F. Strouse**,** NSET Molecular Beacon Analysis of Hammerhead RNA Substrate Binding and Catalysis, *Nano Lett.*, **2006**, *6*, 1318–1324.
26. M P Singh, T L.Jennings and G F. Strouse**,** Tracking Spatial Disorder in an Optical Ruler by Time-Resolved NSET, *J. Phys. Chem. B*, **2009**, *113*, 552–558.
27. Tabor C, Murali R, Mahmoud M and El-Sayed M A, On the Use of Plasmonic Nanoparticle Pairs As a Plasmon Ruler: The Dependence of the Near-Field Dipole Plasmon Coupling on Nanoparticle Size and Shape, , *J. Phys. Chem. A*, **2009**, *113*, 1946–1953.
28. Pons, T., Medintz, I. L., Sapsford, K. E., Higashiya, S., Grimes, A. F., English, D. S., Mattoussi, H., On the Quenching of Semiconductor Quantum Dot Photoluminescence by Proximal Gold Nanoparticles, *Nano Lett.*, **2007**, 7, 3157–3164.
29. Marc Achermann, Melissa A. Petruska ,Simon Kos, Darryl L. Smith, Daniel D. Koleske & Victor I. Klimov, Energy-transfer pumping of semiconductor nanocrystals using an epitaxial quantum well, *Nature* 429, **2004**, 642–646.
30. Griffin, J., Singh, A. K., Senapati, D., Lee, E., Gaylor, K., Jones-Boone, J and Ray, P. C., Sequence Specific HCV-RNA Quantification Using Size Dependent Nonlinear Optical Properties of Gold Nanoparticles, *Small*, **2009**, 5, 839–845.
31. Mirkin, C. A.; Letsinger, R. L.; Mucic, R. C.; Storhoff, J. J. A DNA-based method for rationally assembling nanoparticles into macroscopic materials, *Nature* **1996**, *382*, 607.
32. Alivisatos, A. P.; Johnson, K. P.; Peng, X.; Wislon, T. E.; Bruchez, M. P.; Schultz, P. G. Organization of 'nanocrystal molecules' using DNA, *Nature*, **1996**, *382*, 609.
33. Nanostructures in Biodiagnostics, N. L. Rosi and C. A. Mirkin, *Chem. Rev.* **2005**, 105 1547.
34. Gill, R., Willner, I., Shweky, I., Banin, U., Fluorescence Resonance Energy Transfer in CdSe/ZnS-DNA Conjugates: Probing Hybridization and DNA Cleavage, *J. Phys. Chem. B,* **2005**, 109, 23715 –23719.

35. Zhang, C.-Y., Johnson, L. W., Quantum-Dot-Based Nanosensor for RRE IIB RNA-Rev Peptide Interaction Assay, *J. Am. Chem. Soc.*; **2006**, 128, 5324–5325.
36. Zhuang X, Bartley L E, Babcock H P, Russell R, Ha T, Herschlag D, Chu S, A Single-Molecule Study of RNA Catalysis and Folding, , *Science* 288, **2000**, 2048 – 2051.
37. Seferos, D. S., Giljohann, D. A., Hill, H. D., Prigodich, A. E., Mirkin, C. A., Nano-Flares: Probes for Transfection and mRNA Detection in Living Cells *J. Am. Chem. Soc.*, **2007**, 129, 15477–15479.
38. Bates, A. D., Callen, B. P., Cooper, J. M., Cosstick, R., Geary, C., Glidle, A., Jaeger, L., Pearson, J. L., Proupin-Perez, M., Xu, C., Cumming, D. R. S., Construction and Characterization of a Gold Nanoparticle Wire Assembled Using Mg^{2+}-Dependent RNA-RNA Interactions, *Nano Lett.*, **2006**, 6, 445–448.
39. Aslan K and Geddes C D. Directional Surface Plasmon Coupled Luminescence for Analytical Sensing Applications: Which Metal, What Wavelength, What Observation Angle? *Anal. Chem.*, **2009**, *81*, 6913–6922.
40. Zhang Y, Dragan A and Geddes C D, Wavelength Dependence of Metal-Enhanced Fluorescence, *J. Phys. Chem. C*, **2009**, *113*, 12095–12100.
41. Tabor C, Van Haute D and El-Sayed M A, Effect of Orientation on Plasmonic Coupling between Gold Nanorods, *ACS Nano*, **2009**, *3* , 3670–3678.
42. Mastroianni A J, Claridge S A and Alivisatos P A, Pyramidal and Chiral Groupings of Gold Nanocrystals Assembled Using DNA Scaffolds, *J. Am. Chem. Soc.*, **2009**, *131*,8455–8459.
43. Lal S, Grady N K, Goodrich G P, and Nhalas N J, Profiling the Near Field of a Plasmonic Nanoparticle with Raman-Based Molecular Rulers, *Nano Lett.*, **2006**, *6*, 2338–2343.
44. Funston A M, Novo C, Davis T J and Mulvaney P, Plasmon Coupling of Gold Nanorods at Short Distances and in Different Geometries, *Nano Lett.*, **2009**, *9* 1651–1658.
45. Mayilo S, Kloster M A, Wunderlich M, Lutich A, Klar T A, Nichtl A, Konrad K, Stefani F D and Feldmann J, Long-Range Fluorescence Quenching by Gold Nanoparticles in a Sandwich Immunoassay for Cardiac Troponin T , *Nano Lett.*, **2009**, *9*, 4558–4563.
46. Aslan, K., Lakowicz, J. R., Geddes, C. D., Plasmon light scattering in Biology and Medicine: New sensing approaches, visions and perspectives, *Curr. Opn. Chem. Bio.*, **2005**, 9, 538–544.
47. Zhang, C-Y, Yeh, H-C, Kuroki, M. T., Wang, T-H, Single-quantum-dot-based DNA nanosensor, *Nature Mat.*, **2005**, 4, 826–831.
48. Aslan, K.; Huang, J.; Wilson, G. M., Geddes, C. D., Metal-Enhanced Fluorescence-Based RNA Sensing, *J. Am. Chem. Soc.*; **2006**, 128, 4206–4207.
49. Li, H., Rothberg, L., Detection of Specific Sequences in RNA Using Differential Adsorption of Single-Stranded Oligonucleotides on Gold Nanoparticles, *Anal. Chem.*, **2005**; 77, 6229–6233.
50. Tam, F.; Goodrich, G. P.; Johnson, B. R.; Halas, N. J, Plasmonic Enhancement of Molecular Fluorescence, *Nano Lett.*; **2007**; 7; 496–501.
51. Optical Biosensors, Sergey M. Borisov and Otto S. Wolfbeis, *Chem. Rev.*, **2008**, *108*, 423–461
52. Stewart, M. E.; Anderton, C. R.; Thompson, L. B.; Maria, J.; Gray, S. K.; Rogers, J. A.; Nuzzo, R. G., Nanostructured Plasmonic Sensors, *Chem. Rev.*, **2008**; 108; 494–521.
53. Thoumine, O.; Ewers, H.; Heine, M.; Groc, L.; Frischknecht, R.; Giannone, G.; Poujol, C.; Legros, P.; Lounis, B.; Cognet, L.; Choquet, D., Probing the Dynamics of Protein–Protein Interactions at Neuronal Contacts by Optical Imaging, *Chem. Rev.*; **2008**; 108; 1565–1587.
54. DNA nanomachines, Jonathan Bath, Andrew J. Turberfield, *Nature Nanotechnology* **2**, 2007, 275–284.
55. Gold Nanoparticle Based FRET for DNA Detection, P C Ray, G K Darbha, A Ray, J. Walker, W Hardy and A Perryman, *Plasmonics*, **2007**, 2, 173–183.
56. Dulkeith, E., Morteani, A. C., Niedereichholz, T., Klaar, T. A., Feldmann, J., Levii, S. A., Reinhoudt, D. N., Fluorescence Quenching of Dye Molecules near Gold Nanoparticles: Radiative and Nonradiative Effects, *Phys. Rev. Lett.*, **2002**, 89, 203202–203205.

Chapter 7
Optics and Plasmonics: Fundamental Studies and Applications

Florencio Eloy Hernández

During the last decade, nanotechnology has become one of the major driving forces for basic and applied research [1–3]. The extraordinary physicochemical properties of metal particles as they approach the nanometer scale have increased the interest of physicists and chemists in nanoscience [4]. Matter at this size scale has been demonstrated to outperform in a wide variety of applications, such as catalysis [5], environmental [6], biological labeling and sensing [7, 8], surface Raman scattering [9], nonlinear absorption processes [10], photonics [11], and optoelectronics [12]. Nanoscience has additionally had a strong impact in biophotonics through radiative decay engineering (RDE) of organic dye molecules in the close vicinity of conducting metal surfaces or covalently attached to them [13–15]. Most of the applications explored so far are based on the well-known surface plasmon resonance (SPR) effect [16].

SPR in noble metal nanoparticles is manifested by well-defined absorption bands in the visible region as a result of the interaction of electromagnetic radiation with conduction band electrons [17]. These bands present a remarkable feature that should be underlined, i.e., their spectral position strongly depends on their size, shape, and composition [18]. This dependence has been applied, among other things, to the development of specific optical noses in biology and environmental applications [19, 20]. Furthermore, SPR can also produce a tremendous electric field enhancement at the surface of metal nanoparticles [16]. Such an augmentation has been applied to second harmonic generation [21, 22], SERS [9], multiphoton-assisted excitation, and enhanced fluorescence of chromophores [23–26]. Another effect worth to be highlighted is the modification of the radiative decay rates of emitters close to metal surfaces [27–29]. As a result, RDE has resulted in the design of better fluorophore complexes for bioimaging [30], and has awakened the interest of the

F.E. Hernández (✉)
Department of Chemistry and CREOL/The College of Optics and Photonics,
University of Central Florida, P.O. Box 162366, Orlando, FL 32816-2366, USA
e-mail: florencio.hernandez@ucf.edu

C.D. Geddes (ed.), *Reviews in Plasmonics 2010*, Reviews in Plasmonics,
DOI 10.1007/978-1-4614-0884-0_7,

scientific community to fully develop hybrid systems (metal–chromophores) with potential applications in the biomedical field and environmental sensing, among others [30]. Although many contributions have been made to this exciting field, there is still a great need for the development of more sensitive and reliable sensors, and for a better understanding of the physical, chemical, and optical properties of pure metal nanoparticles that can affect the interaction metal–chromophores.

In this review, we would like to share with you some of our most recent contributions to the field of plasmonics and their applications. First, we present SPR electric field enhancement applied to nonlinear optical processes in a suspension containing gold metal nanoparticles clusters and chromophores. Subsequently, we show the first evidence of fluorescence lifetime augmentation in a dye that has been systematically attached to gold nanoparticles. Afterward, we demonstrate the principles of one of the best Hg sensors at present based on gold nanorods. Following, we display the design of a new approach to the Krechtman geometry that works at a liquid–metal–liquid interface. Finally, we present a fundamental study of the nonlinear optical properties of gold nanoparticles with different shape. It is our hope to reveal through this review the present and future impact that the field of plasmonics can have in our society.

1 Surface Plasmon Enhancement of the Multiphoton Absorption of Dyes in the Presence of Activated Gold Colloidal Solution

Before getting more in depth into our work, we would like to introduce the reader to one of the most common multiphoton absorption process employed nowadays, namely two-photon absorption (2PA) [31]. 2PA is a well-known third-order nonlinear optical phenomenon based on the simultaneous absorption of two photons of less energy (half the energy in the degenerate case) than that required for an electronic transition via one-photon absorption [32]. Because the excitation probability for 2PA scales quadratically with the incident irradiance (I), the excitation is mostly confined to the focal volume of the incident beam. Moreover, since the excitation is induced using longer wavelength radiation, out of the linear absorption spectrum, the linear absorption outside the focal region is usually negligible and scattering is reduced. Consequently, the spatial resolution and the penetration depth using 2PA is increased [32, 33]. This process has been widely used in 3D microfabrication [34–36], two-photon confocal imaging [37–39] and, more recently, in photodynamic therapy (PDT) [40–46]. However, for more practical biomedical applications, there is still a great need of specifically designed molecules exhibiting a stronger 2PA and/or higher order nonlinear absorption responses that afford an even greater penetration depth and spatial resolution.

In 2004, we proposed the use of (3PA) as a possibility in biological imaging and phototherapy treatment [47–53]. The applicability of 3PA processes in these two areas is founded on its inherent improvement in penetration depth when working in

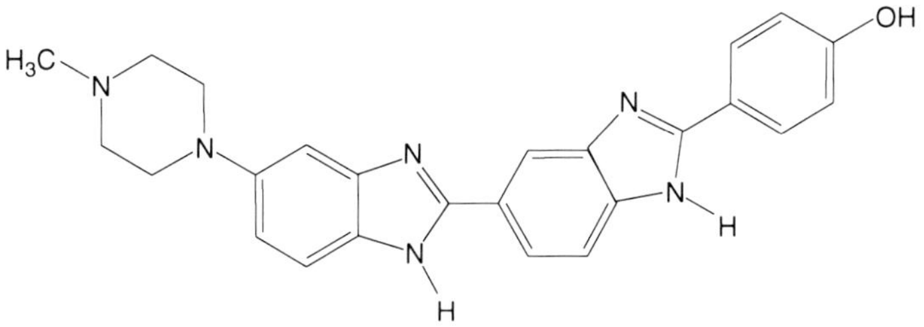

Fig. 7.1 Molecular structure of Hoechst 33258 [63]

the NIR [54], minimization of loses due to scattering when even longer wavelengths are employed [31], and the potential enhancement in the spatial resolution as a result of the I^3 dependence of the process [33]. Although 3PA seems to be promising for biomedical applications, the already mentioned cubic dependence of the rate of energy absorption with the irradiance for this process limits its applicability to real scenarios [55]. Therefore, in order to make a revolutionary contribution to the fields of biophotonics and nanophotonics using 3PA, the development of new materials and processes that can overcome this impediment is extremely necessary.

In order to surmount the existent restriction in 3PA, few years ago we proposed to use the electric field enhancement produced by the quantized oscillation of collective electrons in a plasma surface, i.e., surface plasmon (SP) [56]. Theoretical calculations of the electric field intensity enhancements predict an augmentation of more than two orders of magnitude in noble metal films [54], and even a greater enhancement (>100,000) in noble metal nanoparticles [56]. Therefore, demonstrating the possibility of using SPR enhancement in high order absorption processes would open a new horizon of practical applications of 3PA in the biomedical field. In addition, since this enhancement is larger in metal nanoparticles, working in solutions becomes more attractive. For our sympathy at that time, the application of this effect to 2PA in doped polymeric films deposited on metal nanolayers and nanoparticles was already reported in the literature [57–62]. In fact, fluorescence enhancement factors as high as ≈160,000 were observed on thiol–chromophores attached to silver nanoparticle fractal clusters [62]. According to the authors, the extraordinary enhancement was undeniably due to the stronger electric field obtained via SPR accumulative effects in localized "hot spots" between nanoparticles. Although these results were very impressive and of great relevance in nonlinear optics and its applications, the experimental demonstration using fluorescence needed to consider radiative decay changes due to the presence of the metal. In addition, the application of SPR to higher order nonlinearities was still a puzzle to be solved.

3PA assisted by SPR was first reported by our group in 2005, on Hoechst 33258 in a colloidal aqueous suspension containing gold nanoparticles. This is definitely a very attractive system for biology and medicine since it can be delivered through the blood stream and because of its biocompatibility with living organisms [63]. We also demonstrated 2PA enhancement in the same suspension, as described next (Fig. 7.1).

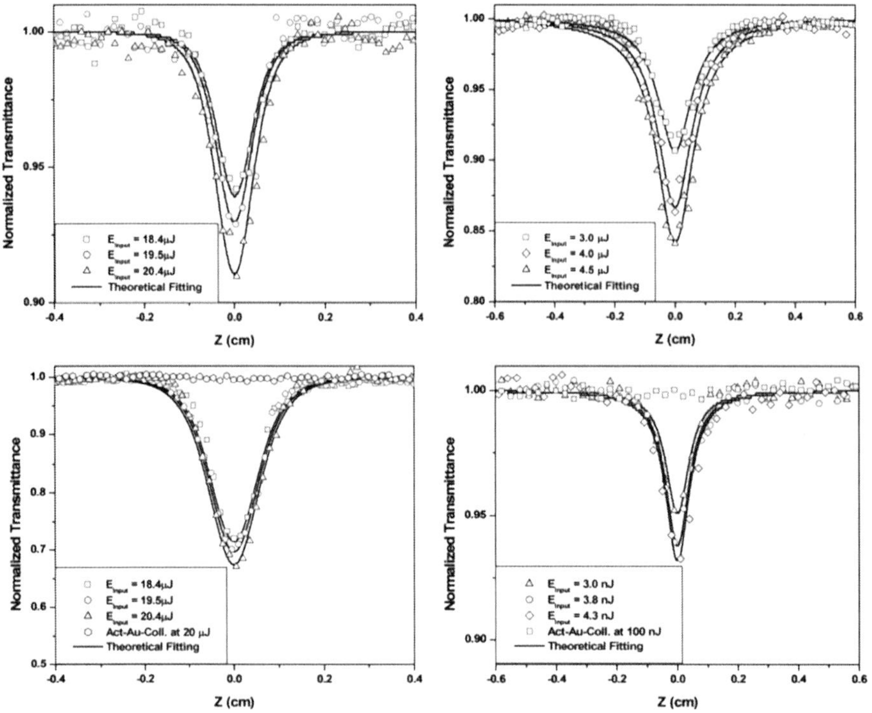

Fig. 7.2 Open aperture Z-scan curves in Hoechst 33258/H_2O solution (2.3×10^{-3} M) pumping at 550 nm (**a**) and 1,064 nm (**b**). *Solid lines* are the best theoretical fittings for the change in transmittance for 2PA (*top*) and 3PA (*bottom*). 2PA and 3PA measurements were done at three different input energies (E_{in}) without (*left*) and with nanoparticles (*right*) [63]

Gold nanospheres of 16 nm average diameter with a size distribution <15% were synthesized with the wet chemical method developed by Turkevich et al. [64, 65]. The nanoparticles size and distribution were determined by high resolution transmission electron microscopy (TEM). In order to ensure multiphoton absorption enhancement using longer excitation wavelengths, i.e., in the NIR region, we activated the metal colloids by electrolyte-induced aggregation [66–68]. Activated gold colloids (hot particles) were prepared by adding NaCl(aq) to the original gold nanoparticle suspension. This method, also commonly used in bulk surface enhanced Raman scattering (SERS) [9], is necessary to induce the formation of hot spots. For this experiment, we excited two- and three-photon process with a tunable OPG pumped by the third harmonic of a mode-locked, 25 ps full-width at half-maximum (FWHM), Nd:YAG laser (EKSPLA), operating at a 10 Hz repetition rate. The 2PA and 3PA coefficients were measured using the well-known open aperture Z-scan [69], which allows to measure pure electric field enhancement effects.

After characterizing the pure dye in solution, we measured the 2PA and 3PA coefficients of the Hoechst 33258/H_2O solutions in the presence of activated gold colloid (α_n(+Au)) and at the same wavelengths, i.e., 550 and 1,064 nm, respectively. Figure 7.2 shows the open aperture Z-scan curves and their

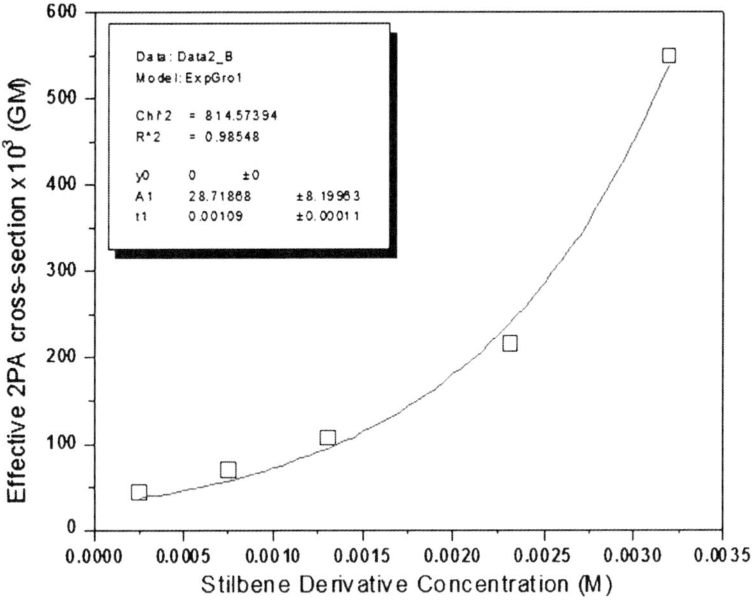

Fig. 7.3 The effective 2PA cross-section of *trans*-4,4′-diaminostilbene as function of dye concentration. The *solid line* represents the theoretical fitting using an exponential function [70]

corresponding theoretical fittings pumping at 550 and 1,064 nm. The measured 2PA and 3PA coefficients $\alpha_2(+\text{Au}) = (2.7 \pm 0.2) \times 10^{-8}\,\text{cm}^2\,\text{W}^{-1}$ and $\alpha_3(+\text{Au}) = (3.5 \pm 0.2) \times 10^{-21}\,\text{cm}^3\,\text{W}^{-2}$ yielded effective $\sigma_2'(+\text{Au}) = (1160 \pm 0.2) \times 10^{-47}\,\text{cm}^4\,\text{sphoton}^{-1}$ and $\sigma_3'(+\text{Au}) = (244 \pm 10) \times 10^{-78}\,\text{cm}^6\,\text{s}^2\,\text{photon}^{-2}$, respectively. $\alpha_n(+\text{Au})$ and $\sigma_n'(+\text{Au})$ refer to the *n*th order absorption coefficients and cross-sections, respectively, in the presence of activated gold colloid. From the experimental data we were able to deduce that pure Hoechst 33258 presented high 2- and 3PA cross-section. Interestingly, in the presence of activated gold colloid, the 2- and 3PA of Hoechst 33258 was enhanced in 480- and 30-folds, respectively. From our work, we concluded that the enhancement of the effective cross-sections was a direct consequence of the electric field enhancement generated by the surface plasmon on activated gold colloid. Throughout our experiments, we determined that activated gold colloid was essential for 2PA and 3PA enhancement as it is for SERS. Of course, the efficiency of multiphoton absorption enhancement via SPR strongly depends on the multiphoton absorption cross-sections of the chromophore. This final remark must be carefully considered when reporting surface plasmon-enhanced multiphoton absorption in molecular probes.

Additional studies of the effect of the concentration of organic dyes on the surface plasmon-enhanced 2PA cross-section on Au nanoparticles covered by an stilbene derivative revealed that metal surface coverage has a crucial effect on the surface plasmon-enhanced 2PA of organic dyes attached to or in the immediate vicinity of metal particles [70]. Figure 7.3 clearly shows this dependence on *trans*-4,4′-diaminostilbene in an activated gold aqueous suspension at different concentrations.

The observed exponential growth of the effective 2PA cross-section ($\sigma_2'(\text{Au})$) of the dye was attributed to the electric field augmentation via SPR between nanoparticles and the molecular density on Au nanospheres. An unprecedented $\sigma_2'(\text{Au}) = 550{,}000\text{GM}$ was found. This unique result is expected to open a new of applications in multiphoton imaging, PDT, telecommunications, optical limiting, and multidimensional data storage using hybrid systems.

In fact, we have found that the enhancement is not due to the molecular density at the surface but to the electric field enhancement via SPR. In addition, we predict a maximum effective enhanced σ_2' of 10^8 for SY183 in suspension. Achieving such a great enhancement would permit the reduction of the maximum excitation intensity 10^4 times. Demonstrating such an enhancement would be of great relevance in biomedical applications.

Thinking about the potential applications of this and other effects in PDT, we considered finding mechanisms to increase the intersystem crossing (ISC) of certain organic molecules. Enhanced ISC could result in a higher singlet oxygen generation. This is what the following section is about.

2 Fluorescence Lifetime Enhancement of Organic Chromophores Attached to Gold Nanoparticles

The field of RDE using noble metal nanoparticles has been mainly promoted by the pioneer work done in Geddes and Lackowicz groups during the last decade [29, 30, 71]. My group, for instance, became interested in this topic after reading some of the distinguished reviews published by these two scientists in the field of plasmonics. Let us review some of the developments in this arena that stimulated us to make a contribution in this field.

It has been demonstrated that fluorescence can be enhanced orders of magnitude when organic dyes are near metal thin films and nanoparticles [29, 72]. This is the outcome of two effects acting synergistically (1) the intensification of the incoming electric field at the SPR, and (2) the increase of the radiative decay rate of the fluorophore due to the presence of the metal surface [29]. Fluorescence can also be quenched if the fluorophores are within the Foster radius with respect to the metal surface, where energy transfer becomes very important [73, 74]. These effects are a consequence of the dipole–dipole interaction between the dye molecules and the metal surface as Barnes has already described theoretically [28]. At present, it is known that the main critical parameters that modify the decay rate of organic dyes near metal surfaces are their location and separation from the metal surface and the molecular dipole moment orientation with respect to the metallic surface. The latter is perhaps the less comprehended effect but currently the most important due to its fundamental implications in the biomedical and solar energy storage fields [28]. In order to fulfill the existent gap and gain control on the decay rate of organic molecules by means of metal surfaces, we proposed a meticulous fundamental study of the effects that gold nanoparticles have on the excitation and relaxation processes of organic molecules with different dipole moment orientation with respect to the metal surface. The proposed study was based on the following theoretical predictions.

Barnes described in 1997 the modification of spontaneous emission produced by a planar metal surface using photonic modes density (PMD) as decay routes [28]. He showed that in the small separation regime (<100 nm) from a planar metal surface, perpendicular orientation of the molecular dipole with respect to the surface tends to be enhanced by image coupling while the parallel one is canceled out. Therefore, the emission rate can be either increased or decreased for perpendicular or parallel molecular dipole orientation, respectively. Ten years before Barnes first calculations, Chew et al. had already calculated the transition rate for the two polarizations of an oscillator atom outside a perfectly conducting sphere of radius r placed at a distance d_r from the center of the sphere [75]. Similar results to those observed with the planar metal surface were obtained [73]. As d_r approaches r, the emission rate either increases or decreases for perpendicular or parallel polarization, respectively. Experiments showing a decrease of fluorescence lifetime of dye molecules in the immediate vicinity of metal surfaces have been extensively reported [29, 72, 74], but there was no experimental evidence of fluorescence lifetime enhancement of organic dyes close to metal surfaces until we published our work in 2004 [76].

Few years ago, we reported the first evidence of fluorescence lifetime enhancement on an organic dye chemically attached to the surface of spherical nanoparticles [76]. The chromophore consisted of a hybrid system (*HS*) containing a modified di-thiol molecular probe attached to the surface of ca. 5 nm diameter gold nanospheres (*Au-Np*) and with its molecular axis parallel to the surface of the spherical nanoparticle. The chromophore was specifically attached with dipole moment parallel to the metal surface and at a distance of approximately 1 nm. In order to perform the experiment, we first generated gold nanospheres of ca. 5 nm of diameter with 10% size distribution, synthesized by the wet chemical method reported by Turkevich et al. [64, 65]. Then, a commercially available molecular probe, 4-acetamido-4′-maleimidylstilbene-2,2′-disulfonic acid, disodium salt was reduced to its corresponding di-thiol homologous (*AMDT*) using phosphorus pentasulfide and $LiAlH_4$ [77]. Thiol groups guaranteed the chemical attachment of AMDT to the metal surface. *AMDT* molecules were incorporated onto *Au-Np* following the thiolate chemistry cape method developed by Brust [78]. The schematic diagram shown in Fig. 7.4 depicts the molecular structure of the fluorescent probe (*AMDT*) and the resulting *HS*.

Figure 7.5 displays the fluorescence wavelength-time matrix (WTM) of AMDT and HS collected with the picosecond laser system and the fiber optic probe describe in the original paper [76]. The inset in Fig. 7.5 shows the normalized decay curves obtained for free AMDT and the probe in the HS obtained at room temperature. As predicted by Barnes [28] and Chew [75], the fluorescence lifetime of AMDT in the HS (HS) 8.73 ± 0.23 ns was two fold longer than the lifetime of the free probe (AMDT) 4.32 ± 0.10 ns. In fact, at lower temperature, i.e., 4.2 K the fluorescence lifetime of AMDT in the HS is almost four times longer. With these experiments,

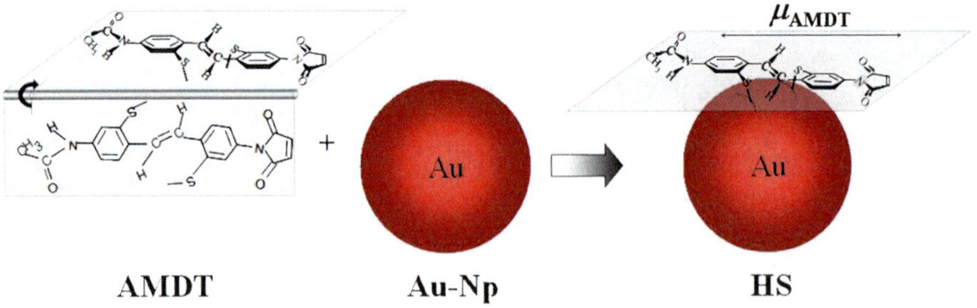

Fig. 7.4 Structure of the 4-acetamido-4′-maleimidylstilbene-2,2′-dithiol (*AMDT*), schematic representation of the 3.5 nm diameter gold spherical nanoparticles (*Au-Np*), and bi-attached *AMDT* onto gold nanoparticle "hybrid system" (*HS*). The *red double side arrow* represents the plane of the molecular dipole moment μ_{AMDT} [76]

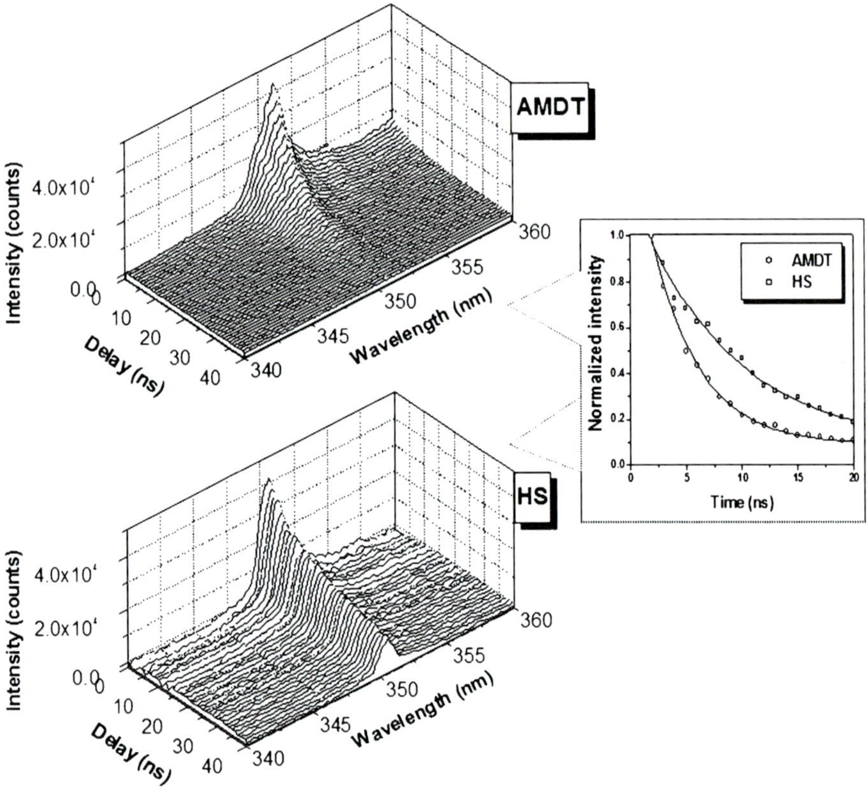

Fig. 7.5 Wavelength–time matrix of *AMDT* (*top*) and *HS* (*bottom*) in octane solutions at 298 K. Initial delay time was 2 ns, gate width 200 ns, and delay step 2 ns. *Inset* normalized intensity vs. delay time (ns) of AMDT (*open circle*) and HS (*open square*). *Solid lines* are the mono-exponential fittings of the experimental data yielding $\tau_{AMDT} = 4.32 \pm 0.10$ ns and $\tau_{HS} = 8.73 \pm 0.23$ ns [76]

we showed that a molecular dipole oriented parallel to a noble metal surface tends to be reduced by the coupling with its image. We also observed that a weak energy transfer between the molecular probe and the nanoparticle inhibits the reduction of the fluorescence quantum yield on the hybrid system. This last statement was validated by the increase of the radiative decay rate in both systems. Further studies are still in great need before achieving the applications. However, we would like to encourage the scientific community to join us in the development of this field.

3 Highly Sensitive and Selective Mercury Sensor Using Gold Nanorods

Mercury (Hg) is a known environmental pollutant routinely released from coal-burning power plants to the air in the USA [79]. Most of the Hg ingested by people comes through the food chain, mainly fresh water fish. The main problem with this metal is that once humans get exposed to it, Hg can never be removed from the organism, accumulating in time, thus harming the brain, heart, kidneys, lungs, and immune system of people of all ages. Consequently, detecting ultra-low concentrations of this pollutant in the ecosystem, primarily in water, is crucial.

Currently, there are several methods employed to monitor concentration levels of Hg in water samples. Established techniques, such as atomic absorption spectroscopy (AAS) [80, 81], gas chromatography-inductively coupled plasma-mass spectrometry (GC-ICP-MS) [82], atomic fluorescence spectrometry (AFS) [83, 84], inductively coupled plasma-atomic emission spectrometry (ICP-AES) [85, 86], and reversed-phase high performance liquid chromatography (HPLC) [87], provide limits of detection at the parts-per-billion level. Their excellent performance, however, is achieved at the expenses of elaborated and time-consuming sample preparation and preconcentration procedures.

As a tentative means of reducing analysis time and cost, on-site sensing approaches capable of providing real-time Hg determination have been actively pursued. These include optical test strips [88], remote electrochemical sensors [89], ion-selective electrodes [90], fluorescence-based sensor membranes [91, 92], and piezoelectric quartz crystals [93]. Although these approaches provide low detection limits and fast response times, they still lack the procedural simplicity for on-site analysis.

In 2006, we developed a new method that takes advantage of the strong affinity between Au and Hg [19]. The interaction between these two elements has been extensively studied by monitoring physical properties of Au upon Hg adsorption. Numerous gravimetric sensors exist based on the mass change when Hg adsorbs onto a gold-coated piezoelectric substrate [94–99]. Au nanofilms have also been employed to monitor Hg upon changes on resistivity [100–102], reflectivity [103–105], and SPR coupling angle change [106]. However, their high limit of detection is still an issue.

Using our novel approach, we can determine Hg upon wavelength changes on absorption spectra of Au nanorods [19]. Only a few reports exist correlating this optical property to Hg–Au interactions but none of them present the required sensitivity [107–109]. Our methodology, nonetheless, demonstrates the analytical potential of Au nanorods for monitoring Hg in water samples. The outstanding selectivity and sensitivity of the method provide a unique way to determine Hg in water samples without previous separation and/or preconcentration of the original sample. In order to understand better how our sensor works, let us get more into the details of the operational mechanisms of this new sensor.

Nanorods of average aspect ratio (AR) (length/diameter) of 1.6 were synthesized according to the photochemical method developed by Kim et al. [110]. The first step of our approach consists of reducing all existing forms of oxidized Hg to Hg(0). This is accomplished by mixing the water sample with $NaBH_4$. This strong reducing agent is also capable of reducing any oxidized form of Au into Au(0). To guarantee the complete reduction of oxidized Hg eventually present in unknown water samples, all studies were performed with an excess of reducing agent.

Based on experimental evidence, El-Sayed and coworkers have provided the theoretical foundation to understand the two absorption bands typically observed in the UV–vis absorption spectra of gold nanorods [111]. According to these authors, the two absorption bands correspond to the transversal and longitudinal modes of SPR. The maximum absorption wavelength of the longitudinal mode – which corresponds to the SPR along the long axis of the rod – presents a linear correlation with the aspect ratio of the nanorod. As the aspect ratio of the nanorod increases, the longitudinal mode band shifts to longer wavelengths. The same behavior is observed as the dielectric constant of the medium increases [111]. The black line in Fig. 7.6 shows a typical absorption spectrum recorded from a nanorod suspension in pure water. The absorption maxima of the transversal and longitudinal modes appear at 520 and 612 nm, respectively. After adding different amounts of mercury to the initial suspension, the observed blue shift of the maximum absorption wavelength of the longitudinal band shows a direct correlation with Hg (II) concentration. The presence of Hg (II) at higher concentration ranges causes an overlap between the longitudinal and the transversal absorption bands of the nanorods (Navy line).

In order to explain the principle of our sensor, we proposed a mechanism (see Fig. 7.7a). In the presence of liquid Hg, our experiments strongly suggest that the main mechanism of action as being through amalgamation at the active sites of Au nanorods which mainly belong to the tips of the nanostructures [112]. Therefore, a decrease in the nanorods' effective aspect ratio is foreseen. Preferential deposition at the tips of the nanorods is also favored by the presence of surfactant (CTAB) on the lateral sides of the nanorods which restricts the amalgamation of Hg on the lateral walls of the nanorods. Experimental evidence of our hypothesis is shown in Fig. 7.7b. TEM images show the shape and the aspect ratio of Au nanorods in the absence of Hg^{2+} and the presence of 1.25×10^{-5} M and 1.57×10^{-4} M Hg^{2+} solutions. As the concentration of Hg^{2+} increases, the aspect ratio indeed decreases to the point at which the shape of the nanoparticles becomes spherical. Additional experimental evidence is provided by EDX analysis. As expected, the Hg content in nanoparticles is inversely proportional to their aspect ratio. Spherical nanoparticles contain more Hg than nanorods.

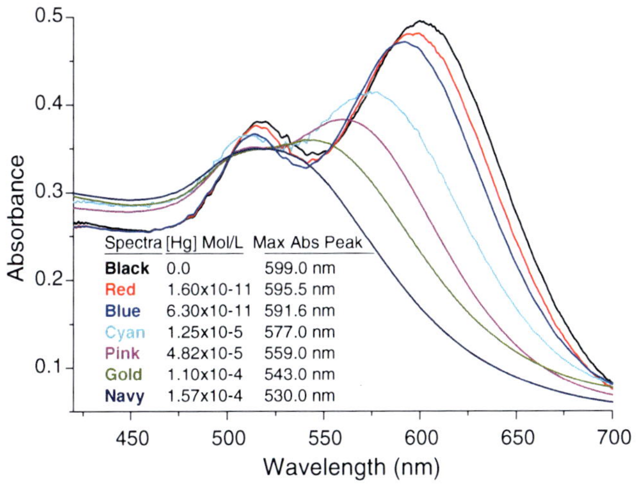

Fig. 7.6 UV–vis absorption spectra showing the spectral shift at several Hg(II) concentrations. The concentration range between 1.6×10^{-11} M and 6.3×10^{-11} M shows the spectral within the linear dynamic range of the calibration curve. The remaining spectra show the overlapping between the longitudinal and transversal absorption bands at higher Hg(II) concentrations [19]

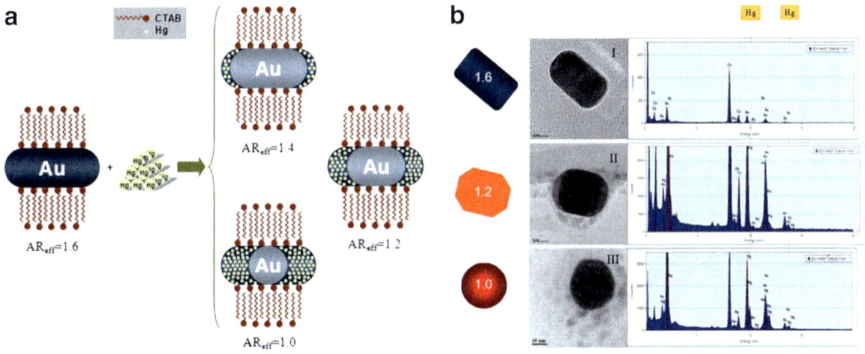

Fig. 7.7 (**a**) Schematic diagram showing the amalgamation of Hg with Au nanorods. (**b**) TEM and EDX analysis of Au nanorods in the absence and the presence of Hg. I = no Hg; II = 1.25×10^{-5} M and III = 1.57×10^{-4} M Hg^{2+}. All solutions were prepared in 1.67×10^{-3} Mol/L $NaBH_4$ [19]

In just few paragraphs, we have presented an exceptional approach to detect Hg that does not need any previous separation and/or sample preconcentration steps. This sensor has a tremendous potential for the determination of mercury in tap water samples, as well as in gas lines. The specificity toward Hg and its low limit of detection is allocated to specific Hg–Au amalgamation and the spectral blue shift of the maximum absorption wavelength of the longitudinal SPR mode band of Au nanorods. In comparison to the existent methodology for mercury analysis, our approach provides an outstanding limit of detection of one order of magnitude better than those previously reported with established techniques [80–87] and three orders of magnitudes better than those reported with the most sensitive sensors [88–99]. In addition, the experimental procedure is quite simple and it takes no longer than 10 min per sample, solution mixing can be done in a test tube, which is well suited for on-site analysis with any portable spectrometer having appropriate spectral resolution. Future studies will apply this approach to the analysis of water samples with high matrix complexity and gases.

4 Surface Novel Kretchman Geometry at a Liquid–Metal–Liquid Interface

Throughout our previous studies, we started thinking about new possibilities using SPR to create a more versatile geometry. Knowing the concept of surface plasmon, introduced for the first time by R. H. Ritchie in June 1957 [113], we came up with a new SPR arrangement that works at a liquid–liquid interfaces.

We have already presented the surface plasmon definition and location. However, we have not described yet how to excite SPR. As we already know, these oscillations are transverse electromagnetic waves with magnetic vector perpendicular to the direction of propagation of the surface plasmon wave and parallel to the plane of interface. Using Maxwell's equations and the continuity relations [56], one can get the dispersion relation of SP propagating along the interface between two media. Because the surface plasmons dispersion relation lies to the right of the light line, SP cannot be excited directly by shining a noble metal at the nanoscale with light. Therefore, the light wave vector has to be increased. This condition can be satisfied using a grating coupler or an attenuated total reflection coupler [114]. The latter is the most utilized method through the employment of a Kretschmann geometry which consists of an equilateral or a right angle quartz prism coated with a noble metal, on one of its faces [115]. If the light is reflected at the quartz–metal surface, its momentum increases and the SP dispersion relation, up to certain frequency, lies to the left of the light line in quartz, allowing for its excitation by light. The resonance excitation of SPR is recognized as a minimum in the totally reflected intensity of the incident light at a specific incident angle θ_{SPR}. The dependence of this angle with the dielectric medium in contact with the metal has pushed the development of new instrumentation based on this principle; some of them have already found important applications in the study of protein binding studies [116]. In addition, as mentioned in previous sections, at SPR, a maximum of the electromagnetic field at the metallic–dielectric interface can be reached making this approach incredibly important in fluorescence enhancement [19, 30, 47, 71]. Conversely, this geometry can only be employed in a solid–liquid interface.

In 2008, we came up with a pseudo-Kretschmann geometry that works at the interface between two immiscible liquids [117]. Pumping a monolayer of Au nanoparticles clusters at the interface of a *p*-xylene–water mixture, we demonstrated fluorescence enhancement in Rose Bengal of three orders of magnitude at resonance excitation. Let us give more experimental details to grasp the principles behind this setup.

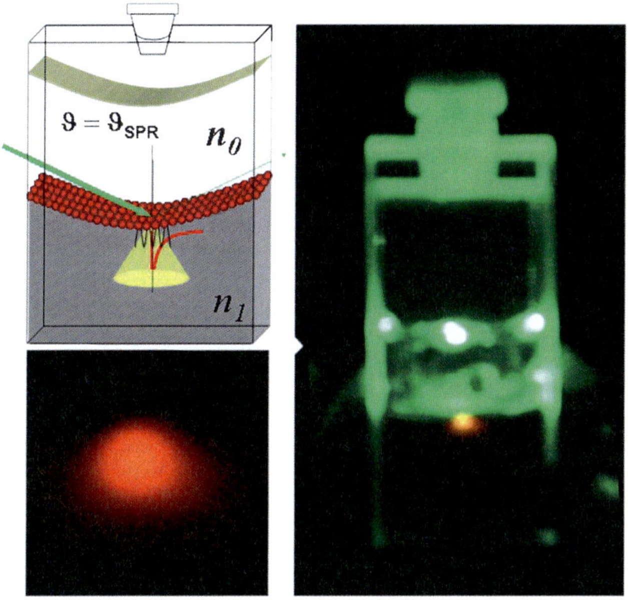

Fig. 7.8 Schematic of the irradiation geometry (*top left*). Picture of RB fluorescence emission in the cell (*right*). Zoom of Rose Bengal emission (*bottom left corner*) [117]

Gold nanospheres of 18 nm average diameter (<15% size distribution), were synthesized with the wet chemical method of Turkevich et al. [64, 65]. The procedure for forming Au nanospheres film at the interface between these two immiscible liquids consisted of a modified form of the Suzuki et al. method to functionalize microscope slides with Au nanorods [118]. Film formation was observed to begin almost immediately by noticing a golden color coming into view at the interface. Figure 7.8 shows the schematic of the unconventional geometry and a picture of the real cell. As in a traditional Krechmann geometry, the coupling takes place by irradiating the interface from the higher refractive index liquid, i.e., the top ($n_0 > n_1$). Once the incoming field has been coupled, the enhanced field can excite the dye molecules dissolved in the other liquid as shown in Fig. 7.8. The observed augmentation was determined to be originated from two different factors: the electric-field enhancement and the reduction of the fluorescence lifetime of dye molecules in the close vicinity of the metal surface [19]. The enhanced fluorescence of Rose Bengal is obvious to the naked eye. Under this new scheme, a continuous film is not generated but instead a layer of activated gold nanoparticles is formed, therefore, the enhanced electric field is higher than in a film [56]. In order to separate the two contributions, we performed the theoretical modeling of the average electric field intensity enhancement of emulated surfaces using the T-Matrix method (TMM). The theory supported our experimental results. With this novel approach, we have opened then a new road for the study of dynamic systems in liquid face.

5 Optical Saturable Absorption in Gold Nanoparticles

Through the previous sections, we have recognized the potential applications of metal nanoparticles of different size and shape in many different fields, being SPR one of the most interesting phenomena to understand. However, a better comprehension of the physical, chemical, and optical properties of pure metal nanoparticles, that can affect the interaction metal–chromophores, is still in great need. Thus, it is necessary to exceed the current level of knowledge of the photophysical properties of pure nanoparticles by performing experiment through the whole SPR bands and in nanoparticles of different size, shape, and composition.

In an effort to gain a superior insight of the optical properties of pure metal nanoparticles, Elim et al. have studied the nonlinear absorption effects in these systems [119]. Using discrete wavelengths, within a narrow spectral band, they observed induced bleaching in the plasmon absorption, i.e., saturable absorption (SA). Similar type of behavior has been observed by Link and coworkers using transient absorption measurements pumping in the interband region [18]. These authors have demonstrated that the time of such a nonlinear optical process can be correlated with the electron–phonon (~3 ps) and phonon–phonon (~90 ps) relaxation processes.

In 2009, we reported broadband SA effects in gold nanospheres and nanorods with different aspect ratios within their corresponding full SPR bands, using a white-light supercontinuum (SC) beam (20 ps) generated in water. We determined, as shown next, that the apparent reverse saturable absorption (RSA) observed for input energies greater than 9 μJ is the result of photo-degradation of the nanoparticles. This encouraging result insinuates that pure gold nanoparticles could be used as saturable absorber for mode-locked lasers, or materials for optical limiters in the high intensity range.

Au nanorods, of aspect ratio 2.2 and 2.4, length ca. 28 nm and size distribution <15% were synthesized using the modified seed-mediated growth method proposed by Pelton et al. [120]. In our SC Z-scan, a pulse of white-light supercontinuum, between 450 and 750 nm, was used to measure the nonlinear spectrum at all wavelengths simultaneously [121]. SC was generated with a 25 ps (FWHM) pulses, from an Nd:YAG laser (1,064 nm) working at a 10 Hz repetition rate, were focused with a 25 cm focal distance lens (L1) into a 10 cm path length cylindrical cell containing nanopure water.

Figure 7.9 shows the normalized transmittance (NT) as a function of wavelength for suspensions containing Au nanorods of aspect ratio 2.2 and 2.4, and also for nanospheres. The plotted NT corresponds to the maximum value obtained at the focal plane in the Z-scan geometry [69] reveal a saturable absorption effect within the SPR bands. One can notice that for Au nanospheres there is only one band centered at approximately the same wavelength as the SPR band. Yet, two bands can be observed in Au nanorods, which correspond to the SPR transversal and longitudinal modes. The observed SA effect in Au nanoparticles was first considered as the result of the saturation of the collective oscillation of electrons in the conduction band driven by the incoming electric field. This effect was found to have a dependence on

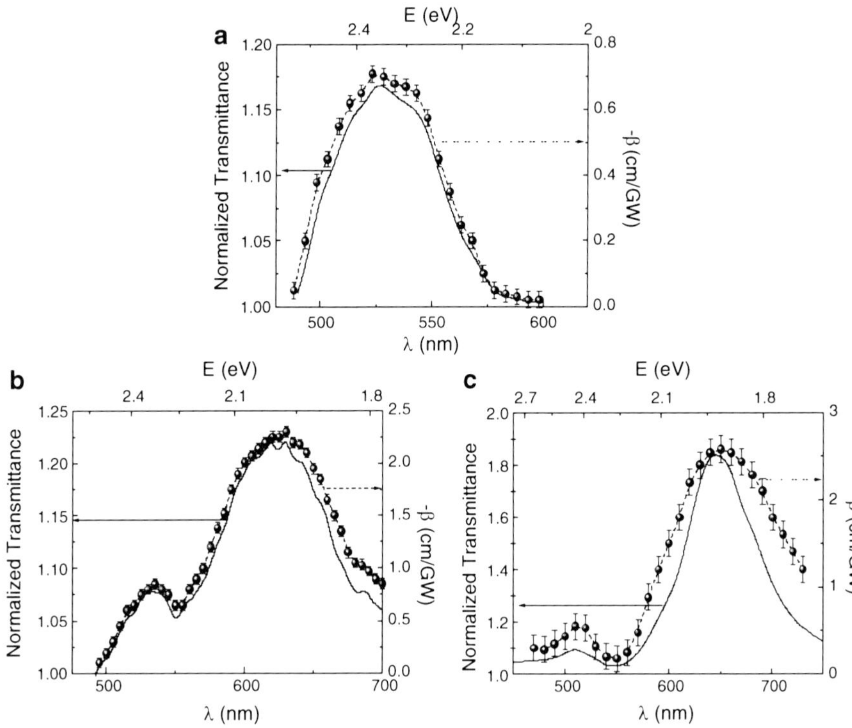

Fig. 7.9 Normalized transmittance (*solid line*) as a function of wavelength are shown for nanosphere (**a**) and different nanorods aspect ratio, 2.2 (**b**) and 2.4 (**c**). *Solid circles* represent the nonlinear absorption coefficient obtained by fitting the SCZS signatures [120]

the input energy of the SC. During this experiment, we noticed that NT increased up to 9 μJ. For energies greater than this value, the Z-scan signal gradually decreases in transmittance at the focal position. This behavior was thought to be originated by the presence of a reversible saturable absorption processes at high energies. However, further experiments demonstrated that it was due to photo-degradation. The latter was confirmed by the observation of a change in color of the suspension from blue (nanorods) or red (nanospheres) to brownish. The linear absorption spectra taken after irradiation with energies above 9 μJ exhibited an increase of the transversal SPR band at expenses of a decrease in amplitude of the longitudinal SPR band in the nanorods. These results were slightly different from those reported by Elim et al. [119] on RSA in Au nanorods measured with femtosecond laser pulses. Perhaps, using femtosecond pulses the possibility of photodegradation is less likely than with picoseconds.

So far there has been an enormous amount of contributions to the field of plasmonics. Many applications have been explored in different domains of chemistry, physics, biology, medicine, optics and material sciences. However, there is still much more to do from the fundamental point of view as well as from the applicational. Therefore, further studies are in great need, mainly multidisciplinary approaches. The combination of skills and knowledge can only turn into more relevant discoveries in this prosperous and relatively new field. In order to invite young scientists to join this exciting field, I would like to recall Richard P. Feynman's classic talk, on 29 Dec 1959, at the annual meeting of the American Physical Society at the California Institute of Technology, where he stated for the very first time in public that "There's Plenty of Room at the Bottom," referring to an emerging field in sciences, i.e., nanoscience and nanotechnology.

Acknowledgments I would like to acknowledge all the members of my research group and collaborators for their wonderful contribution to the work reported in this review. Also, I am very grateful with the University of Central Florida for its unconditional support through all these years of my career.

References

1. J. He, H. F. VanBrocklin, B. L. Franc, Y. Seo, F. F. Jones, Current Nanoscience **4**, 17–29 (2008)
2. B. Bhushan, "*Springer Handbook of Nanotechnology*", New York, Springer (2006)
3. M. R. Mohamadi, L. Mahmoudian, N. Kaji, M. Tokeshi, H. Chuman, Y. Baba, Nano Today **1**,38-45 (2006)
4. P. V. Kamat, Photophysical, J. Phys. Chem. B **106**, 7729–7744 (2002)
5. M. Valden, X. Lai, D. W. Goodman, Science **281**, 1647–1650 (1998)
6. M. Minunni, M. Mascini, Anal. Lett. **26**, 1441–1460 (1993)
7. S. R. Nicewarner-Peña, R. G. Freeman, B. D. Reiss, L. He, D. J. Peña, I. D. Walton, R. Cromer, C. D. Keating, M. J. Natan, Science **294**, 137–141 (2001)
8. M. Jr. Bruchez, M. Moronne, P. Gin, S. Weiss, A. P. Alivisatos, Science **281**, 2013–2016 (1998)
9. S. Nie, S. R. Emory, Science **275**, 1102–1106 (1997)
10. H. Kano, S. Kawata, Opt. Lett. **21**, 1848–1850 (1996)
11. S. A. Maier, M. L. Brongersma, P. G. Kik, S. Meltzer, A. A. G. Requicha, H. A. Atwater, Adv. Mater. **13**, 1501–1505 (2001)
12. G. K. Wertheim, S. B. Dicenzo, S. E. Youngquist, Phys. Rev. Lett. **51**, 2310–2313 (1983)
13. J. R. Lakowicz, Anal. Biochem. **298**, 1–24 (2001)
14. B. Dubertret, M. Calame, A. J. Libchaber, Nature Biotech. **19**, 365–370 (2001)
15. K. G. Thomas, P. V. Kamat, Acc. Chem. Res. **36**, 888–898 (2003)
16. H. Raether, "*Surface Plasmons on Smooth and Rough Surfaces and Grattings*", New York, Springer-Verlag (1988)
17. U. Kreibig, M. Vollmer, "*Optical Properties of Metal Clusters*", Berlin, Springer (1995)
18. S. Link, M. A. El-Sayed, J. Phys. Chem. B **103**, 8410–8426 (1999)
19. Matt Rex, Florencio E. Hernandez, and Andres Campiglia, Anal. Chem. **78**, 445–451 (2006)
20. S. Westcott, S. Oldenburg, T. R. Lee, and N. J. Halas, Langmuir **14**, 5396 (1998)
21. M. Zavelani-Rossi, M. Celebrano, P. Biagioni, D. Polli, M. Finazzi, L. Duo, G. Cerullo, M. Labardi, M. Allegrini, J. Grand, P. M. Adam, Appl. Phys. Lett. **92**, 0931191–0931192 (2008)
22. M. Finazzi, P. Biagioni, M. Celebrano, L. Duo, Phys. Rev. B **76**, 1254141–1254148 (2007)
23. Chis D. Geddes, "*Metal-Enhanced Fluorescence*", Wiley, 1st ed. (2010)
24. Y. Z. Shen, J. Swiatkiewicz, T. C. Lin, P. Markowicz, P. N. Prasad, J. Phys. Chem. B **106**, 4040–4042 (2002)

25. H. Kano, S. Kawata, Opt. Lett. **21**, 1848–1850 (1996)
26. Y. X. Zhang, K. Aslan, M. J. R. Previte, C. D. Geddes, Appl. Phys. Lett. **90**, 0531071–0531073 (2007)
27. F. E. Hernández, S. Yu, M. García, A. D. Campiglia, J. Phys. Chem. B **109**, 9499–9504 (2005)
28. W. L. Barnes, J. Mod. Opt., **45**, 661–699 (1998)
29. J. R. Lakowicz, Y. B. Shen, S. D'Auria, J. Malicka, J. Fang, Z. Gryczynski, I. Gryczynski, Anal. Biochem. **301**, 261–277 (2002)
30. Joseph R. Lakowicz, Plasmonics **1**, 5–33 (2006)
31. Boyd, R. W. "*Nonlinear Optics*". Academic Press, eds. (San Diego, CA, 1992), Chap. 4 and 8
32. M. Goeppert-Mayer, Ann. Physik **9**, 273–294 (1931)
33. Min Gun, Opt. Lett. **21**, 988–990 (1996)
34. W. Zhou, S. M. Kuebler, K. L. Braun, T. Yu, J. K. Cammack, C. K. Ober, J. W. Perry, S. R. Marder, Science **296**, 1106–1109 (2002)
35. B. H. Cumpston, S. P. Ananthavel, S. Barlow, D. L. Dyer, J. E. Ehrlich, L. L. Erskine, A. A. Heikal, S. M. Kuebler, I.-Y. S. Lee, D. McCord-Maughon, J. Qin, J.; M. Rumi, X.-L.Wu, S. R. Marder, J. W. Perry, Nature **398**, 51–54 (1999)
36. Claudia C. Corredor, Kevin D. Belfield, Mykhailo V. Bondar, Olga V. Przhonska, Florencio E. Hernandez, Oleksiy D. Kachkovsky, Journal of Photochemistry and Photobiology A: Chemistry. **184**, 177 (2006)
37. W. Denk, J. H. Strickler, W. W. Webb, Science **248**, 73–76 (1990)
38. D. R. Larson, W. R. Zipfel, R. M. Williams, S. W. Clark, M. P. Bruchez, F. W. Wise, Webb, W. W., Science **300**, 1434–1436 (2003)
39. J. M. Squirel, D. L. Wokosin, J. G. White, B. D. Bavister, Nature Biotechnology **17**, 763–767 (1999)
40. A. Karotki, M. A. Drobizhev, M. Kurk, A. Rebane, E. Nickel, C. W. Spangler, Proc. SPIE-Int. Soc. Opt. Eng. **4612**, 143–151 (2002)
41. S. J. Madsen, C. Sun, B. J. Tomberg, V. P. Wallace, H. Hirschberg, Photochem. Photobiol. **72**, 128–134 (2000)
42. E. A. Wachter, W. P. Partridge, W. G. Fisher, H. C. Dees, M. G. Petersen, Proc. SPIE-Int. Soc. Opt. Eng. **3269**, 68–75 (1998)
43. P. K. Frederiksen, M. Jorgensen, P. R. Ogilby, J. Am. Chem. Soc. **123**, 1215–1221 (2001)
44. W. G. Fisher, W. P. Partridge, C. Jr. Dees, E. A. Wachter, Photochem. Photobiol. **66**, 141–155 (1997)
45. T. D. Poulsen, P. K. Frederiksen, M. Jorgensen, K. V. Mikkelsen, P. R. Ogilby, J. Phys. Chem. **105**, 11488–11495 (2001)
46. D. Bhawalkar, N. D. Kumar, C. F. Zhao, P. N. Prasad, J. Clin. Laser Med. Surg. **15**, 201–204 (1997)
47. I. Gryczynski, H. Gryczynski, M. Malak, P. Schrader, H. Engelhardt, H. Kano, S. W. Hell, Biophys. J. **72**, 567–578 (1997)
48. S. W. Hell, K. Bahlmann, M. Schrader, A. Soini, H. Malak, I. Gryczynski, J. R. Lakowicz, J. Biomed. Opt. **1**, 71–74 (1996).
49. H. Szmacinski, I. Gryczynski, H. Gryczynski, Biophys. J. **70**, 547–555 (1996)
50. S. Maiti, J. B. Shear, R. M. Williams, W. R. Zipfel, W. W. Webb, Science **275**, 530–532 (1997)
51. Florencio E. Hernández, Kevin D. Belfield, Ion Cohanoschi, Chem. Phys. Lett. **391**, 22–26 (2004)
52. Florencio E. Hernández, Kevin D. Belfield, Ion Cohanoschi, Mihaela Balu and Katherine. J. Schafer, Appl. Opt. **43**, 5394–5398 (2004)
53. Ion Cohanoschi, Kevin D. Belfield, Florencio E. Hernández, Chem. Phys.Lett. **406**, 462–466 (2005)
54. E. M. Attas, M. G. Sowa, T. B. Posthumus, B. J. Schattka, H. H. Mantsch, S. L. L. Zhang, Biopolymers **67**, 96–106 (2002)

55. D. M. J. Friedrich, J. Chem. Phys. **75**, 3258–3268 (1981).
56. Heinz Raether, "*Surface Plasmons on smooth and rough Surfaces and on Grattings*", Springer-Verlag, eds. (New York, NY, 1988)
57. H. Kano, S. Kawata, Opt. Lett. **21**, 1848–1850 (1996)
58. E. Z. Kretschmann, Phys. **241**, 313–324 (1971)
59. E. J. Sánchez, L. Novotny, X. S. Xie, Phys. Rev. Lett. **82**, 4014–4017 (1999)
60. I. Gryczynski, J. Malicka, Y. Shen, Z. Gryczynski, J. R. Lakowicz, J. Phys. Chem. B **106**, 2191–2195 (2002)
61. Y. Shen, J. Swiatkiewicz, T.-C. Lin, P. Markowicz, P. N. Prasad, J. Phys. Chem. B **106**, 4040–4042 (2002)
62. W. Wenseleers, F. Stellacci, T. Meyer-Friedrichsen, T. Mangel, C. A. Bauer, S. J. K. Pond, S. R. Marder, J. W. Perry, J. Phys. Chem. B **106**, 6853–6863 (2002)
63. Ion Cohanoschi, Florencio E. Hernández, J. Phys. Chem. B 109, 14506–14512 (2005)
64. J. Turkevich, P. C. Stevenson, J. Hiller, Discuss. Faraday Soc. **11**, 55–75 (1951)
65. G. Frens, Nature Physical Science **241**, 20–22 (1973)
66. G. Schatz, Acc. Chem. Res. **17**, 370–376 (1984)
67. M. Moskovits, Rev. Mod. Phys. **57**, 783–826 (1985)
68. V. M. Shalaev, Phys. Rep. **272**, 61–137 (1996)
69. M. Sheik-Bahae, A. A. Said, T. Wei, D. J. Hagan, E. W. Van Stryland, IEEE J. Quantum Elect. **26**, 760–769 (1990)
70. Ion Cohanoschi, Sheng Yao, Kevin D. Belfield, Florencio E. Hernández, J. Appl. Phys. **101**, 861121–861123 (2007)
71. J. R. Lakowicz, Analytical Biochemistry **298**, 1–24 (2001).
72. K. H. Drexhage, Prog. Optics **12**, 165–232 (1974)
73. A. Camplon, A. R. Gallo, C. B. Harris, H. J. Robota, P. M. Whutemore, Chem. Phys. Lett. **73**, 447–450 (1980)
74. R. R. Chance, A. Prock, R. Silbey, Adv. Chem. Phys. **37**, 1–65 (1978)
75. H. J. Chew, *Chem. Phys.* **87**, 1355–1360 (1987)
76. Florencio E. Hernández, Shenjiang Yu, Marisol García, Andrés D. Campiglia, J. Phys. Chem. B **109**, 9499–9504 (2005)
77. S. Oae, H. Togo, Tetrahedron Lett. **23**, 4701–4704 (1982)
78. M. Brust, M. Walker, D. Bethell, D. J. Schiffrin, R. J. Whyman, Chem. Soc., Chem. Commun. **104**, 801–802 (1994)
79. http://www.epa.gov/mercury/about.htm
80. Y. A. Vil'pan, I. L. Grinshtein, A. A. Akatove, S. J. Gucer, J. Anal. Chem. **60**, 45–51 (2005)
81. E. Kopysc, K. Pyrzynska, S. Garbos, E. Bulska, Anal. Sci. **16**, 1309–1312 (2000)
82. M. J. Bloxham, S. J. Hill, P. J. Worsfold, J. Anal. Atom. Spec. **11**, 511–514 (1996)
83. D. Karunasagar, J. Arunachalam, S. Gangadharan, J. Anal. Atomic Spec. **13**, 679–682 (1998)
84. L Yu, X. Yan, Atomic Spec. **25**, 145–153 (2004)
85. C. A. Trimble, R. W. Hoenstine, A. B. Highley, J. F. Donoghue, P. C. Ragland, Marine Geores. & Geotech. **17**, 187–197 (1999)
86. T. Smigelski, K. O'Brien, J. Fusco, J. C. Schaumloffel, J. Tausta, *226th ACS National Meeting* , September 7–11 (2003)
87. X. Yin, Q. Xu, X. Xu, Fenxi Huaxue **23**, 1168–1171 (1995)
88. L. F. Capitan-Vallvey, C. Cano Raya, E. Lopez Lopez, M. D. Fernandez Ramos, Anal. Chim. Acta **524**, 365–372 (2004)
89. J. Wang, B. Tian, J. Lu, D. Luo, D. MacDonald, Electroanalysis **10**, 399–402 (1998)
90. Y. G. Vlasov, Y. E. Ermolenko, V. V. Kolodnikov, A. V. Ipatov, S. Al-Marok, S. Sens. Actuators B **24–25**, 317–319 (1995)
91. I. Murkovic, O. S. Wolfbeis, Sens. Actuators B **38–39**, 246–251 (1997)
92. W. H. Chan, R. H. Yang, K. M. Wang, Anal. Chim. Acta **444**, 261–269 (2001)
93. B. Palenzuela, L. Manganiello, A. Riso, M. Valcarcel, M. Anal. Chim Acta **511**, 289–294 (2004)

94. L. Manganiello, A. Rios, M. Valcarcel, Anal. Chem. **74**, 921–925 (2002)
95. D. P. Ruys, J. F. Andrade, O. M. Guimaraes, Anal. Chim. Acta **404**, 95–100 (2000)
96. S. Yao, S. Tan, L. Nie, Fenxi Huaxue **14**, 729–734 (1986)
97. S. Casilli, C. Malitesta, S. Conoci, S. Petralia, S. Sortino, L. Valli, Biosens. & Bioelec. **20**, 1190–1195 (2004)
98. B. Rogers, C. A. Bauer, J. D. Adams, Micro-electro-mechanical Systems **5**, 663–666 (2003)
99. M. Gomes, S. R. Teresa, E. V. Morgado, Joao A. B. P. Oliveira, Anal. Lett. **32**, 2715–2723 (1999)
100. M. Skreblin, A. R. Byrne, Vestnik Slovenskega Kemijiskega Drustva **38**, 521–536 (1991)
101. B. Mazzolai, V. Mattoli, V. Raffa, G. Tripoli, D. Accoto, A. Menciassi, P. Dario, Sens. Microsystems **12–14**, 369–375 (2003)
102. M. A. George, W. S. Glaunsinger, Thin Solid Films **245**, 215–224 (1994)
103. T. Morris, G. Szulczewski, Langmuir **18**, 5823–5829 (2002)
104. E. DiMasi, H. Tostmann, B. M. Ocko, P. Huber, O. G. Shpyrko, P. S. Pershan, M. Deutsch, L. E. Berman, Materials Research Society Symposium Proceedings **590**, 183–188 (2000)
105. M. A. Butler, A. J. Ricco, R. J. Baughman, J. App. Physics **67**, 4320–4326 (1990)
106. S. Chah, J. Yi, R. N. Zare, Sens. Actuators B **99**, 216–222 (2004)
107. A. Henglein, M. Giersig, J. Phys. Chem. B **104**, 5056–5060 (2000)
108. T. Morris, H. Copeland, E. McLinden, S. Wilson, G. Szulczewiski, Langmuir **18**, 7261–7264 (2002)
109. T. Morris, K. Kloepper, S. Wilson, G. Szulczewiski, J. Coll. Interf. Sci. **254**, 49–55 (2002)
110. F. Kim, J. H. Song, P. Yang, J. Am. Chem. Soc. **124**, 14316–14317 (2002)
111. S. Link, M. B. Mohamed, M. A. El-Sayed, J. Phys. Chem. B **103**, 3073–3077 (1999)
112. J. Chang, H. Wu, H. Chen, Y. Ling, W. Tan, Chem. Comm. **8**, 1092–1094 (2005)
113. R. H. Ritchie, Phys. Rev. ***106***, 874–881 (1957)
114. A. Otto, Z. Phys. **216**, 398–410 (1968)
115. E. Z. Kretschmann, Phys. **241**, 313–324 (1971)
116. Biacore,http://www.rci.rutgers.edu/~longhu/Biacore/pdf_files/SPR_Technology_Brochure.pdf
117. Ion Cohanoschi, Carlos Toro, Athur Thibert, Shengli Zou, Florencio E. Hernández, Plasmonics **2**, 89–94 (2007)
118. M. Suzuki, Y. Niidome, N. Terazaki, Y. Kuwahara, K. Inoue, S. Yamada, Jpn. J. Appl. Phys. **43**, L554–L556 (2004)
119. H. I. Elim, J. Yang, J. Y. Lee, J. Mi, W. Ji, Appl. Phys. Lett. **88**, 0831071–0831073 (2006)
120. Leonardo De Boni, Carlos Toro, Erin Leigh Wood, Florencio E. Hernández, Plasmonics **3**, 171 (2008)
121. M. Pelton, M. Liu, H. Y. Kim, G. Smith, P. Guyot-Sionnest, N. F. Scherer, Opt. Lett. **31**, 2075–2077 (2006)
122. Leonardo De Boni, Carlos Toro, and Florencio E. Hernández, Opt. Exp. **16**, 957 (2008)

Chapter 8
Optical Properties and Applications of Shape-Controlled Metal Nanostructures

Rebecca J. Newhouse and Jin Z. Zhang

1 Introduction

Research on the properties, synthesis, and applications of metal nanoparticles has surged in the last two decades. Yet metal nanoparticles have a long and rich history. Gold nanoparticles were used as decorative pigments in stained glass dating back to the fourth century [1]. However, it was not until 1857 when Michael Faraday performed systematic experiments on the synthesis and resultant absorption of these solutions that the vivid ruby color was correctly associated with gold "in an excessively subdivided condition" [2, 3].

The bright red and yellow color of gold and silver nanoparticles is due to strong absorption in the visible region of the electromagnetic (EM) spectrum from their surface plasmon resonance (SPR). SPR describes the collective oscillation of the metal's conduction band electrons, induced through interaction with an incident EM field. The resonant frequency of the SPR is determined by the metal's shape, structure, and the dielectric of the surrounding medium. Many nanoparticle syntheses result in nearly spherical particles, and their SPR can be modeled to a good approximation with Mie theory [4, 5]. However, for nonspherical particle shapes there exists no exact solution, and numerical methods such as the discrete dipole approximation (DDA) are used to study the SPR [5]. Although spherical metal nanoparticles are useful in many applications, the ability to predict absorption and scattering profiles for other particle shapes is becoming essential as more complex structures are constantly developed for diverse applications. The combination of different surfactants and seed-mediated growth methods has given wet-chemical synthesis a high

R.J. Newhouse • J.Z. Zhang (✉)
Department of Chemistry and Biochemistry, University of California,
Santa Cruz, CA 95060, USA
e-mail: zhang@chemistry.ucsc.edu

C.D. Geddes (ed.), *Reviews in Plasmonics 2010*, Reviews in Plasmonics,
DOI 10.1007/978-1-4614-0884-0_8, © Springer Science+Business Media, LLC 2012

degree of control and reproducibility in recent years. Other novel structures can be engineered by templating materials on others, which has been used to synthesize hollow gold spheres with tunability of the SPR absorption band over the visible and NIR range [6].

The small size of gold and silver nanoparticles gives rise to unique optical and electronic properties utilized in numerous applications, including photovoltaics [7], catalysis [8–11], plasmon-assisted catalysis [12], surface-enhanced Raman scattering (SERS) [13], metal-enhanced fluorescence [14], imaging [15], orientation sensors [16], SPR spectroscopy and in vitro cancer therapeutics [17, 18].

Metal nanoparticle research extends to a variety of different metals; however, here we focus on silver and gold metal nanomaterials that have attracted the most attention. We aim to provide a basic understanding of the origin of their unique optical properties, emphasizing their dependence on particle shape. We begin by discussing methods of colloidal synthesis and common techniques used to characterize metal nanoparticles. Unique structures of gold and silver nanomaterials and their optical properties will be examined in detail. We conclude with an overview of a few select applications of nanomaterials, including SPR spectroscopy, SERS, photothermal imaging and ablation therapy, drug delivery, and solar energy conversion.

2 Synthesis and Characterization

2.1 Synthetic Methods

The most common and versatile method for synthesizing metal nanomaterials is colloidal synthesis based on wet chemistry. Colloidal synthesis of metal nanoparticles requires a metal precursor, a reducing agent, and a stabilizing molecule. The most popular method of preparation for silver and gold nanoparticles is the citrate-reduction technique first published by Turkevich [19]. This simple and robust method employs citrate as the reducing agent and the stabilizing molecule (capping ligand). Frens improved upon the synthesis by controlling the ultimate size of the nanoparticles with citrate ion concentration [20]. A simplified explanation for this behavior follows. For a given concentration of metal ion precursors, increased citrate results in more nucleation sites during the initial stages of the reaction, yielding a higher concentration of smaller gold nanoparticles. Conversely, decreased citrate produces fewer nucleation sites, so that for the same concentration of metal ions the nanoparticles will be larger but lower in concentration. Typically, lower temperatures enable more synthetic control in colloidal synthesis, but citrate is a weak reductant and requires elevated temperatures to reduce gold ions. Nevertheless, gold ion reduction at 100°C produces spherical particles with a surprisingly narrow size distribution [19, 20]. Small particle sizes made with higher concentrations of citrate are optically transparent, ruby-colored solutions with a characteristic absorption peak at 520 nm [20]. As particle size increases, the solutions scatter visible light much more efficiently, becoming opaque dispersions [20]. This "well-behaved" characteristic of gold is one of the reasons that gold nanoparticles have received so much attention and have found use in a variety of applications. For aqueous syntheses, the metal ion precursor is commonly $HAuCl_4$ or $AuCl_3$. Weak reducing agents such as sodium citrate are popular, although stronger reducing agents such as $NaBH_4$ can be used at lower temperatures to gain additional control of reaction processes.

Molecules that bind to the surface of nanoparticles are critical to long-term stability of the colloidal suspension. In the case of citrate-capped nanoparticles, aggregation is prevented by electrostatic repulsion between particles since this repulsive force is greater than the attractive van der Waals force between particles [21]. Instability is introduced when additional salts are added, since excess ions in solution work to screen the repulsive force between particles and flocculation is induced. More robust gold nanoparticle systems can be created by using different synthetic methods.

One method of gold nanoparticles synthesis with broad application is the Brust–Schiffrin method, which produces monodispersed, small (~1–3 nm) gold particles [22, 23]. This technique is popular due to the robust nature of the ligand-stabilized particles that are produced. Indeed, the solvent can be removed, the particles dried, and subsequently redispersed in solvent without the irreversible aggregation seen for particles made with the citrate-reduction method. In the original paper published in 1994, a two-phase liquid–liquid system is used where $AuCl_4^-$ is first transferred from an aqueous phase to toluene using tetraoctylammonium bromide. Aqueous $NaBH_4$ is introduced to reduce the gold precursor in the presence of dodecanethiol, or in general, a long chain alkane thiol [23]. Expanding on this work, Leff et al. showed that by varying the initial concentration of $AuCl_4^-$/alkane-thiol ratio, the size of the gold nanocrystals can be controlled from 1.5 to 20 nm [24]. This basic synthesis has been the starting point for others who have increased the flexibility of the procedure to include many other thiol-functionalized ligands as particle stabilizers as well as a variety of surfactants for micelles and reverse-micelle syntheses [25–27].

Lee and Meisel explored silver and gold nanoparticle synthesis and developed a silver particle synthesis technique that again uses citrate as the reducing agent [28]. Unlike gold, silver nanoparticles produced by this method are not completely uniform in particle shape or size, but do preferentially synthesize spherical or spheroid nanoparticles [28].

Additional control of colloid syntheses can be obtained using a seed-mediated growth method which has been used successfully in developing gold and silver nanoparticles of well-defined shapes [29–34]. The first step in the process is the development of 3.5–4 nm spherical "seeds" of the gold or silver (or other metals, such as platinum [34]) by the reduction of silver or gold salt with a strong reducing agent. A solution of additional metal ions, a weak reducing agent, and a shape-controlling agent, such as cetyltrimethylammonium bromide (CTAB), is added to the seed solution. The weak reducing agent, often ascorbic acid in the case of Au^{3+}, is only able to reduce the species to Au^+, not the elemental metal. However, the ascorbic acid works to reduce the monocation gold ion to elemental gold on the surface of the seed [35]. It should also be noted that the addition of the counter-ion, bromide in the case of CTAB, is crucial to the synthesis of a particular shape such as a nanorods. Spherical particles are produced when cetyltrimethylammonium chloride is used and cetyltrimethylammonium iodide produces a mixture of shapes [35]. Additional ions are useful during this synthetic method for further improving shape control and yields such as Ag^+ for gold nanorods [36] and HCl for silver nanocubes [37].

Highly controlled silver (and gold) nanoparticles have been synthesized using the polyol method [32, 38]. A seed-mediated growth mechanism is also used; however, the seeds are reduced by ethylene glycol at temperatures >100°C. Additional metal precursors are introduced into the system to further reduce onto the structures and grow them to the appropriate size. In the presence of poly(vinyl pyrrolidone) or PVP, metal nanowires [39, 40], nanocubes [32, 37, 41], nanocages [41], and bipyramids [33] can be fabricated. The synthetic factors that ultimately determine the final geometric shape include the initial concentration of metal ion precursor, the ratio of metal precursor to PVP, and the interaction strength of PVP with the various crystalline facets [30]. Figure 8.1 shows representative EM images of highly controlled silver nanostructures synthesized using a seed-mediated growth method, PVP, and other shape-directing anions.

Layered nanostructures are commonly synthesized by templating a metal on an existing structure of a particular size and shape. To make the structures hollow, the original template material can be dissolved, or, "used up" as in the case with a galvanic replacement. Galvanic replacement in this context uses the interior metal structure as a sacrificial template. A hollow structure, instead of simply a solid structure of alternate metallic composition, can be obtained when the balanced electrochemical equation generates fewer M^0 atoms than template atoms are oxidized. This technique has been used with successfully to create symmetric hollow gold nanostructures out of essentially any silver nanostructure. Examples are gold nanocages and nanotubes resulting from sacrificial silver nanocube and nanorod templates [42]. Recently, asymmetric hollow nanorods have been synthesized using partial galvanic replacement [43]. In this work, a Ag–Au–Ag nanorod was synthesized by growing Ag metal on a gold decahedron particle to create a solid rod structure with a Au core and silver elongation from both ends. By carefully controlling the amount of $AuCl_4^-$, Seo and Song hollowed out both ends of the rod structure, leaving the gold core intact, with an Au/Ag alloy shell that maintained the framework of the nanorod [43]. Extremely monodispersed hollow gold nanospheres (HGNs) can also be synthesized by using a sacrificial cobalt template [6]. Cobalt nanoparticles are first synthesized using air-exclusion Schlenk-line techniques. $AuCl_4^-$ is then added and for every Au^{3+} ion that is reduced, three Co^+ ions are oxidized. Since cobalt is air-sensitive, any elemental cobalt that remains is oxidized when the solution is exposed to O_2 and leaves a pure gold shell [6, 44]. In addition, silver can be deposited onto these structures using a seed-mediated growth method to reduce silver precursors on HGNs. This results in extremely monodispersed Ag–Au double shell structures with applications in SERS [45].

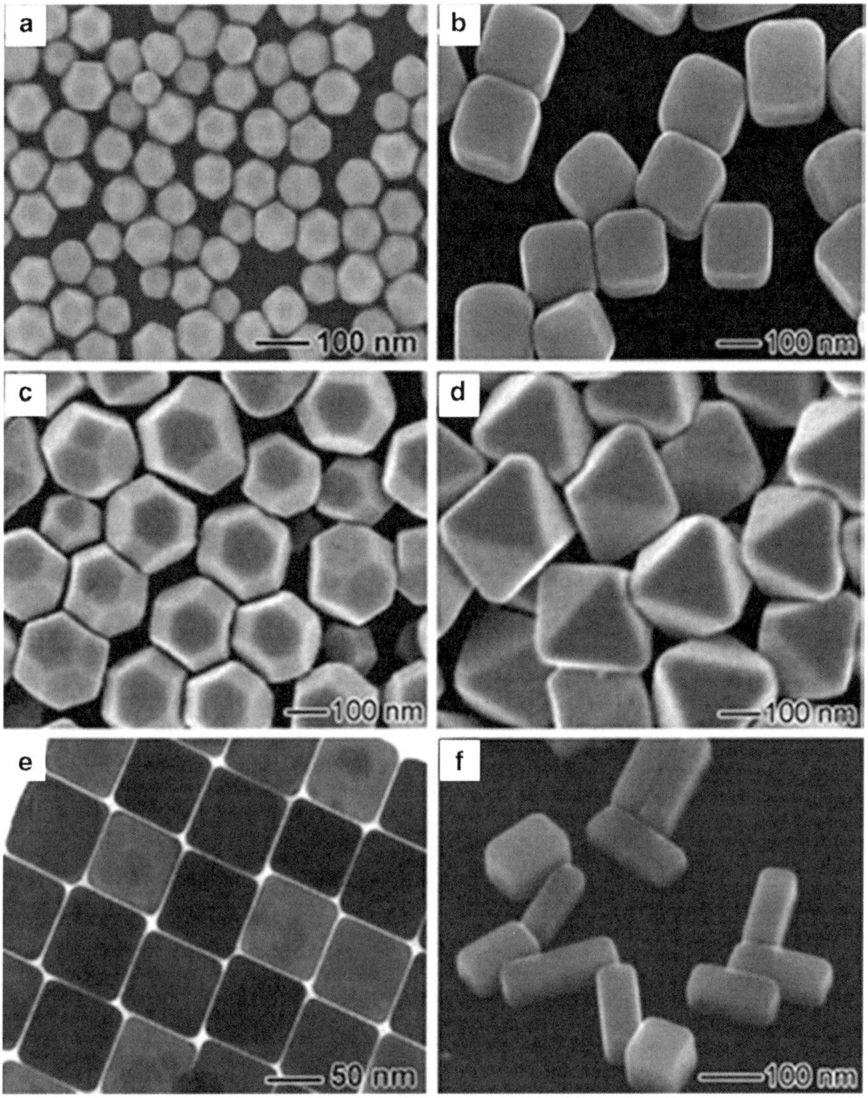

Fig. 8.1 Electron microscopy images of single-crystal Ag nanocrystals that can be synthesized using a seed-mediated growth method, PVP, and other shape-directing anions including (**a**) cuboctahedrons, (**b**) nanocubes, (**c**) truncated octahedrons, (**d**) octahedrons, (**e**) nanocubes, and (**f**) nanobars. Reproduced with permission from [34]

2.2 Structural Characterization

Structural characterization is an essential aspect of nanomaterial research. The choice of technique depends on the desired information. For example, electron microscopy (EM) provides information about size, shape, and morphology of nanomaterials, while X-ray diffraction (XRD) probes the crystal structure. Scanning probe microscopy (SPM), including atomic force microscopy (AFM), affords information about surface and topological properties.

Electron microscopy (EM) is a powerful tool for the structural characterization of nanomaterials. Transmission electron microscopy (TEM) accelerates electrons through a sample, and the interaction of the electrons with the electron-transparent sample creates an image. EM exploits the wave-particle duality of electrons by accelerating them with a positive potential to increase their momentum and decrease their de Broglie wavelength, overcoming the resolution limit of visible light (~300 nm) [46]. Subnanometer resolution can be obtained with high-resolution TEM (HRTEM). At this resolution, lattice fringe spacing in crystalline samples is observed and can be used to determine which facets are expressed [46]. An advantage of this technique is that secondary electrons are emitted from the surface of the sample, which can provide additional information. X-ray energy dispersive spectroscopy (XEDS) and electron energy loss spectrometry (EELS) both give characteristic peaks from these secondary electrons, which can give quantitative elemental distribution in the sample [46]. Electron diffraction (ED) is also a useful technique for determining crystal structure with TEM. Electrons diffract through crystalline or semicrystalline solids, creating patterns that reveal information on the crystal structure and lattice repeat distances [46].

Optical microscopy is not commonly used to characterize the structure of metal nanoparticles since the resolution is limited to approximately half the wavelength of light. However, techniques that avoid this constraint have been developed and are in use, such as near-field scanning optical microscopy (NSOM) [47] and scattering near-field scanning optical microscopy (s-NSOM) [48, 49]. This technique exploits the phase shift of the incident field and the field scattered by the particle and ultimately images the particles as circular interference patterns. Particles smaller than 100 nm can be visualized by these methods (better resolution can be obtained with s-NSOM) [47]. However, the use of this optical technique for nanomaterial structural characterization is limited since the resolution is insufficient to provide any information regarding crystallinity or surface morphology.

Scanning electron microscopy (SEM) and AFM are popular surface imaging techniques with nanometer resolution; AFM can even yield subangstrom vertical resolution [50]. AFM is part of a category of techniques termed SPM. In these techniques, a sharp tip is scanned across the sample surface and the surface–tip interaction is detected [51]. Scanning tunneling microscopy (STM) measures the current from the sample through the tip to provide a 3D image of the sample [52]. AFM is more versatile since it can also image nonconducting samples by raster scanning an atomically sharp tip attached to a flexible cantilever along the surface of the material and measuring the force on the tip from the surface. Atomic resolution is obtained by using a piezoelectric stage. Advantages include little or no sample preparation or coatings and the ability to work in nonvacuum or liquid environments [50]. There are some disadvantages, though, including the slow scanning speed and the limited area that can be examined. SEM, like TEM, takes advantage of the small de Broglie wavelength of electrons, but rather than detecting transmitted electrons, SEM detects scattered electrons from the surface of the material and uses them to construct an image. This provides accurate shape and size information, as well as surface topologies depending on the feature size, but unlike AFM, does not provide true 3D maps. As in TEM, secondary electrons can be used to provide additional information regarding elemental composition [50].

X-ray techniques such as XRD, X-ray photoelectron spectroscopy (XPS), and X-ray absorption spectroscopy (XAS) can provide additional information about the sample [53, 54]. XRD probes the spacing between crystallographic planes, providing a fingerprint for the material type and crystal phase. Moreover, information about the crystallite grain sizes (under 100 nm) can be obtained from the full width at half maximum of reflections using the Debye–Sherrer equation. XPS can also be used to measure the elemental composition of the material and varieties of XAS exist that serve to probe the local geometric and electronic structure of the material [53, 54].

3 Surface Plasmon Resonance

To first-order approximation, metals can be described by the "free-electron" or "electron gas" model, in which electrons are considered to be delocalized, resulting in high electrical conductivity [55]. An incident electric field can perturb the electron gas from equilibrium and create a nonuniform distribution of charge, producing an electric field that attempts to restore charge neutrality. A collective oscillation results since the electrons repeatedly overshoot the equilibrium position because of their acquired momentum [56]. The term plasmon succinctly encompasses this phenomenon of *plasma* oscillation. Electronic motion is confined as the size approaches the electron mean free path (about 10–100 nm at room temperature for most metals) and as the size shrinks, the oscillating charges appear localized on the surface of the particle [15, 55, 57]. Thus, the term "surface plasmon" is used to describe this phenomenon in metals on the nanoscale [55, 56, 58]. At small particle sizes, only dipolar surface plasmons need be considered, but larger particle sizes can excite higher multipoles, such as quadruple oscillations [5, 59]. The resonant frequency of oscillation depends on a combination of factors including shape and size of the nanoparticles, metal type, and dielectric of the embedding medium. When the frequency of the incident light matches the intrinsic electron oscillation frequency, the particles absorb the light extremely efficiently, resulting in the observed SPR absorption. For silver and gold spheroid nanoparticles, there is a single resonant wavelength in the visible region of the electromagnetic spectrum. For silver particles, this resonance is ~400 nm producing a vibrant yellow solution, whereas spherical gold particle dispersions absorb at 520 nm and are seen as a brilliant ruby solution. Similar to the frequency of the SPR, the intensity also depends on the particle size, shape, and surrounding environment.

3.1 Particle Shape and SPR

3.1.1 Spherical Au and Ag Nanoparticles

Spheroid or near-spherical gold and silver particles of various sizes are readily synthesized with standard colloidal preparation techniques. Their single resonance can be modeled with good accuracy using Mie theory, which is the exact solution to Maxwell's equations for spherical particles of arbitrary sizes and describes the scattering and absorption spectra [4]. In general, as a spheroid particle gets larger, the resonant frequency is lowered [60, 61]. This is predicted from Mie theory since, as particle size increases, the extinction spectra have contributions from multipole absorption as well as scattering which shifts the absorption to longer wavelengths [4, 5, 60, 61]. Qualitatively, this red-shift can be explained by considering the distance that the electrons must travel in order to generate the oscillating dipole. For the same material, a larger nanoparticle will result in a reduced plasmon frequency since there is a longer path, and the particle absorption is observed to red-shift. For gold nanoparticles, as they increase in diameter from 9 to 99 nm, there is a 48 nm SPR shift from the original value of 517 nm [60]. The width of the plasmon does not follow the same monotonic trend. It is observed to decrease when the size is increased from 9 to 22 nm and then increase at larger particle diameters [60]. Although there is some tunability with spheroid gold particles, much larger variations of the SPR wavelength can be achieved by changing the nanoparticle shape.

3.1.2 Nanorods

As the nanoparticle aspect ratio increases, e.g., into a nanorod, the SPR splits into two distinct bands: a transverse mode corresponding to electron oscillation perpendicular to the long axis of the rod and longitudinal mode corresponding to electron oscillation along the long axis of the rod. This is in direct contrast to the single SPR for spherical nanoparticles for which all the dipolar modes are degenerate. The anisotropy of the metal nanorod breaks the symmetry and allows electron oscillation in the direction perpendicular and parallel to the long axis of the nanorod [15]. For gold, the transverse SPR mode is typically around 520 nm (for silver, ~400 nm is observed) and the longitudinal SPR is at longer wavelengths for both gold and silver with the specific peak position dependent on the length of the nanorod. Yu et al. in 1997 and later Chang et al. used electrochemical means to synthesize aqueous dispersion of high-quality Au nanorods of varying aspect ratios [62, 63]. With increasing nanorod length, the SPR of the longitudinal mode for Au nanorods can be tuned from 520 to ~1,100 nm as the aspect ratio is increased from 1 to 11 [63]. As the aspect ratio is increased, the transverse mode also shifts, although the magnitude of this shift is small relative to that observed for the longitudinal SPR [62]. Link et al. modeled this optical behavior and found good agreement between calculation and experiment [64]. Recent theoretical work on elongated metal structures has revealed that the end shape of the nanoparticles, in addition to the aspect ratio, also has an effect on the longitudinal and transverse SPR modes [65, 66]. Since the optical properties are directly tied to the nanorod shape, there has been a tremendous effort to better understand the growth mechanism and develop improved synthetic routes with increased control over the nanorod aspect ratio [36, 64, 67–76]. For example, gold nanowires with an aspect ratio >200 have been synthesized recently by reducing the seed concentration and growing the structures in highly acidic solution [77]. Silver nanorods also exhibit a transverse and longitudinal SPR and have been synthesized with excellent control over the rod's aspect ratio and thus the position of the longitudinal SPR [78–80]. These silver nanostructures have also been studied using theoretical methods to gain insight into their fundamental optical properties [81, 82].

3.1.3 Plasmon Coupling

Another interesting optical property of metal nanostructures is the ability of their individual plasmons to couple when the nanoparticles are physically near each other, e.g., in aggregates. Weak or moderate interaction or coupling between particles usually results in shift of the SPR, while strong coupling can result in new SPR bands. For gold nanoparticles, this effect can manifest visually as a vibrant color change from a ruby red solution to darker red, magenta, and purple solutions as a growing number of particles interact and increase the size of the gold nanoparticle aggregate [83]. With the appropriate choice of surface functionalization and specificity toward a particular analyte, aggregation can be induced in the presence of a particular analyte. This effect was exploited to create a variety of colorimetric sensors for alkali ions [84], heavy metals [85], and single strand oligonucleotides [83, 86, 87].

As discussed previously, the strength of particle interaction can affect the SPR signal. When the interparticle interaction is strong, the original, single SPR band from isolated nanoparticles can split into multiple bands, similar to the case of gold nanorods. The new SPR bands can be significantly broadened due to multiple dipolar modes of each aggregate and inhomogeneous broadening resulting from a distribution of sizes and shapes of aggregates [88]. For example, strongly interacting gold aggregates can be obtained from the reduction of $HAuCl_4$ by Na_2S [89, 90]. As in the case of the nanorods, the interacting nanoparticles retain the transverse SPR close to 520 nm, but their close proximity to each other creates aggregates that have multiple dipolar modes. Hole-burning experiments are able to clearly show inhomogeneous broadening in the aggregate absorption [91].

Plasmon coupling has been studied recently with the so-called "gold plasmon rulers" and the efficiency was observed to increase as the interparticle distance for gold nanoparticles decreased. This is monitored as an exponentially decreasing redshift of the EPB as the interparticle spacing is increased (this occurs until a spacing of ~2.5 times the short axis of the particle) [92–94]. A "plasmon ruler equation" was proposed that uses the plasmon shift to estimate interparticle Au NP (nanoparticle) distance which is of interest for biological applications as well as a possible alternative to fluorescence resonance energy transfer (FRET) where fluorescent probe have limitations in terms of lifetime and distance measuring capabilities [94, 95]. Analogous studies on silver nanoparticles revealed that silver plasmon rulers are more sensitive to small distance changes than gold NPs [96]. In addition, two distinct regimes of coupling behavior were observed where the EPB red-shifts until the ratio of the particle separation and the particle diameter is ~1.05. As the distance continues to decrease so that the particles are nearly touching, the EPB no longer shifts, which contrasts the coupling in Au NPs [96].

3.1.4 Hollow Au and Ag/Au Nanospheres

HGNs are another interesting example of how shape can dramatically affect the optical properties of metal nanomaterials. These structures have been synthesized using a variety of techniques, including reducing gold ions onto a silica bead and using a sacrificial template such as a silver or cobalt nanoparticle [6, 44, 97–99]. In particular, monodispersed HGNs made in our laboratory by the reduction of Au^{3+} by Co^{0} as described in the previous synthetic methods section show tunability from 550 nm (solid gold nanoparticles) to ~820 nm, which is in the near-infrared (NIR) region of the spectrum (Fig. 8.2a) [6]. Applications of this NIR absorption are discussed later in this article. HGNs fabricated in this method are monodispersed, with narrow SPR in the visible to NIR region [6]. The tunability is controlled by adjusting the core diameter and the thickness of the shell. Hao et al. used Mie theory and DDA to predict a red-shift of the SPR for hollow gold particles with decreasing shell thickness and constant core diameter as well as increasing core diameter with constant shell thickness [100]. The HGN synthesis by Schwartzberg et al. produced monodispersed particles with a high degree of control over their shell thickness and particle diameter. By varying the particle's aspect ratio, the tunable extinction spectra for these particles were obtained, which agreed very well with the theoretical prediction that the most red-shifted particles have both large core diameters and thin shells.

Further tunability of the SPR can be obtained by depositing a silver shell on the HGN nanoparticles [45]. Figure 8.2b shows the UV–vis spectra as increasing aliquots of $AgNO_3$ are added to a solution of HGNs. With the continued addition of silver ions to form a silver shell of a few nanometers thickness, the gold plasmon becomes significantly blue-shifted from 610 to ~480 nm and the silver plasmon is red-shifted from 410 to 425 nm. The shift in the gold plasmon band can be explained by the refractive index change from water to silver [45]. The red-shift for the silver plasmon occurs since the silver is essentially forming a shell and the plasmon shifts accordingly (as in the case of the red-shift for HGNs compared to solid gold particles) [45].

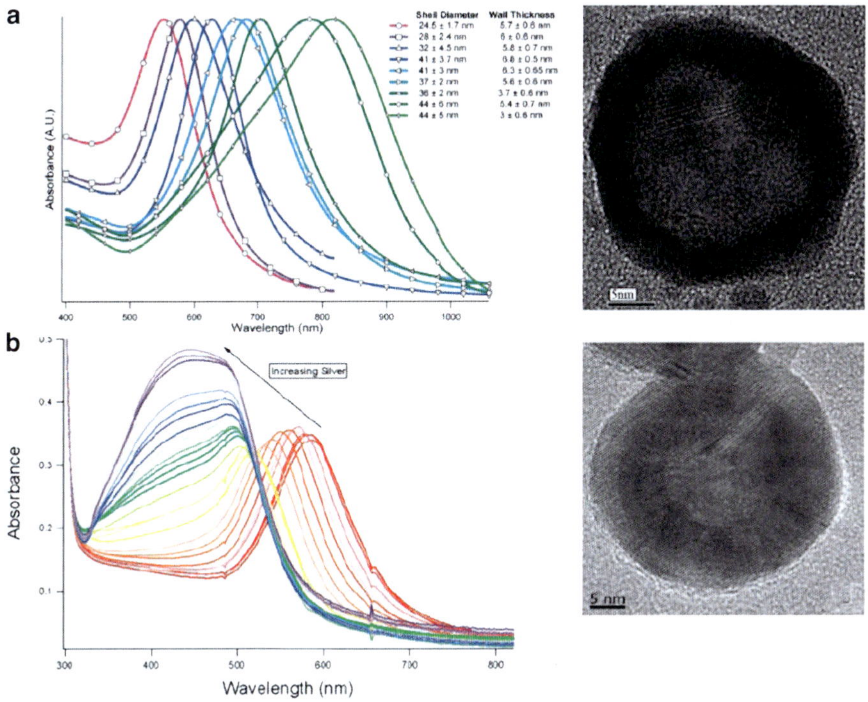

Fig. 8.2 (**a**) UV–visible absorption spectra of HGN samples with varying diameters and wall thicknesses. The image to the right is a representative TEM image of an HGN. Reproduced with permission from [6]. (**b**) UV–vis spectra of AuHNS with incremental addition of silver ions during silver shell growth. The image to the *right* is a representative TEM image of a silver-coated hollow gold nanosphere. The scale bar in both images is 5 nm. Reproduced with permission from [45]

3.1.5 Other Shapes

Increased understanding of crystal growth and reaction mechanisms has resulted in the ability to synthesize nearly any shape or size of gold and silver nanoparticles, including nanorods [43, 62, 71, 72, 76, 79, 101–103], nanowires [31, 77, 78], hollow gold and silver nanoparticles [6, 41, 43–45, 97–99, 104], nanocubes [37, 41], nanocages [41], nanoframes [105], bipyramids [33], triangular nanoprisms [29, 106], pentagons [107], tetrahexahedra [108], nanoantenna [93], nanostars [109], and many other well-controlled and defined shapes. In general, as a nanomaterial becomes less spherical with an increasing number of vertices, the extinction spectrum becomes more complex [81]. As in the case of nanorods, increased anisotropy can lead to additional dipole modes along the various symmetry axes. Further complexity can result from spectral broadening over the visible region and increased *multipole* contribution from less spherical shapes and larger dimensions of the

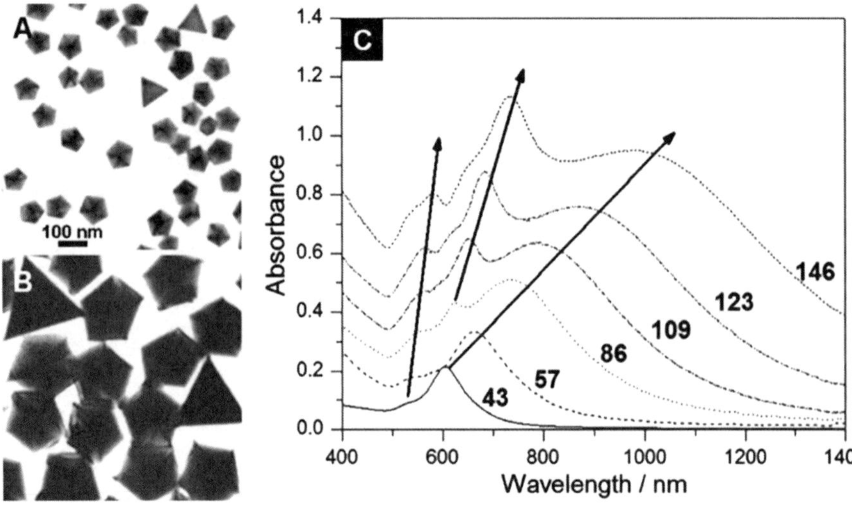

Fig. 8.3 Representative TEM micrographs of gold decahedra grown by using the ultrasound-induced reduction of $HAuCl_4$ in DMF, in the presence of poly(vinylpyrrolidone) (PVP) on preformed 2–3 nm seeds (**a**) and on preformed gold decahedra (**b**). The average decahedra edge lengths are 54 (**a**) and 146 nm (**b**). The scale is the same for both images. (**c**) UV–vis–NIR spectra of dispersions of Au decahedra in DMF with varying side lengths (as indicated). The *arrows* indicate the size dependence of the different plasmonic mode. Reproduced with permission from [113]

nanomaterial. Multipole resonances are generally blue-shifted with respect to the dipole excitation wavelength [5, 81, 109]. For gold triangular nanoprisms, a quadrupole band was observed at ~800 nm for gold which was blue-shifted from the in-plane dipole band at ~1,300 nm and the observed spectrum was in excellent agreement with DDA calculations [110]. In terms of the general morphology and the observed optical response, increasing resonances and broadness appear as the polyhedral metal nanoparticle shows fewer faces and with increasing sharpness of the vertices [111]. As the polyhedral nanoparticles are truncated, the primary resonance is blue-shifted and can overlap with higher-order modes to broaden the spectrum [112].

In particular, gold decahedra, otherwise known as pentagonal bipyramids, are highly faceted metal nanoparticles that have ten triangular faces and seven vertices. Figure 8.3a, b shows representative TEM images of the particles. Two distinct, although broad, dipole bands are observed when the particle size is relatively small (a side length of ~42 nm) in Fig. 8.3c [113, 114]. One band is due to oscillations along the fivefold symmetry axis, and the other arises from oscillations in the equatorial plane of the nanoparticle [113, 114]. As the size increases to a side length of ~60 nm, a blue-shifted quadrupole band develops. All the bands are red-shifted as the particle size increases, as expected [113, 114]. In Fig. 8.4, changes in morphology of a single particle are directly related to the optical properties for individual decahedron particles [113]. SEM images compare decahedra particles which are

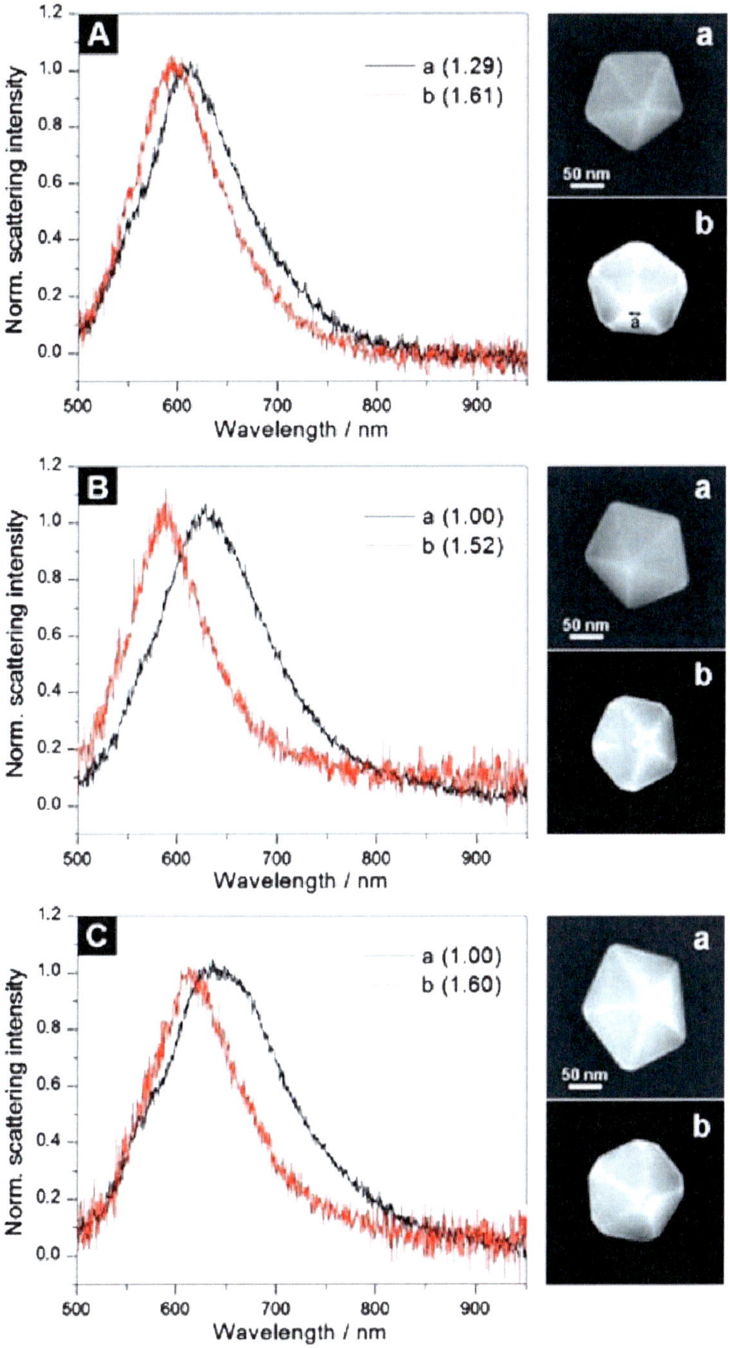

Fig. 8.4 (**A**–**C**) Correlated scattering spectra and SEM micrographs pairs of Au decahedra with increasing tip truncation (indicated in the SEM *inset*), from (**a**) to (**b**), where (**a**) and (**b**) have the same effective side length *L*. *Red* and *black spectra* are from truncated and nontruncated decahedra. Reproduced with permission from [113]

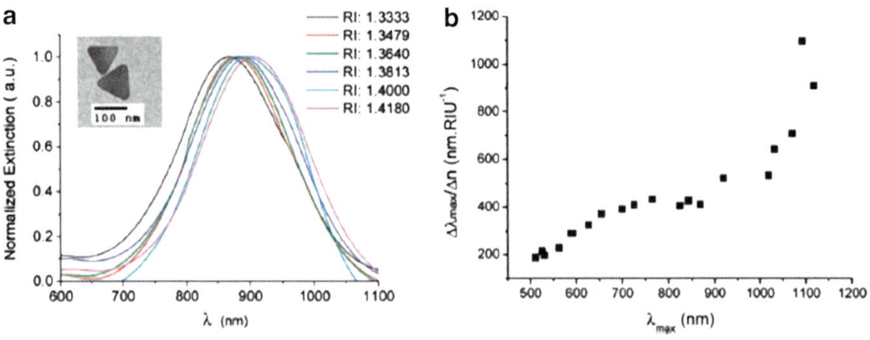

Fig. 8.5 (**a**) Example of spectral shift observed for a specific TSNP ensemble with original peak wavelength of 868 nm, mean edge length 82 nm, mean thickness 11 nm, in varying sucrose concentrations. (**b**) Peak LSPR wavelength of the 20 samples plotted against the corresponding LSPR sensitivity of the ensemble to the sucrose refractive index analysis. Reproduced with permission from [118]

partially truncated to those with the same side length, but with sharp tips. The scattering spectrum in each case is shown in the three left panels (a, b, and c each compare a different particle size) and in general, the peak becomes less broad and blue-shifts for the partially truncated particle in comparison to the decahedron with sharp tips [113]. This can be explained by the truncated particles more closely resembling spherical particles (i.e., higher symmetry), and therefore have a narrowed, blue-shifted plasmon.

3.2 *Effect of Surrounding Environment*

The localized surface plasmon resonance (LSPR) shifts in response to changes in local refractive index surrounding the metal nanoparticles. The magnitude of the shift is dependent on multiple components including a sensitivity factor, the difference between the refractive index at the particle surface and in solution, the thickness of the absorbed layer, and the electromagnetic field decay length [115]. Higher electromagnetic fields resulting from sharp features of nanomaterials increase the sensitivity [116]. This results in varying degrees of sensitivity for nanoparticles of different shapes [117, 118]. Increasing the aspect ratio of metal nanoparticles typically results in an increased sensitivity factor. In fact, hollow gold nanoparticles have been shown to have a much larger response to environmental changes than solid gold nanoparticles and serve to increase the sensitivity of the detection system [99]. Triangular silver nanoplates (TSNPs) with varying edge lengths have also been shown to be highly sensitive SPR sensors (Fig. 8.5a) [118]. SPR sensitivities of these tunable structures was shown to increase linearly with a lambda max up to 800 nm, after which the shift was not linear (Fig. 8.5b) [118].

4 Applications

The unique optical properties of noble metal nanostructures have found application in many technologies, ranging from sensing and photocatalysis to biomedical detection and therapy. We will highlight several applications including SPR spectroscopy, SERS, photothermal imaging and ablation therapy, drug delivery, and solar energy conversion.

4.1 SPR Spectroscopy

As previously discussed, the SPR of metal nanoparticles is sensitive to changes in size, shape, and embedding media of the particles as well as interparticle plasmon coupling. The ease of measurement, the sensitivity of the SPR to particle aggregation and molecular adsorption, and the versatility of surface functionalization make SPR-based sensors ideal for detecting a variety of analytes. Specifically, the change of refractive index when molecules bind to a metal surface has been exploited to study the binding affinity and obtain kinetic constants for interacting biomolecules such as antibody–antigen interactions and complementary strands of DNA [119, 120]. In a typical SPR spectroscopy experiment, molecules are immobilized on a thin metal film. Light is focused through a prism on the opposite surface and the incidence angle, polarization or wavelength is modulated until the SPR condition is achieved. This is observed as a dip in reflectance. The resonance condition is modified by the binding of analytes to bound surface molecules, resulting in a change in reflectivity [115, 121]. SPR spectroscopy is advantageous because it allows real-time and label-free determination of kinetic parameters [122]. Arrays of SPR platforms enable high-throughput screening for drug discovery [121].

Sensors based on the LSPR of silver and gold nanoparticles have also been examined [123–130]. Advantages of nanomaterials over typical thin-film substrates include less interference from the bulk refractive index and increased lateral resolution [115, 128]. The sensitivity of LSPR is improved by employing single metal nanoparticle arrays rather than immobilized ensembles of nanoparticles, since single nanoparticle LSPR detection serves to increase the detection limit as well as improve the lateral resolution [115, 123, 130, 131]. Suspension-based nanomaterials are promising for in vivo sensing applications where they are able to penetrate tissue and cells, in contrast to array-based LSPR sensors [129]. One challenge to this goal is false LSPR shifting that can result from nonspecific particle aggregation. To address this, Wu et al. developed a colloidal gold nanorod sensor, surrounded by a mesoporous silica shell which produced a linear spectral shift with increasing glutathione concentration [129]. The mesoporous nature of the shell allowed the nanorod to be exposed to the solution while also providing increased stability to prevent nonspecific particle aggregation [129].

The specific aggregation of nanoparticles can also be exploited to create sensitive, simple, and inexpensive sensors based on LSPR spectral shifts. To accomplish this, nanoparticles must be functionalized with tightly bound species, which serve to "link" individual particles when the analyte of interest is introduced. The extent of plasmon coupling or LSPR shift increases with the increasing size of aggregate structure and the magnitude of the shift is proportional to the concentration of analyte. Gold nanoparticles have been used to detect ultra-low concentrations of polynucleotide sequences by functionalizing the surface of the nanoparticles with sulfur-terminated polynucleotides, which results in plasmon coupling and an observable (often by eye as a colorimetric detection) LSPR red-shift when the complementary oligonucleotide is present [83, 132]. The extent of aggregation can be visualized for increased practicality, but the high sensitivity of this simple technique is obtained by monitoring the absorbance at the SPR wavelength as a function of temperature, which simply dissolves the aggregate by overcoming the attractive forces involved in binding [83, 132]. In this way, even single base-pair mismatches can be discriminated from complementary strands [87]. Hg^{2+} sensors were created by Lee et al. using a variation of this method by designing complementary oligonucleotides with a thymine–thymine mismatch [85]. The presence of Hg^{2+} can selectively bridge two thymine bases, essentially eliminating the base-pair mismatch and creating aggregates that melt at higher temperatures. The ultimate Hg^{2+} detection limit is 100 nM, which ranks the sensitivity of this technique among the highest for colorimetric-based sensors of mercury ions [85].

4.2 *Surface-Enhanced Raman Scattering*

Raman spectroscopy is based on detection of inelastically scattered light that differs in frequency from the elastically scattered Rayleigh line by a characteristic vibrational mode of the molecule [133]. A particular combination of Raman-active modes is unique to a molecule and can be used as a molecular fingerprint. This quality allows molecules that can be identified based on their corresponding peaks to act as probes for detection applications. A significant limitation of Raman spectroscopy, however, is that the Raman cross-section is very low, ~10^{-30} cm^2/molecule, which severely limits the sensitivity and applicability of this technique [134].

Fortunately, Raman signal can be dramatically enhanced when the analyte molecules are placed at or near a metal substrate surface via the SERS process. SERS was first observed in 1974 [135] but properly identified as an anomalously high enhancement from surface effects in 1977 [136, 137]. The SERS effect is primarily due to an electromagnetic field enhancement effect that occurs upon SPR excitation. Molecules that are close to the surface experience the field generated by the localized plasmon resonance and thus, form a much larger electromagnetic field. The scattering efficiency is increased with a typical enhancement factor of 10^6 [138]. Even higher enhancements of 10^9 are observed when molecules are trapped in junctions between aggregated nanoparticles; it has been suggested that enhancements from these so-called "hot-spots" may allow for single-molecule detection [139, 140].

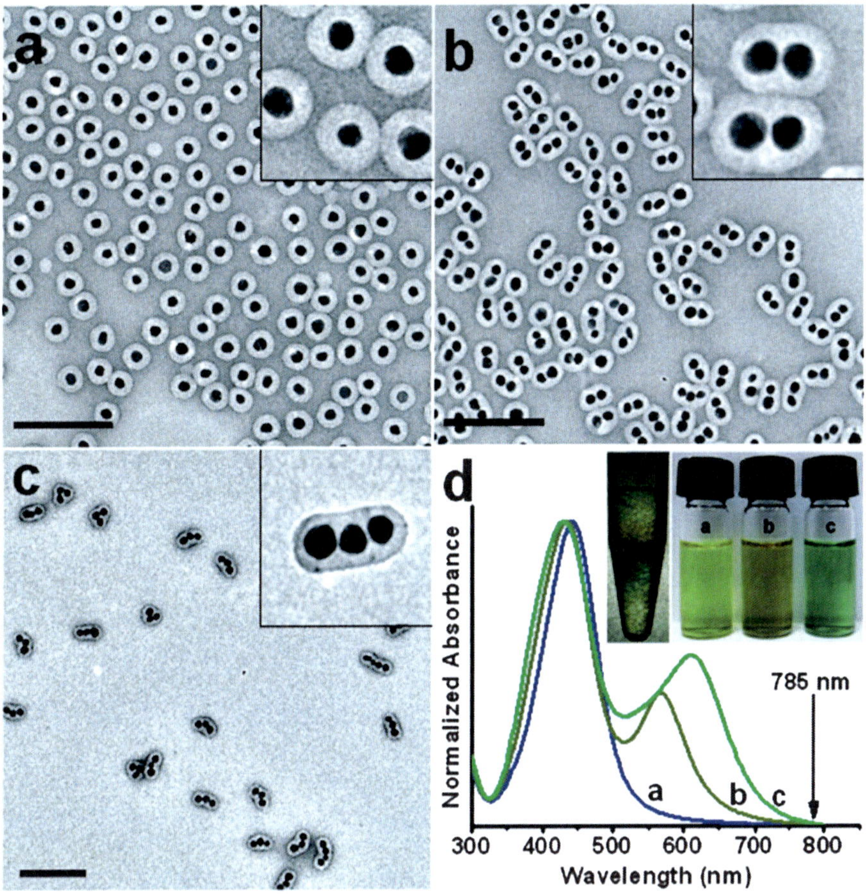

Fig. 8.6 (**a–c**) TEM images and (**d**) UV–vis spectra of the samples enriched with Au at Ag monomers (**a**), dimers (**b**), and trimers (**c**). Though often in a bent conformation, the trimers mostly contained only two hot-spots. Clear gaps existed within the nanoclusters, and no fusion was observed. The *inset* in (**d**) shows a typical outcome of the differential centrifugation, where monomers, dimers, and trimers were enriched in distinct *yellow*, *brown*, and *green bands*, respectively. Scale bars: 200 nm. Reproduced with permission from [141]

The enhancement factors (EFs) due to hot-spots has mainly been studied using single nanoparticle aggregates. This makes it difficult to get statistically relevant correlations between the SERS EF and a particular nanostructure morphology. To overcome this challenge, Chen et al. have engineered spatially isolated colloidal nanostructures (Fig. 8.6) with hot-spots of uniform morphology and size to more accurately interpret statistically significant ensemble enhancement factors [141].

One problem that arises when trying to use SERS as an analytical technique is that typical enhancements are achieved with induced colloidal nanoparticle aggregation to force molecules into nanoparticles junctions and increase the number of "hot-spots" [142]. There is an inherent disorder to this aggregation which leads to aggregates of various shape and sizes and ultimately results in inconsistencies in enhancement factors with repeated measurements. Work to improve the reproducibility and consistency of SERS sensors has been done by using extremely monodispersed structures, such as HGNs, that can enhance at the single particle level, thereby avoiding aggregation that necessarily leads to inconsistencies [142].

Another technique involves the rational design of nanoparticle junctions to take advantage of huge enhancement factors of nanoparticles with sharp corners, such as nanocrescents, and the scattering enhancements seen at the junctions between particles [143, 144]. In general, to combine huge SERS efficiencies with reproducibility, it would be desirable to "put" molecules at the junction of two nanoparticles. One approach in this direction is the synthesis and application of gold nanoframes for SERS [105, 145]. Gold nanoframes are essentially gold nanocages where the side wall thickness is significantly reduced and eventually leads to a gold metal frame [145]. Particles that diffuse into the structure experience a very large electromagnetic field, which is desirable for SERS applications [105]. An interesting feature of these unique structures is that the particle aggregation, which usually intensifies SERS signal, actually leads to a decrease in SERS intensity [105]. This is explained by using DDA modeling which found that although the interparticle field is significantly enhanced between particles as they are brought close together, the intraparticle electromagnetic field is substantially reduced [105].

Because SERS combines molecular specificity with high sensitivity, there is great potential for its wide-spread use as a sensing platform. Colloidal and single use film substrates have been the most well-studied platforms for SERS analysis, however, practical and remote sensing applications require a robust, reusable, and convenient substrates. End-polished optical fibers modified with silver films as SERS probes have stability, compactness, low cost, and portability combined with specificity to make strides toward an ideal sensor. A significant limitation of these end-polished fiber probes is the weak SERS signal that arises from the small number of nanoparticle substrates located around the cross-section of the fiber core [146]. In these cases, high laser powers and long integration times need to be used to acquire SERS spectra with reasonable intensities [146]. To increase the interaction volume, Zhang et al. proposed and successfully developed a D-shaped fiber by polishing the side of the fiber (producing a "D" shaped cross-section). This results in a significant increase in the active surface area over using the fiber tip [146]. Another way to increase the interaction volume is by using a hollow core photonic crystal fiber (HCPCF). He et al. demonstrated that HCPCF probes were a convenient, robust, and reliable method for SERS detection [147]. HCPCFs were modified by first coating the inner walls of the air channels with silver. Liquid sample was introduced into the fiber by capillary action through the air holes and analyte molecules in close proximity to the coated air channels experienced an increased SERS effect over solid fibers [147]. In addition to increasing the interaction volume with HCPCFs, signal can be improved by engineering fiber probes to maximize the number of SERS hot-spots. This has been done by coating a solid multimode fiber with silver and then introducing the probe into a solution mixture of silver nanoparticles and analyte [148]. The combination of two substrates to trap analyte molecules in nanoparticle junctions between nanoparticles on the fiber and those in solution to create a "sandwich-structure" resulted in a tenfold increase in signal over using either substrate individually (colloidal nanoparticles or silver-coated fiber) for SERS measurements [148]. This structure was recently improved by combing enhancements observed for "sandwich-structure" arrangements as well as those obtained for SERS with hollow core fibers. By maximizing SERS hot-spots and increasing the interaction volume by using a hollow fiber, Shi et al. observed a 100-fold enhancement over simple colloidal SERS measurements [149]. These studies demonstrate that optical fibers are indeed promising for SERS sensing applications.

4.3 Photothermal Imaging and Photothermal Ablation Therapy

Metal nanomaterials are increasingly being used for imaging applications. Common problems with typical fluorophore detection, such as photobleaching and blinking, could be circumvented by using metal nanoparticles as optical probes [150]. Small gold and silver nanoparticles have very high extinction coefficients dominated by absorption rather than scattering at wavelengths matching their SPRs [150]. Excitation at the resonant frequency is followed by relaxation, and for metals energy is released nonradiatively as heat which increases the temperature of the environment around the particle, and therefore creates a gradient in the index of refraction surrounding the particle [150]. This phenomenon is called the photothermal effect. Metal nanoparticles can be optically detected by using photothermal interference contrast (PIC) where an additional light source is able to detect the varying index of refraction around the metal nanoparticle with a resolution of 2.5 nm [150, 151]. This technique provides a robust and sensitive imaging technique that can be used to detect single nanoparticles–protein conjugates inside live cells [150, 151].

Nanorods have also been used for photothermal imaging and have the advantage of anisotropy which allows them to function as orientation sensors as well (Fig. 8.7) [16]. A thermal profile of the nanorod as the polarization of the excitation source is changed produces thermal maxima and minima at different polarization angles (Fig. 8.8). Since the longitudinal SPR mode of a nanorod can only be excited with light polarized parallel to the long axis of the rod, the orientation of the nanoparticles can be uniquely determined from this type of analysis [16].

The photothermal effect can also be used as a way to damage biological material in close proximity to the surface of a nanoparticle. Functionalizing the nanoparticles appropriately ensures that the metal nanomaterial is directed to a target and subsequent irradiation of the particle at the SPR frequency can give off enough heat to damage surrounding cells. This application is called photo-ablation therapy.

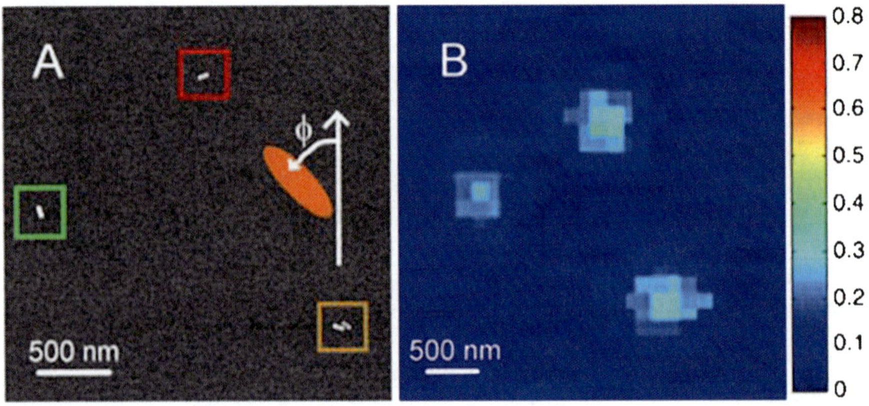

Fig. 8.7 SEM (**a**) and photothermal (**b**) images of the same area showing two single Au NR and one NR dimer. The photothermal image was recorded with 675 nm excitation. Reproduced with permission from [16]

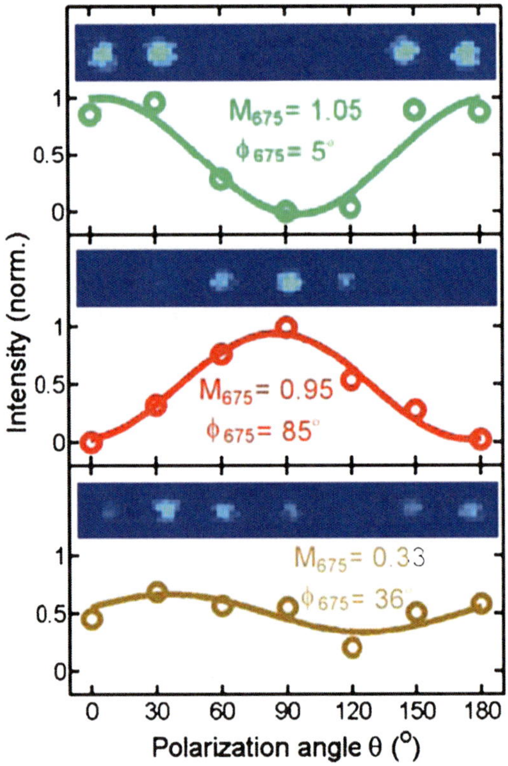

Fig. 8.8 Polarization dependence of the photothermal intensity for the NRs is highlighted in Fig. 8.6. Note that the colors of the traces correspond to those of the *boxes*. The *insets* show the photothermal signal of the individual particles as a function of excitation polarization angle θ. Reproduced with permission from [16]

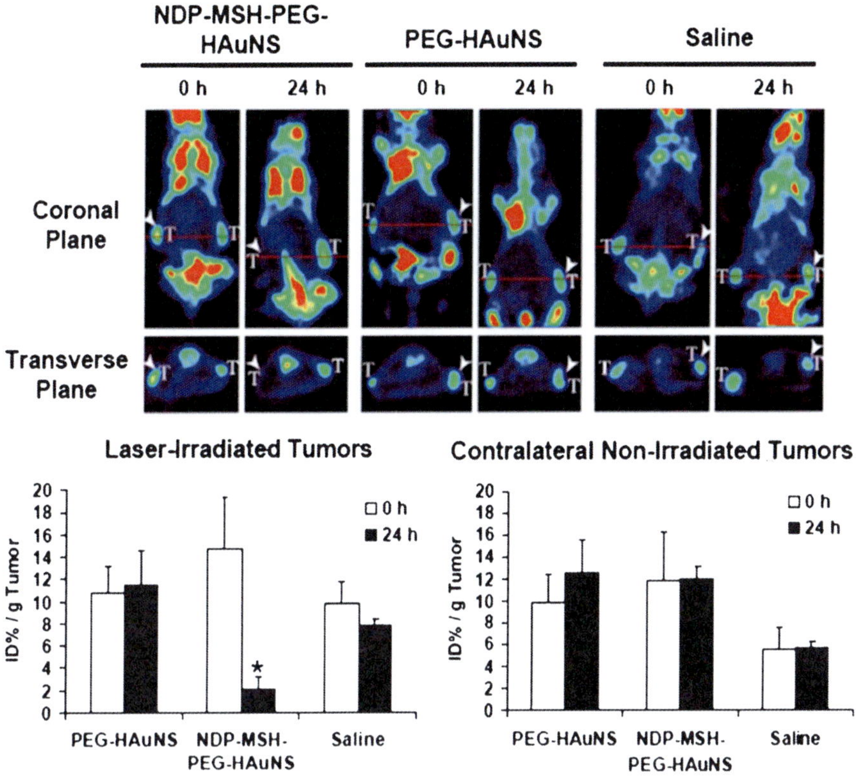

Fig. 8.9 In vivo photothermal ablation with targeted NDP–MSH–PEG–HGNs induced selective destruction of B16/F10 melanoma in nude mice. [^{18}F]fluorodeoxyglucose positron emission tomography (PET) imaging shows significantly reduced metabolic activity in tumors after photothermal ablation in mice that were pretreated with NDP–MSH–PEG–HGNs, but not in mice pretreated with PEG–HGNs or saline. [^{18}F]fluorodeoxyglucose PET was conducted before (0 h) and 24 h after near-IR laser irradiation (0.5 W/cm^2 at 808 nm for 1 min), which was commenced 4 h after i.v. injection of HGNs or saline (T for tumor). *Arrowheads* in the figure indicate tumors irradiated with near-IR light. [^{18}F]fluorodeoxyglucose uptakes (%ID/g) before and after laser treatment are shown graphically at the bottom. *Bars*, SD ($n=3$). *$P<0.01$ for %ID/g posttreatment versus %ID/g pretreatment. HAuNS or HGNs termed in this original work is the same as HGN. Reproduced with permission from [17]

HGNs (HGNs or HAuNS) have been functionalized to selectively target and destroy tissue in vivo [17, 18]. HAuNS are uniquely suited to this purpose for several reasons. Most importantly, they have a tunable absorption spectrum that extends to the NIR. This is required for in vivo work where light must penetrate tissue (which is more transparent to NIR light) in order to excite the nanomaterials [17, 18]. Specifically, Lu et al. modified the surface of HAuNS with a low molecular weight melanocyte-stimulating hormone which binds to the melanocortin type-1 receptor typically overexpressed in melanoma cells [17]. In Fig. 8.9, [^{18}F] fluorodeoxyglucose positron emission topography (PET) images are shown for mice treated with targeted HAuNS (NDP–MSH–PEG–HAuNS), nontargeted HAuNS and with saline solution. Arrowheads indicate tumors irradiated with NIR light. Only NDP–MSH–PEG–HAuNS resulted in significant reduction of [^{18}F] fluorodeoxyglucose uptake 24 h after irradiation with NIR laser light [17].

4.4 *Drug Delivery*

One of the emerging applications of metal nanostructures is drug delivery sometimes implemented in conjunction with light for remotely controlled delivery and release [152–157]. The underlying principle is based on photothermal conversion of the metal nanostructures upon light irradiation in the SPR region, usually chosen to be in the near-infrared (NIR) for deep tissue penetration. Two general approaches have been used for loading target drug molecules of interest. In the first approach, drug molecules, including small organic molecules and oligonucleotides, are attached to the surface of metal nanostructures, e.g., gold nanorods or nanoprisms, and are released upon NIR illumination [153–155, 158, 159]. In the second approach, metal nanoparticles, primarily gold, are combined with liposomes loaded with drug molecules and used to destabilize the liposomes upon NIR light absorption to release the content [156, 157, 160]. A limitation with these approaches is inevitable leakage or degradation of cargo molecules [161–165].

Very recently, a new nanostructure, or so-called nanocontainer, has been reported to be leakage-free, spectrally programmable, and capable of monitoring drug release [166]. As shown in Fig. 8.10, cargo-loaded liposomes are encapsulated with a thin shell of gold, which is biocompatible but nonbiodegradable. The spatially confined gold shell growth is directed by a layer of poly-L-histidine (PLH) on the liposome surface, which is capable of chelating metal ions. In contrast to liposome–gold nanoparticle clusters, the integrity of the nanocontainer is determined by the gold shell instead of the original liposome, thus rendering the nanocontainer stable and the embedded cargo unavailable at the ambient temperature. The nanocontainer's SPR spectra can be tuned by varying the overall particle size and shell thickness. Photothermal heating based on gold nanostructures under pulsed laser irradiation can dramatically increase *local* temperature or even melt nanoparticles [167, 168], which can be used to control content release. With this design, drug release can be spatially confined to targeted sites and help reduce toxicity to other organs. Furthermore, the release profile can be controlled for optimal dosage using laser illumination at various powers.

4.5 *Solar Energy Conversion*

Solar hydrogen generation from water splitting using photoelectrochemical (PEC) cells is one of the most promising approaches for energy production [169–171]. Semiconductor metal oxides, such as TiO_2, ZnO, and WO_3, with various film morphologies have been investigated for PEC water splitting [172–176]. Despite the fact that these metal oxides are inexpensive and chemically stable, most of them have

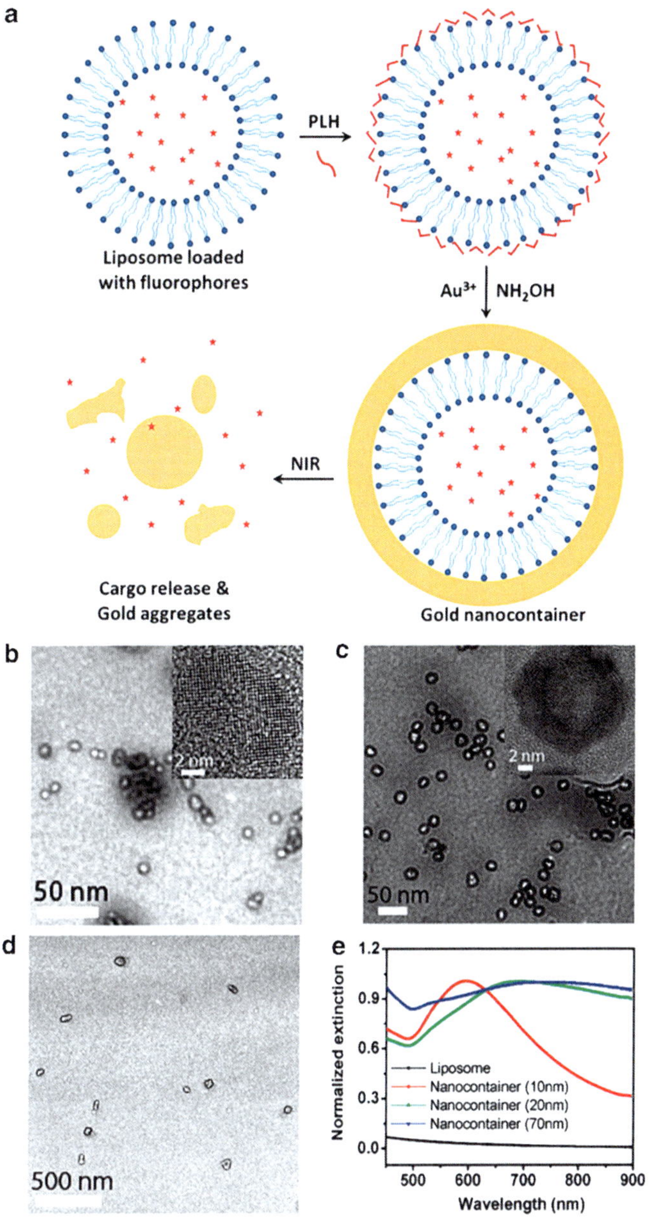

Fig. 8.10 Nanocontainer preparation and characterization. (**a**) Schematic plot of gold nanocontainer (not drawn to scale) and its cargo release mechanism. Cargo (*red stars*)-loaded liposomes are first coated with a layer of PLH for gold ion immobilization. Reduction with NH_2OH leads to formation of nanocontainers directed by the liposome scaffold. When irradiated with NIR light at their SPR, nanocontainers deform and release the contents. (**b–d**) TEM and HRTEM (*inset*) images of gold nanocontainers with mean diameters of 10 ± 3, 20 ± 5, and 70 ± 10 nm. Gold crystalline structure is seen in some small nanocontainers such as the *inset image* in panel **b**. (**e**) Corresponding extinction spectra of the three gold nanocontainers. Compared with that of the original liposomes, the intensities of the three extinction peaks centered at ~600, 690, and 750 nm are dramatically enhanced. Reproduced from [166] with permission

large bandgaps and thus limited visible light absorption. In order to enhance their visible light absorption for solar energy conversion, various approaches have been taken including dye and quantum dot (QD) sensitization as well as elemental doping. For instance, transition metals as well as nitrogen and carbon have been used as dopants to reduce their bandgap energy for enhancing visible light absorption [174, 177–181]. Alternatively, organic dyes and inorganic QDs have been utilized as visible light absorbers to sensitize the metal oxides [182, 183].

It has been reported that metal nanoparticles (NPs), such as gold and silver, can enhance the photoresponse of semiconductors in PEC. For example, enhancement of PEC performance was reported for nanostructured semiconductor–metal composite films [184–187], although the mechanism for the observed enhancement was not fully understood. Tatsuma et al. suggested that the gold or silver nanoparticles serve as light sensitizers similar to organic dyes since the photoresponse of these TiO_2–metal composite films was consistent with the absorption spectra of Au or Ag nanoparticles [187, 188]. Chen et al. also observed similar enhancement in photocurrent in ZnO nanowires incorporated with gold nanoparticles [182]. It was believed that the gold nanoparticles act as photosensitizers and photo-generated electrons can be transferred from gold nanoparticles to the connected ZnO nanowires by the force of Schottky Au–ZnO contact.

Alternatively, Kamat et al. has proposed that another role of noble metal nanoparticles in semiconductor–metal composite films, e.g., Au–TiO_2 nanoparticles, was to help separate the photo-generated charges and thus improve the interfacial charge transfer kinetics [186, 189, 190]. The noble metal nanoparticles act as electron sinks or traps to accumulate the photo-generated electrons, which can minimize charge recombination in the semiconductor films. The IPCE data showed that there is no photoresponse in the absorption region of gold nanoparticles. In this case, the gold nanoparticles did not act as light sensitizer, and increasing their amount led to a gradual decrease of the photocurrent in the PEC cells [182, 190].

Besides acting as photosensitizers for charge injection or as electron traps for facilitating charge separation, noble metal nanoparticles could also act as the light trapping or scattering agent in semiconductor–metal structures [191–195]. Metal nanoparticles of silver and gold have strong absorption in the visible region due to SPR absorption as a consequence of collective electron oscillation in response to electromagnetic field (EM) radiation [196]. Metal nanoparticles can exhibit significant Rayleigh scattering, especially when their size is comparable or larger than the wavelength of light. While the scattering can occur at any wavelength for a given sized particle, one may expect enhanced scattering as well as absorption in the SPR region. As an example, Liu et al. has very recently found a strong synergistic effect in CdSe QD-sensitized Au/TiO_2 hybrid structures in which the gold nanoparticles strongly enhance the photoresponse and IPCE CdSe QD-sensitized TiO_2 NPs [197]. This synergistic enhancement is proportional to the amount of gold nanoparticles embedded in the range of 0–5% (weight %), which cannot be simply explained by the notions of gold nanoparticles acting as photosensitizers or electron traps. Instead, it was suggested that the synergistic enhancement be due to increased light absorption of the CdSe QDs as sensitizers for TiO_2 resulting from enhanced light scattering caused by Au NPs. Thus, the specific function of the metal nanoparticles seems to depend strongly on the structural or morphological details of the nanocomposite films and devices. This observation provides useful fundamental insights for developing new nanocomposite structures with metal nanomaterials as a key component tailored for PEC and possibly other applications including photovoltaic solar cells.

5 Concluding Remarks

In this chapter, we reviewed several important aspects of metal nanomaterial synthesis with emphasis on shape and size control as well as techniques for their structural characterization. Their optical properties, especially SPR, were discussed in detail. The relationship between particle shape and SPR was examined for nanorods, hollow Au, and Ag–Au double sphere structures, as well as aggregates and other more complicated structures. The effect of surrounding media on SPR was discussed as well. In addition, we reviewed several examples of applications of optical properties of metal nanomaterials that are currently being explored as active areas of research, such as SPR spectroscopy and localized SPR-based sensors, SERS, photothermal imaging and photothermal ablation therapy (PTA), drug delivery, and solar energy conversion. These examples illustrate the diversity and usefulness of metal nanostructures in emerging technologies.

Despite remarkable progress in synthesis and understanding of metal nanostructures, there is significant room for further research to advance the field. Challenges and future directions in the study of metal nanomaterials include better control and characterization of shape, size, and surface at the atomic scale, the study of single nanoparticles to remove inhomogeneities in size, shape, and surface properties for better understanding of fundamental properties, new and expanded applications in energy and medicine, and better understanding and control of SERS substrates for surface-enhanced applications. Combined experiments and theoretical or simulation-based approaches could be the key to future developments and advances in metal nanomaterials and their applications.

Acknowledgment We are grateful to financial support from the US NSF, US DOE, US DoD, and NASA UARC.

References

1. H. Z. Cummins, S. J. Williams (1983) *Light and Color in Nature and Art*, John Wiley and Sons.
2. M. Faraday, (1857), On the color of colloidal gold, Phil. Trans. R. Soc. London. *147*, 145–181.
3. M. Faraday, (1857), The Bakerian lecture: experimental relations of gold (and other metals) to light, Philosophical Transactions of the Royal Society of London. *147*, 145–181.
4. G. Mie, (1908), Optical Properties of Colloidal Gold Solutions, Ann. Phys. *25*, 329.

5. K. L. Kelly, E. Coronado, L. L. Zhao, G. C. Schatz, (2003), The optical properties of metal nanoparticles: the influence of size, shape, and dielectric environment, J. Phys. Chem. B. *107*, 668–677.
6. A. M. Schwartzberg, T. Y. Olson, C. E. Talley, J. Z. Zhang, (2006), Synthesis, Characterization, and Tunable Optical Properties of Hollow Gold Nanospheres†, J. Phys. Chem. B. *110*, 19935–19944.
7. B. P. Rand, P. Peumans, S. R. Forrest, (2004), Long-range absorption enhancement in organic tandem thin-film solar cells containing silver nanoclusters, Journal Of Applied Physics. *96*, 7519–7526.
8. T. Mitsudome, A. Noujima, T. Mizugaki, K. Jitsukawa, K. Kaneda, (2009), Supported gold nanoparticle catalyst for the selective oxidation of silanes to silanols in water, Chemical Communications. 5302–5304.
9. H. Hartshorn, C. J. Pursell, B. D. Chandler, (2009), Adsorption of CO on Supported Gold Nanoparticle Catalysts: A Comparative Study, Journal Of Physical Chemistry C. *113*, 10718–10725.
10. T. Mitsudome, Y. Mikami, H. Mori, S. Arita, T. Mizugaki, K. Jitsukawa, K. Kaneda, (2009), Supported silver nanoparticle catalyst for selective hydration of nitriles to amides in water, Chemical Communications. 3258–3260.
11. L. Guczi, G. Peto, A. Beck, K. Frey, O. Geszti, G. Molnar, C. Daroczi, (2003), Gold nanoparticles deposited on SiO2/Si(100): Correlation between size, electron structure, and activity in CO oxidation, Journal Of The American Chemical Society. *125*, 4332–4337.
12. J. R. Adleman, D. A. Boyd, D. G. Goodwin, D. Psaltis, (2009), Heterogenous Catalysis Mediated by Plasmon Heating, Nano Letters. *9*, 4417–4423.
13. K. Kneipp, H. Kneipp, I. Itzkan, R. R. Dasari, M. S. Feld, (1999), Ultrasensitive chemical analysis by Raman spectroscopy, Chem. Rev. *99*, 2957–2976.
14. C. D. Geddes, J. R. Lakowicz, (2002), Metal-enhanced fluorescence, Journal Of Fluorescence. *12*, 121–129.
15. C. J. Murphy, A. M. Gole, S. E. Hunyadi, J. W. Stone, P. N. Sisco, A. Alkilany, B. E. Kinard, P. Hankins, (2008), Chemical sensing and imaging with metallic nanorods, Chemical Communications. *2008*, 544–557.
16. W. S. Chang, J. W. Ha, L. S. Slaughter, S. Link, (2010), Plasmonic nanorod absorbers as orientation sensors, Proceedings Of The National Academy Of Sciences Of The United States Of America. *107*, 2781–2786.
17. W. Lu, C. Xiong, G. Zhang, Q. Huang, R. Zhang, J. Z. Zhang, C. Li, (2009), Targeted Photothermal Ablation of Murine Melanomas with Melanocyte-Stimulating Hormone Analog–Conjugated Hollow Gold Nanospheres, Clinical Cancer Research. *15*, 876.
18. M. P. Melancon, W. Lu, Z. Yang, R. Zhang, Z. Cheng, A. M. Elliot, J. Stafford, T. Olson, J. Z. Zhang, C. Li, (2008), In vitro and in vivo targeting of hollow gold nanoshells directed at epidermal growth factor receptor for photothermal ablation therapy, Molecular cancer therapeutics. *7*, 1730.
19. J. Turkevitch, P. C. Stevenson, (1951), Hillier,"A study on the nucleation and growth processes in the synthesis of colloidal gold", J. Discuss. Faraday Soc. *11*, 55–75.
20. G. Frens, (1973), Controlled nucleation for the regulation of the particle size in monodisperse gold solutions, Nat. Phys. Sci. *241*, 20–22.
21. C. Burda, X. Chen, R. Narayanan, M. A. El-Sayed, (2005), Chemistry and properties of nanocrystals of different shapes, Chem. Rev. *105*, 1025–1102.
22. M. Brust, J. Fink, D. Bethell, D. J. Schiffrin, C. Kiely, (1995), Synthesis And Reactions Of Functionalized Gold Nanoparticles, Journal Of The Chemical Society-Chemical Communications. 1655–1656.
23. M. Brust, M. Walker, D. Bethell, D. J. Schiffrin, R. Whyman, (1994), Synthesis Of Thiol-Derivatized Gold Nanoparticles In A 2-Phase Liquid-Liquid System, Journal Of The Chemical Society-Chemical Communications. 801–802.
24. D. V. Leff, P. C. Ohara, J. R. Heath, W. M. Gelbart, (1995), Thermodynamic Control of Gold Nanocrystal Size: Experiment and Theory, The Journal of Physical Chemistry. *99*, 7036.

25. K. Sun, J. Qiu, J. Liu, Y. Miao, (2009), Preparation and characterization of gold nanoparticles using ascorbic acid as reducing agent in reverse micelles, Journal of Materials Science. *44*, 754–758.
26. A. B. Smetana, J. S. Wang, J. Boeckl, G. J. Brown, C. M. Wai, (2007), Fine-tuning size of gold nanoparticles by cooling during reverse micelle synthesis, Langmuir. *23*, 10429–10432.
27. A. P. Herrera, O. Resto, J. G. Briano, C. Rinaldi, (2005), Synthesis and agglomeration of gold nanoparticles in reverse micelles, Nanotechnology. *16*, S618–S625.
28. P. C. Lee, D. Meisel, (1982), Adsorption and surface-enhanced Raman of dyes on silver and gold sols, The Journal of Physical Chemistry. *86*, 3391.
29. J. E. Millstone, S. J. Hurst, G. S. Metraux, J. I. Cutler, C. A. Mirkin, (2009), Colloidal Gold and Silver Triangular Nanoprisms, Small. *5*, 646–664.
30. W. Benjamin, S. Yugang, M. Brian, X. Younan, (2005), Shape-Controlled Synthesis of Metal Nanostructures: The Case of Silver, Chemistry - A European Journal. *11*, 454–463.
31. Y. Sun, B. Mayers, T. Herricks, Y. Xia, (2003), Polyol synthesis of uniform silver nanowires: a plausible growth mechanism and the supporting evidence, Nano Letters. *3*, 955–960.
32. Y. G. Sun, Y. N. Xia, (2002), Shape-controlled synthesis of gold and silver nanoparticles, Science. *298*, 2176–2179.
33. B. J. Wiley, Y. J. Xiong, Z. Y. Li, Y. D. Yin, Y. A. Xia, (2006), Right bipyramids of silver: A new shape derived from single twinned seeds, Nano Letters. *6*, 765–768.
34. Y. Xia, Y. J. Xiong, B. Lim, S. E. Skrabalak, (2009), Shape-Controlled Synthesis of Metal Nanocrystals: Simple Chemistry Meets Complex Physics? Angewandte Chemie-International Edition. *48*, 60–103.
35. C. J. Murphy, T. K. San, A. M. Gole, C. J. Orendorff, J. X. Gao, L. Gou, S. E. Hunyadi, T. Li, (2005), Anisotropic metal nanoparticles: Synthesis, assembly, and optical applications, Journal Of Physical Chemistry B. *109*, 13857–13870.
36. T. K. Sau, C. J. Murphy, (2004), Seeded high yield synthesis of short Au nanorods in aqueous solution, Langmuir. *20*, 6414–6420.
37. S. H. Im, Y. T. Lee, B. Wiley, Y. N. Xia, (2005), Large-scale synthesis of silver nanocubes: The role of HCl in promoting cube perfection and monodispersity, Angewandte Chemie-International Edition. *44*, 2154–2157.
38. F. Fievet, J. P. Lagier, B. Blin, B. Beaudoin, M. Figlarz, (1989), Homogeneous and heterogeneous nucleations in the polyol process for the preparation of micron and submicron size metal particles, Solid State Ionics. *32*, 198–205.
39. Y. Sun, Y. Yin, B. T. Mayers, T. Herricks, Y. Xia, (2002), Uniform Silver Nanowires Synthesis by Reducing AgNO3 with Ethylene Glycol in the Presence of Seeds and Poly(Vinyl Pyrrolidone), Chemistry of Materials. *14*, 4736.
40. Y. G. Sun, B. Gates, B. Mayers, Y. N. Xia, (2002), Crystalline silver nanowires by soft solution processing, Nano Letters. *2*, 165–168.
41. S. E. Skrabalak, L. Au, X. D. Li, Y. Xia, (2007), Facile synthesis of Ag nanocubes and Au nanocages, Nature Protocols. *2*, 2182–2190.
42. Y. G. Sun, B. Mayers, Y. N. Xia, (2003), Metal nanostructures with hollow interiors, Advanced Materials. *15*, 641–646.
43. D. Seo, H. Song, (2009), Asymmetric Hollow Nanorod Formation through a Partial Galvanic Replacement Reaction, Journal Of The American Chemical Society. *131*, 18210–18211.
44. H.-P. Liang, L.-J. Wan, C.-L. Bai, L. Jiang, (2005), Gold Hollow Nanospheres: Tunable Surface Plasmon Resonance Controlled by Interior-Cavity Sizes, The Journal of Physical Chemistry B. *109*, 7795.
45. T. Y. Olson, A. M. Schwartzberg, C. A. Orme, C. E. Talley, B. O'Connell, J. Z. Zhang, (2008), Hollow gold-silver double-shell nanospheres: Structure, optical absorption, and surface-enhanced Raman scattering, Journal Of Physical Chemistry C. *112*, 6319–6329.
46. D. B. Williams, C. B. Carter (1996) *The Transmission Electron Microscope: Basics*, Plenum Press, New York.
47. J. Prikulis, H. Xu, L. Gunnarsson, M. Käll, H. Olin, (2002), Phase-sensitive near-field imaging of metal nanoparticles, Journal of Applied Physics. *92*, 6211.

48. S. Benrezzak, P. M. Adam, J. L. Bijeon, P. Royer, (2001), Observation of nanometric metallic particles with an apertureless scanning near-field optical microscope, Surface science. *491*.
49. R. Hillenbrand, F. Keilmann, (2001), Optical oscillation modes of plasmon particles observed in direct space by phase-contrast near-field microscopy, Applied Physics B: Lasers and Optics. *73*, 239–243.
50. P. Russell, D. Batchelor, J. Thornton, (2004), SEM and AFM: complementary techniques for high resolution surface investigations, Veeco Instruments Inc., AN46, Rev A. *1*.
51. H. J. Güntherodt, D. Anselmetti, E. Meyer (1995) *Forces in scanning probe methods*, Kluwer Academic in cooperation with NATO Scientific Affairs Division.
52. R. Wiesendanger, H. J. Güntherodt (1993) *Scanning tunneling microscopy III: theory of STM and related scanning probe methods*, Springer-Verlag.
53. T. L. Barr (1994) *Modern ESCA: The Principles and Practice of X-ray Photoelectron Spectroscopy Vol. 384*, CRC Press, New York.
54. D. C. Koningsberger, R. Prins (1988) *X-Ray Absorption: Principles, Applications, Techniques of EXAFS, SEXAFS and XANES*, Wiley-Interscience, New York.
55. M. A. El-Sayed, (2001), Some interesting properties of metals confined in time and nanometer space of different shapes, Acc. Chem. Res. *34*, 257–264.
56. C. F. Bohren, (1983), How can a particle absorb more than the light incident on it? Am. J. Phys. *51*, 323–327.
57. M. C. Daniel, D. Astruc, (2004), Gold nanoparticles: assembly, supramolecular chemistry, quantum-size-related properties, and applications toward biology, catalysis, and nanotechnology, Chemical reviews. *104*, 293–346.
58. J. A. Creighton, D. G. Eadon, (1991), Ultraviolet-visible absorption spectra of the colloidal elements, J Chem Soc Faraday Trans. *87*, 3.
59. J. RodrÃguez-FernÃ¡ndez, J. PÃ©rez-Juste, F. J. GarcÃa de Abajo, L. M. Liz-MarzÃ¡n, (2006), Seeded Growth of Submicron Au Colloids with Quadrupole Plasmon Resonance Modes, Langmuir. *22*, 7007.
60. S. Link, M. A. El-Sayed, (1999), Size and temperature dependence of the plasmon absorption of colloidal gold nanoparticles, J. Phys. Chem. B. *103*, 4212–4217.
61. S. Link, M. A. El-Sayed, (1999), Spectral properties and relaxation dynamics of surface plasmon electronic oscillations in gold and silver nanodots and nanorods, Journal Of Physical Chemistry B. *103*, 8410–8426.
62. S.-S. Chang, C.-W. Shih, C.-D. Chen, W.-C. Lai, C. R. C. Wang, (1998), The Shape Transition of Gold Nanorods, Langmuir. *15*, 701.
63. Yu, S.-S. Chang, C.-L. Lee, C. R. C. Wang, (1997), Gold Nanorods: Electrochemical Synthesis and Optical Properties, The Journal of Physical Chemistry B. *101*, 6661.
64. S. Link, M. B. Mohamed, M. A. El-Sayed, (1999), Simulation of the optical absorption spectra of gold nanorods as a function of their aspect ratio and the effect of the medium dielectric constant, Journal Of Physical Chemistry B. *103*, 3073–3077.
65. A. L. Gonzalez, J. A. Reyes-Esqueda, C. Noguez, (2008), Optical properties of elongated noble metal nanoparticles, Journal Of Physical Chemistry C. *112*, 7356–7362.
66. C. Pecharroman, J. Perez-Juste, G. Mata-Osoro, L. M. Liz-Marzan, P. Mulvaney, (2008), Redshift of surface plasmon modes of small gold rods due to their atomic roughness and end-cap geometry, Physical Review B. *77*.
67. K.-S. Lee, M. A. El-Sayed, (2006), Gold and Silver Nanoparticles in Sensing and Imaging: Sensitivity of Plasmon Response to Size, Shape, and Metal Composition, The Journal of Physical Chemistry B. *110*, 19220.
68. N. R. Jana, L. Gearheart, S. O. Obare, C. J. Murphy, (2002), Anisotropic chemical reactivity of gold spheroids and nanorods, Langmuir. *18*, 922–927.
69. K. K. Caswell, J. N. Wilson, U. H. F. Bunz, C. J. Murphy, (2003), Preferential end-to-end assembly of gold nanorods by biotin-streptavidin connectors, Journal Of The American Chemical Society. *125*, 13914–13915.
70. J. X. Gao, C. M. Bender, C. J. Murphy, (2003), Dependence of the gold nanorod aspect ratio on the nature of the directing surfactant in aqueous solution, Langmuir. *19*, 9065–9070.

71. B. D. Busbee, S. O. Obare, C. J. Murphy, (2003), An improved synthesis of high-aspect-ratio gold nanorods, Advanced Materials. *15*, 414–416.
72. C. J. Johnson, E. Dujardin, S. A. Davis, C. J. Murphy, S. Mann, (2002), Growth and form of gold nanorods prepared by seed-mediated, surfactant-directed synthesis, Journal Of Materials Chemistry. *12*, 1765–1770.
73. C. J. Orendorff, P. L. Hankins, C. J. Murphy, (2005), pH-triggered assembly of gold nanorods, Langmuir. *21*, 2022–2026.
74. T. K. Sau, C. J. Murphy, (2004), Room temperature, high-yield synthesis of multiple shapes of gold nanoparticles in aqueous solution, Journal Of The American Chemical Society. *126*, 8648–8649.
75. Z. Q. Li, J. Tao, X. M. Lu, Y. M. Zhu, Y. N. Xia, (2008), Facile synthesis of ultrathin Au nanorods by aging the AuCl(oleylamine) complex with amorphous Fe nanoparticles in chloroform, Nano Letters. *8*, 3052–3055.
76. B. P. Khanal, E. R. Zubarev, (2008), Purification of high aspect ratio gold nanorods: Complete removal of platelets, Journal Of The American Chemical Society. *130*, 12634–12635.
77. F. Kim, K. Sohn, J. Wu, J. Huang, (2008), Chemical Synthesis of Gold Nanowires in Acidic Solutions, Journal of the American Chemical Society. *130*, 14442.
78. Y. G. Sun, Y. D. Yin, B. T. Mayers, T. Herricks, Y. N. Xia, (2002), Uniform silver nanowires synthesis by reducing AgNO3 with ethylene glycol in the presence of seeds and poly(vinyl pyrrolidone), Chemistry Of Materials. *14*, 4736–4745.
79. K. K. Caswell, C. M. Bender, C. J. Murphy, (2003), Seedless, surfactantless wet chemical synthesis of silver nanowires, Nano Letters. *3*, 667–669.
80. N. R. Jana, L. Gearheart, C. J. Murphy, (2001), Wet chemical synthesis of silver nanorods and nanowires of controllable aspect ratio, Chemical Communications. 617–618.
81. I. O. Sosa, C. Noguez, R. G. Barrera, (2003), Optical properties of metal nanoparticles with arbitrary shapes, Journal Of Physical Chemistry B. *107*, 6269–6275.
82. J. P. Kottmann, O. J. F. Martin, D. R. Smith, S. Schultz, (2001), Dramatic localized electromagnetic enhancement in plasmon resonant nanowires, Chemical Physics Letters. *341*, 1–6.
83. R. Jin, G. Wu, Z. Li, C. A. Mirkin, G. C. Schatz, (1997), What Controls the Melting Properties of DNA-Linked Gold Nanoparticle Assemblies? Science. *277*, 1078.
84. S. Y. Lin, C. H. Chen, M. C. Lin, H. F. Hsu, (2005), A cooperative effect of bifunctionalized nanoparticles on recognition: Sensing alkali ions by crown and carboxylate moieties in aqueous media, Analytical Chemistry. *77*, 4821–4828.
85. M. S. H. C. A. M. Jae-Seung Lee, (2007), Colorimetric Detection of Mercuric Ion (Hg^{2+}) in Aqueous Media using DNA-Functionalized Gold Nanoparticles, Angewandte Chemie International Edition. *46*, 4093–4096.
86. R. Elghanian, J. J. Storhoff, R. C. Mucic, R. L. Letsinger, C. A. Mirkin, (1997), Selective Colorimetric Detection of Polynucleotides Based on the Distance-Dependent Optical Properties of Gold Nanoparticles, Science. *277*, 1078.
87. J. J. Storhoff, R. Elghanian, R. C. Mucic, C. A. Mirkin, R. L. Letsinger, (1998), One-pot colorimetric differentiation of polynucleotides with single base imperfections using gold nanoparticle probes, Journal Of The American Chemical Society. *120*, 1959–1964.
88. A. N. Shipway, M. Lahav, R. Gabai, I. Willner, (2000), Investigations into the Electrostatically Induced Aggregation of Au Nanoparticlesâ€, Langmuir. *16*, 8789.
89. T. J. Norman, C. D. Grant, D. Magana, J. Z. Zhang, J. Liu, D. L. Cao, F. Bridges, A. Van Buuren, (2002), Near infrared optical absorption of gold nanoparticle aggregates, Journal Of Physical Chemistry B. *106*, 7005–7012.
90. T. J. Norman, C. D. Grant, A. M. Schwartzberg, J. Z. Zhang, (2005), Structural correlations with shifts in the extended plasma resonance of gold nanoparticle aggregates, Optical Materials. *27*, 1197–1203.
91. C. D. Grant, A. M. Schwartzberg, T. J. Norman, J. Z. Zhang, (2002), Ultrafast Electronic Relaxation and Coherent Vibrational Oscillation of Strongly Coupled Gold Nanoparticle Aggregates, Journal of the American Chemical Society. *125*, 549.

92. K. H. Su, Q. H. Wei, X. Zhang, J. J. Mock, D. R. Smith, S. Schultz, (2003), Interparticle Coupling Effects on Plasmon Resonances of Nanogold Particles, Nano Letters. *3*, 1087.
93. D. P. Fromm, A. Sundaramurthy, P. J. Schuck, G. Kino, W. E. Moerner, (2004), Gap-Dependent Optical Coupling of Single "Bowtie" Nanoantennas Resonant in the Visible, Nano Letters. *4*, 957.
94. B. r. M. Reinhard, M. Siu, H. Agarwal, A. P. Alivisatos, J. Liphardt, (2005), Calibration of Dynamic Molecular Rulers Based on Plasmon Coupling between Gold Nanoparticles, Nano Letters. *5*, 2246.
95. P. K. Jain, W. Y. Huang, M. A. El-Sayed, (2007), On the universal scaling behavior of the distance decay of plasmon coupling in metal nanoparticle pairs: A plasmon ruler equation, Nano Letters. *7*, 2080–2088.
96. L. Yang, H. Wang, B. Yan, B. r. M. Reinhard, (2009), Calibration of Silver Plasmon Rulers in the 1–25 nm Separation Range: Experimental Indications of Distinct Plasmon Coupling Regimes, The Journal of Physical Chemistry C. *114*, 4901.
97. C. Graf, A. van Blaaderen, (2001), Metallodielectric Colloidal Core-Shell Particles for Photonic Applications, Langmuir. *18*, 524.
98. S. J. Oldenburg, R. D. Averitt, S. L. Westcott, N. J. Halas, (1998), Nanoengineering of optical resonances, Chemical Physics Letters. *288*, 243–247.
99. Y. G. Sun, Y. N. Xia, (2002), Increased sensitivity of surface plasmon resonance of gold nanoshells compared to that of gold solid colloids in response to environmental changes, Analytical Chemistry. *74*, 5297–5305.
100. E. Hao, S. Y. Li, R. C. Bailey, S. L. Zou, G. C. Schatz, J. T. Hupp, (2004), Optical properties of metal nanoshells, Journal Of Physical Chemistry B. *108*, 1224–1229.
101. N. R. Jana, L. Gearheart, C. J. Murphy, (2001), Evidence for Seed-Mediated Nucleation in the Chemical Reduction of Gold Salts to Gold Nanoparticles, Chemistry of Materials. *13*, 2313.
102. N. R. Jana, L. Gearheart, C. J. Murphy, (2001), Wet chemical synthesis of high aspect ratio cylindrical gold nanorods, Journal Of Physical Chemistry B. *105*, 4065–4067.
103. S. Kim, S. K. Kim, S. Park, (2009), Bimetallic Gold-Silver Nanorods Produce Multiple Surface Plasmon Bands, Journal Of The American Chemical Society. *131*, 8380–8381.
104. E. Prodan, P. Nordlander, (2003), Structural tunability of the plasmon resonances in metallic nanoshells, Nano Letters. *3*, 543–547.
105. M. A. Mahmoud, M. A. El-Sayed, (2009), Aggregation of Gold Nanoframes Reduces, Rather Than Enhances, SERS Efficiency Due to the Trade-Off of the Inter- and Intraparticle Plasmonic Fields, Nano Letters. *9*, 3025–3031.
106. V. Bastys, I. Pastoriza-Santos, B. Rodríguez-González, R. Vaisnoras, L. M. Liz-Marzán, (2006), Formation of Silver Nanoprisms with Surface Plasmons at Communication Wavelengths, Advanced Functional Materials. *16*, 766–773.
107. J. J. Mock, M. Barbic, D. R. Smith, D. A. Schultz, S. Schultz, (2002), Shape effects in plasmon resonance of individual colloidal silver nanoparticles, Journal Of Chemical Physics. *116*, 6755–6759.
108. T. Ming, W. Feng, Q. Tang, F. Wang, L. Sun, J. Wang, C. Yan, (2009), Growth of Tetrahexahedral Gold Nanocrystals with High-Index Facets, Journal of the American Chemical Society. *131*, 16350.
109. P. S. Kumar, I. Pastoriza-Santos, B. Rodriguez-Gonzalez, F. J. Garcia de Abajo, L. M. Liz-Marzan, (2008), High-yield synthesis and optical response of gold nanostars, Nanotechnology. *19*.
110. J. E. Millstone, S. Park, K. L. Shuford, L. Qin, G. C. Schatz, C. A. Mirkin, (2005), Observation of a Quadrupole Plasmon Mode for a Colloidal Solution of Gold Nanoprisms, Journal of the American Chemical Society. *127*, 5312.
111. A. L. Gonzalez, C. Noguez, (2007), Influence of morphology on the optical properties of metal nanoparticles, Journal Of Computational And Theoretical Nanoscience. *4*, 231–238.
112. C. Noguez, (2007), Surface plasmons on metal nanoparticles: The influence of shape and physical environment, Journal Of Physical Chemistry C. *111*, 3806–3819.
113. J. Rodriguez-Fernandez, C. Novo, V. Myroshnychenko, A. M. Funston, A. Sanchez-Iglesias, I. Pastoriza-Santos, J. Perez-Juste, F. J. Garcia de Abajo, L. M. Liz-Marzan, P. Mulvaney,

(2009), Spectroscopy, Imaging, and Modeling of Individual Gold Decahedra, Journal Of Physical Chemistry C. *113*, 18623–18631.
114. I. Pastoriza-Santos, A. Sanchez-Iglesias, F. J. G. de Abajo, L. M. Liz-Marzan, (2007), Environmental optical sensitivity of gold nanodecahedra, Advanced Functional Materials. *17*, 1443–1450.
115. J. N. Anker, W. P. Hall, O. Lyandres, N. C. Shah, J. Zhao, R. P. Van Duyne, (2008), Biosensing with plasmonic nanosensors, Nat Mater. *7*, 442.
116. A. J. Haes, S. Zou, G. C. Schatz, R. P. Van Duyne, (2004), Nanoscale optical biosensor: short range distance dependence of the localized surface plasmon resonance of noble metal nanoparticles, J. Phys. Chem. B. *108*, 6961–6968.
117. J. J. Mock, D. R. Smith, S. Schultz, (2003), Local refractive index dependence of plasmon resonance spectra from individual nanoparticles, Nano Letters. *3*, 485–492.
118. D. E. Charles, D. Aherne, M. Gara, D. M. Ledwith, Y. K. Gun'ko, J. M. Kelly, W. J. Blau, M. E. Brennan-Fournet, (2010), Versatile Solution Phase Triangular Silver Nanoplates for Highly Sensitive Plasmon Resonance Sensing, ACS Nano. *4*, 55–64.
119. M. Malmqvist, (1993), Biospecific Interaction Analysis Using Biosensor Technology, Nature. *361*, 186–187.
120. A. Madeira, E. Ohman, A. Nilsson, B. Sjogren, P. E. Andren, P. Svenningsson, (2009), Coupling surface plasmon resonance to mass spectrometry to discover novel protein-protein interactions, Nature Protocols. *4*, 1023–1037.
121. C. T. Campbell, G. Kim, (2007), SPR microscopy and its applications to high-throughput analyses of biomolecular binding events and their kinetics, Biomaterials. *28*, 2380–2392.
122. S. Paul, P. Vadgama, A. K. Ray, (2009), Surface plasmon resonance imaging for biosensing, Iet Nanobiotechnology. *3*, 71–80.
123. G. Raschke, S. Brogl, A. S. Susha, A. L. Rogach, T. A. Klar, J. Feldmann, B. Fieres, N. Petkov, T. Bein, A. Nichtl, K. Kurzinger, (2004), Gold nanoshells improve single nanoparticle molecular sensors, Nano Letters. *4*, 1853–1857.
124. A. J. Haes, L. Chang, W. L. Klein, R. P. Van Duyne, (2005), Detection of a biomarker for Alzheimer's disease from synthetic and clinical samples using a nanoscale optical biosensor, Journal Of The American Chemical Society. *127*, 2264–2271.
125. M. D. Malinsky, K. L. Kelly, G. C. Schatz, R. P. Van Duyne, (2001), Chain Length Dependence and Sensing Capabilities of the Localized Surface Plasmon Resonance of Silver Nanoparticles Chemically Modified with Alkanethiol Self-Assembled Monolayers, Journal of the American Chemical Society. *123*, 1471.
126. A. B. Dahlin, J. O. Tegenfeldt, F. Hook, (2006), Improving the instrumental resolution of sensors based on localized surface plasmon resonance, Analytical Chemistry. *78*, 4416–4423.
127. C. R. Yonzon, E. Jeoungf, S. L. Zou, G. C. Schatz, M. Mrksich, R. P. Van Duyne, (2004), A comparative analysis of localized and propagating surface plasmon resonance sensors: The binding of concanavalin a to a monosaccharide functionalized self-assembled monolayer, Journal Of The American Chemical Society. *126*, 12669–12676.
128. W. P. Hall, J. N. Anker, Y. Lin, J. Modica, M. Mrksich, R. P. Van Duyne, (2008), A calcium-modulated plasmonic switch, Journal Of The American Chemical Society. *130*, 5836–5837.
129. C. Wu, Q.-H. Xu, (2009), Stable and Functionable Mesoporous Silica-Coated Gold Nanorods as Sensitive Localized Surface Plasmon Resonance (LSPR) Nanosensors, Langmuir. *25*, 9441.
130. A. D. McFarland, R. P. Van Duyne, (2003), Single silver nanoparticles as real-time optical sensors with zeptomole sensitivity, Nano Letters. *3*, 1057–1062.
131. O. L. Muskens, P. Billaud, M. Broyer, N. Fatti, F. Vallee, (2008), Optical extinction spectrum of a single metal nanoparticle: Quantitative characterization of a particle and of its local environment, Physical Review B. *78*.
132. R. Elghanian, J. J. Storhoff, R. C. Mucic, R. L. Letsinger, C. A. Mirkin, (1997), Selective Colorimetric Detection of Polynucleotides Based on the Distance-Dependent Optical Properties of Gold Nanoparticles, Science. *277*, 1078–1081.
133. B. B. Schrader, D. (1995) *Infrared and Raman spectroscopy: methods and applications*, VCH: Weinheim, New York.

134. S. Nie, S. R. Emory, (1997), Probing Single Molecules and Single Nanoparticles by Surface-Enhanced Raman Scattering, Science. *275*, 1102.
135. M. Fleischmann, P. J. Hendra, A. J. McQuillan, (1974), Raman spectra of pyridine adsorbed at a silver electrode, Chemical Physics Letters. *26*, 163–166.
136. D. L. Jeanmaire, R. P. Vanduyne, (1977), Surface Raman Spectroelectrochemistry.1. Heterocyclic, Aromatic, And Aliphatic-Amines Adsorbed On Anodized Silver Electrode, Journal Of Electroanalytical Chemistry. *84*, 1–20.
137. M. G. Albrecht, J. A. Creighton, (1977), Anomalously Intense Raman-Spectra Of Pyridine At A Silver Electrode, Journal Of The American Chemical Society. *99*, 5215–5217.
138. Y. Fang, N.-H. Seong, D. D. Dlott, (2008), Measurement of the Distribution of Site Enhancements in Surface-Enhanced Raman Scattering, Science. *321*, 388–392.
139. K. Kneipp, H. Kneipp, I. Itzkan, R. R. Dasari, M. S. Feld, (1999), Surface-enhanced non-linear Raman scattering at the single-molecule level, Chemical Physics. *247*, 155.
140. J. T. Krug, G. D. Wang, S. R. Emory, S. Nie, (1999), Efficient Raman Enhancement and Intermittent Light Emission Observed in Single Gold Nanocrystals, Journal of the American Chemical Society. *121*, 9208.
141. G. Chen, Y. Wang, M. Yang, J. Xu, S. J. Goh, M. Pan, H. Chen, Measuring Ensemble-Averaged Surface-Enhanced Raman Scattering in the Hotspots of Colloidal Nanoparticle Dimers and Trimers, Journal of the American Chemical Society. *132*, 3644.
142. A. M. Schwartzberg, T. Y. Oshiro, J. Z. Zhang, T. Huser, C. E. Talley, (2006), Improving nanoprobes using surface-enhanced Raman scattering from 30-nm hollow gold particles, Analytical Chemistry. *78*, 4732–4736.
143. Y. Lu, G. L. Liu, J. Kim, Y. X. Mejia, L. P. Lee, (2004), Nanophotonic Crescent Moon Structures with Sharp Edge for Ultrasensitive Biomolecular Detection by Local Electromagnetic Field Enhancement Effect, Nano Letters. *5*, 119.
144. A. M. Schwartzberg, C. D. Grant, A. Wolcott, C. E. Talley, T. R. Huser, R. Bogomolni, J. Z. Zhang, (2004), Unique Gold Nanoparticle Aggregates as a Highly Active Surface-Enhanced Raman Scattering Substrate, The Journal of Physical Chemistry B. *108*, 19191.
145. X. M. Lu, L. Au, J. McLellan, Z. Y. Li, M. Marquez, Y. N. Xia, (2007), Fabrication of cubic nanocages and nanoframes by dealloying Au/Ag alloy nanoboxes with an aqueous etchant based on Fe(NO3)(3) or NH4OH, Nano Letters. *7*, 1764–1769.
146. Y. Zhang, C. Gu, A. M. Schwartzberg, J. Z. Zhang, (2005), Surface-enhanced Raman scattering sensor based on D-shaped fiber, Applied Physics Letters. *87*.
147. H. Yan, C. Gua, C. X. Yang, J. Liu, G. F. Jin, J. T. Zhang, L. T. Hou, Y. Yao, (2006), Hollow core photonic crystal fiber surface-enhanced Raman probe, Applied Physics Letters. *89*.
148. C. Shi, H. Yan, C. Gu, D. Ghosh, L. Seballos, S. W. Chen, J. Z. Zhang, B. Chen, (2008), A double substrate "sandwich" structure for fiber surface enhanced Raman scattering detection, Applied Physics Letters. *92*.
149. C. Shi, C. Lu, C. Gu, L. Tian, R. Newhouse, S. W. Chen, J. Z. Zhang, (2008), Inner wall coated hollow core waveguide sensor based on double substrate surface enhanced Raman scattering, Applied Physics Letters. *93*.
150. D. Boyer, P. Tamarat, A. Maali, B. Lounis, M. Orrit, (2002), Photothermal Imaging of Nanometer-Sized Metal Particles Among Scatterers, Science. *297*, 1160–1163.
151. L. Cognet, C. Tardin, D. Boyer, D. Choquet, P. Tamarat, B. Lounis, (2003), Single metallic nanoparticle imaging for protein detection in cells, Proceedings of the National Academy of Sciences. *100*, 11350.
152. W. A. Zhao, J. M. Karp, (2009), Nanoantennas heat up, Nature Materials. *8*, 453–454.
153. M. R. Jones, J. E. Millstone, D. A. Giljohann, D. S. Seferos, K. L. Young, C. A. Mirkin, (2009), Plasmonically Controlled Nucleic Acid Dehybridization with Gold Nanoprisms, Chemphyschem. *10*, 1461–1465.
154. S. E. Lee, G. L. Liu, F. Kim, L. P. Lee, (2009), Remote optical switch for localized and selective control of gene interference, Nano letters. *9*, 562.
155. G. B. Braun, A. Pallaoro, G. H. Wu, D. Missirlis, J. A. Zasadzinski, M. Tirrell, N. O. Reich, (2009), Laser-Activated Gene Silencing via Gold Nanoshell-siRNA Conjugates, ACS Nano. *3*, 2007–2015.

156. D. V. Volodkin, A. G. Skirtach, H. Mohwald, (2009), Near-IR Remote Release from Assemblies of Liposomes and Nanoparticles, Angewandte Chemie-International Edition. *48*, 1807–1809.
157. T. S. Troutman, S. J. Leung, M. Romanowski, (2009), Light-Induced Content Release from Plasmon-Resonant Liposomes, Advanced Materials. *21*, 2334–2338.
158. C. C. Chen, Y. P. Lin, C. W. Wang, H. C. Tzeng, C. H. Wu, Y. C. Chen, C. P. Chen, L. C. Chen, Y. C. Wu, (2006), DNA-gold nanorod conjugates for remote control of localized gene expression by near infrared irradiation, Journal Of The American Chemical Society. *128*, 3709–3715.
159. H. Takahashi, Y. Niidome, S. Yamada, (2005), Controlled release of plasmid DNA from gold nanorods induced by pulsed near-infrared light, Chemical communications. *2005*, 2247–2249.
160. G. H. Wu, A. Milkhailovsky, H. A. Khant, C. Fu, W. Chiu, J. A. Zasadzinski, (2008), Remotely triggered liposome release by near-infrared light absorption via hollow gold nanoshells, Journal Of The American Chemical Society. *130*, 8175–8177.
161. T. K. Jain, M. K. Reddy, M. A. Morales, D. L. Leslie-Pelecky, V. Labhasetwar, (2008), Biodistribution, clearance, and biocompatibility of iron oxide magnetic nanoparticles in rats, Molecular Pharmaceutics. *5*, 316–327.
162. L. Qi, X. Gao, (2008), Quantum Dot-Amphipol Nanocomplex for Intracellular Delivery and Realtime Imaging of siRNA, ACS Nano. *2*, 1403.
163. V. P. Torchilin, (2005), Recent advances with liposomes as pharmaceutical carriers, Nature Reviews Drug Discovery. *4*, 145–160.
164. H. Chen, S. Kim, W. He, H. Wang, P. S. Low, K. Park, J. X. Cheng, (2008), Fast release of lipophilic agents from circulating PEG-PDLLA micelles revealed by in vivo Forster resonance energy transfer imaging, Langmuir. *24*, 5213–5217.
165. Y. D. Jin, X. H. Gao, (2009), Plasmonic fluorescent quantum dots, Nat. Nanotechnol. *4*, 571–576.
166. Y. Jin, X. Gao, (2009), Spectrally Tunable Leakage-Free Gold Nanocontainers, J. Am. Chem. Soc. *131*, 17774–17776.
167. V. S. Kalambur, E. K. Longmire, J. C. Bischof, (2007), Cellular level loading and heating of superparamagnetic iron oxide nanoparticles, Langmuir. *23*, 12329–12336.
168. A. N. Volkov, C. Sevilla, L. V. Zhigilei, (2007), Numerical modeling of short pulse laser interaction with Au nanoparticle surrounded by water, Applied Surface Science. *253*, 6394–6399.
169. A. Fujishima, K. Honda, (1972), Electrochemical Photolysis Of Water At A Semiconductor Electrode, Nature. *238*, 37–38.
170. O. Khaselev, J. A. Turner, (1998), A monolithic photovoltaic-photoelectrochemical device for hydrogen production via water splitting, Science. *280*, 425.
171. T. Bak, J. Nowotny, M. Rekas, C. C. Sorrell, (2002), Photo-electrochemical hydrogen generation from water using solar energy. Materials-related aspects, International Journal of Hydrogen Energy. *27*, 991–1022.
172. C. Santato, M. Odziemkowski, M. Ulmann, J. Augustynski, (2001), Crystallographically oriented Mesoporous WO3 films: Synthesis, characterization, and applications, Journal Of The American Chemical Society. *123*, 10639–10649.
173. A. Wolcott, T. R. Kuykendall, W. Chen, S. Chen, J. Z. Zhang, (2006), Synthesis and Characterization of Ultrathin WO3 Nanodisks Utilizing Long-Chain Poly (ethylene glycol)†, J. Phys. Chem. B. *110*, 25288–25296.
174. J. H. Park, S. Kim, A. J. Bard, (2006), Novel carbon-doped TiO2 nanotube arrays with high aspect ratios for efficient solar water splitting, Nano Lett. *6*, 24–28.
175. K. S. Ahn, Y. F. Yan, S. H. Lee, T. Deutsch, J. Turner, C. E. Tracy, C. L. Perkins, M. Al-Jassim, (2007), Photoelectrochemical properties of n-incorporated ZnO films deposited by reactive RF magnetron sputtering, Journal Of The Electrochemical Society. *154*, B956-B959.
176. A. Wolcott, W. A. Smith, T. R. Kuykendall, Y. P. Zhao, J. Z. Zhang, (2009), Photoelectrochemical Water Splitting Using Dense and Aligned TiO2 Nanorod Arrays, Small. *5*, 104–111.
177. W. Choi, A. Termin, M. R. Hoffmann, (1994), The role of metal ion dopants in quantum-sized TiO2: correlation between photoreactivity and charge carrier recombination dynamics, The Journal of Physical Chemistry. *98*, 13669–13679.

178. R. Asahi, T. Morikawa, T. Ohwaki, K. Aoki, Y. Taga, (2001), Visible-light photocatalysis in nitrogen-doped titanium oxides, Science. *293*, 269.
179. G. R. Torres, T. Lindgren, J. Lu, C. G. Granqvist, S. E. Lindquist, (2004), Photoelectrochemical study of nitrogen-doped titanium dioxide for water oxidation, J. Phys. Chem. B. *108*, 5995–6003.
180. X. F. Qiu, Y. X. Zhao, C. Burda, (2007), Synthesis and characterization of nitrogen-doped group IVB visible-light-photoactive metal oxide nanoparticles, Advanced Materials. *19*, 3995–3999.
181. J. Hensel, G. M. Wang, Y. Li, J. Z. Zhang, (2010), Synergistic Effect of CdSe Quantum Dot Sensitization and Nitrogen Doping of TiO2 Nanostructures for Photoelectrochemical Solar Hydrogen Generation, Nano Letters. *10*, 478–483.
182. Z. H. Chen, Y. B. Tang, C. P. Liu, Y. H. Leung, G. D. Yuan, L. M. Chen, Y. Q. Wang, I. Bello, J. A. Zapien, W. J. Zhang, C. S. Lee, S. T. Lee, (2009), Vertically Aligned ZnO Nanorod Arrays Sensitized with Gold Nanoparticles for Schottky Barrier Photovoltaic Cells, Journal Of Physical Chemistry C. *113*, 13433–13437.
183. Y. Tak, S. J. Hong, J. S. Lee, K. Yong, (2009), Fabrication of ZnO/CdS core/shell nanowire arrays for efficient solar energy conversion, Journal Of Materials Chemistry. *19*, 5945–5951.
184. Y. Nakato, M. Shioji, H. Tsubomura, (1982), Photoeffects on the potentials of thin metal films on a n-TiO2 crystal wafer. The mechanism of semiconductor photocatalysts, Chemical Physics Letters. *90*, 453–456.
185. G. L. Zhao, H. Kozuka, T. Yoko, (1996), Photoelectrochemical properties of dye-sensitized TiO2 films containing dispersed gold metal particles prepared by sol-gel method, Journal Of The Ceramic Society Of Japan. *104*, 164–168.
186. N. Chandrasekharan, P. V. Kamat, (2000), Improving the photoelectrochemical performance of nanostructured TiO2 films by adsorption of gold nanoparticles, Journal Of Physical Chemistry B. *104*, 10851–10857.
187. Y. Tian, T. Tatsuma, (2005), Mechanisms and applications of plasmon-induced charge separation at TiO2 films loaded with gold nanoparticles, Journal Of The American Chemical Society. *127*, 7632–7637.
188. Y. Tian, T. Tatsuma, (2004), Plasmon-induced photoelectrochemistry at metal nanoparticles supported on nanoporous TiO 2, Chemical Communications. *2004*, 1810–1811.
189. A. Dawson, P. V. Kamat, (2001), Semiconductor-metal nanocomposites. Photoinduced fusion and photocatalysis of gold-capped TiO2 (TiO2/Gold) nanoparticles, Journal Of Physical Chemistry B. *105*, 960–966.
190. V. Subramanian, E. Wolf, P. V. Kamat, (2001), Semiconductor-metal composite nanostructures. To what extent do metal nanoparticles improve the photocatalytic activity of TiO2 films? Journal Of Physical Chemistry B. *105*, 11439–11446.
191. D. Derkacs, S. H. Lim, P. Matheu, W. Mar, E. T. Yu, (2006), Improved performance of amorphous silicon solar cells via scattering from surface plasmon polaritons in nearby metallic nanoparticles, Applied Physics Letters. *89*.
192. D. M. Schaadt, B. Feng, E. T. Yu, (2005), Enhanced semiconductor optical absorption via surface plasmon excitation in metal nanoparticles, Applied Physics Letters. *86*, 063106.
193. K. R. Catchpole, A. Polman, (2008), Design principles for particle plasmon enhanced solar cells, Applied Physics Letters. *93*.
194. K. Nakayama, K. Tanabe, H. A. Atwater, (2008), Plasmonic nanoparticle enhanced light absorption in GaAs solar cells, Applied Physics Letters. *93*, 121904.
195. W. Smith, S. Mao, G. H. Lu, A. Catlett, J. H. Chen, Y. P. Zhao, The effect of Ag nanoparticle loading on the photocatalytic activity of TiO2 nanorod arrays, Chemical Physics Letters. *485*, 171–175.
196. J. Z. Zhang, C. Noguez, (2008), Plasmonic optical properties and applications of metal nanostructures, Plasmonics. *3*, 127–150.
197. L. Liu, G. Wang, Y. Li, Y. Li, J. Z. Zhang, (2011), Enhanced photoresponse of CdSe quantum dot sensitized TiO2 nanoparticles by plasmonic gold nanoparticles, Nano Research. *4*, 249–258.

Chapter 9
Enhanced Optical Transmission Through Annular Aperture Arrays: Role of the Plasmonic Guided Modes

Fadi Baida and Jérôme Salvi

1 Introduction

Let us recall what the enhanced or extraordinary optical transmission (EOT) is. This phraseology was firstly used by T. W. Ebbesen's team in 1998 to qualify the far-field light transmission obtained through an array of cylindrical apertures engraved into an opaque metallic film [1–5]. It was noticed that the normalized measured transmission per aperture is very large compared to the transmission of a single aperture. Thus, the collective response of the whole structure is at the origin of this extraordinary effect. Nevertheless, this phenomenon becomes more usual for holes of big diameters compared to the illumination wavelength. So, the extraordinary character is directly linked to the ratio λ/p because the diameter of the aperture cannot exceed the period value (p).

Since the publication of T. W. Ebbesen et al., many experimental and theoretical studies were carried out in order to determine the physical origin of the extraordinary optical transmission (EOT) reported in that paper. A cylindrical aperture array drilled in opaque but thin metallic films can lead to a large transmission (see Fig. 9.1) comparing to the hollow-surface to total-surface ratio (factor 5).

This is fascinating because the sum of the intensity transmitted by each of the apertures is very small compared to the total transmission due to the whole array. Moreover, the wavelength for which the transmission peak is located corresponds exactly to the excitation of the surface plasmon wave of the unstructured interface (without apertures). Consequently, surface plasmon resonances were firstly suggested as the origin of this EOT [6–10], despite the existence of other assumptions such cavity hole resonances [11–14]. Since its first demonstration, the EOT has

F. Baida (✉) • J. Salvi
Institut FEMTO-ST, CNRS UMR 6174, Université de Franche-Comté, Besançon, France
email: fbaida@univ-fcomte.fr

C.D. Geddes (ed.), *Reviews in Plasmonics 2010*, Reviews in Plasmonics,
DOI 10.1007/978-1-4614-0884-0_9, © Springer Science+Business Media, LLC 2012

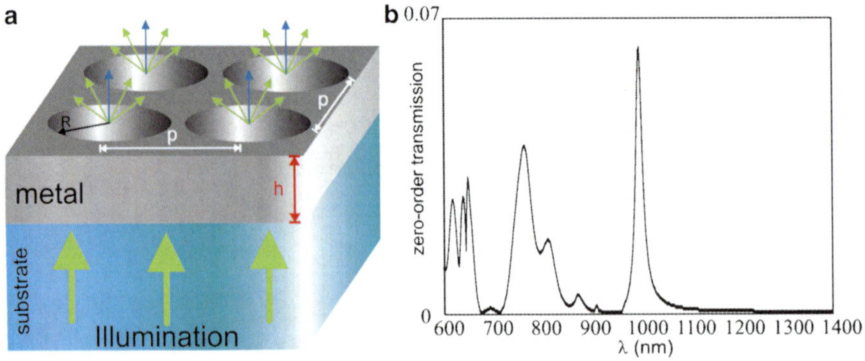

Fig. 9.1 (**a**) Schematic of a hole aperture array and (**b**) the corresponding transmission spectrum. The metal is gold and the geometrical parameters of the structure are: $p = 600$ nm, $h = 250$ nm, $R = 100$ nm: both substrate and superstrate are dielectrics with $\varepsilon = 2.34$. Dispersion of gold is described by a Drude model with a plasma frequency $\omega_p = 1.236 \times 1{,}016$ rad/s and a dumping term $\Gamma = 1.3 \times 10^{14}$ rad/s

attracted much interest from not only experimenters but also theoreticians because such structures require the development of new methods or, at least, the optimization and the improvement of existing numerical methods. The first calculation of the transmission through a 2D grating was performed by Popov et al. [14] by means of a Fourier modal method that was extended to crossed gratings case. In [15] and [9], the transmission spectra of a 2D hole grating are calculated by using a modal expansion of the fields. In those papers, a very simple and efficient minimal model is developed which allows the authors to conclude that "the holes behave like subwavelength cavities for the evanescent waves coupling the surface plasmon on either sides of the films." By using a differential method, Salomon et al. [16] numerically studied the transmission of a very thin (20 nm) 2D hole grating. References [11–13] attribute the transmission enhancement to cavity resonances into the holes. In [17], Vigoureux analyses the Ebbesen experiment in terms of short range diffraction of evanescent waves. Moreover, it is clear that the EOT phenomenon can be compared to what is obtained with frequency selective surface (FSS) components in the spectral region from near-infrared to microwave wavelengths ([18, 19] and references therein). Those components are designed for a broad domain of applications such as Fabry–Perot interferometers, filters, couplers for laser cavity output, or simply as polarizers. Their spectral responses are directly connected to their geometrical parameters, i.e., their thickness, size, period, and especially the shape of the aperture.

A review paper published by Garcia de Abajo [20] in 2007 is a very good discussion on the EOT through cylindrical aperture arrays. Catrysse and Fan also published an interesting paper that clearly elucidates the mechanism of the EOT in the case of Ebbesen's experiments in term of different regimes [21]. They give an explanation that incorporates both surface plasmon waves with propagation plasmonic guided modes. This seems to be a compromise between the two major explanations

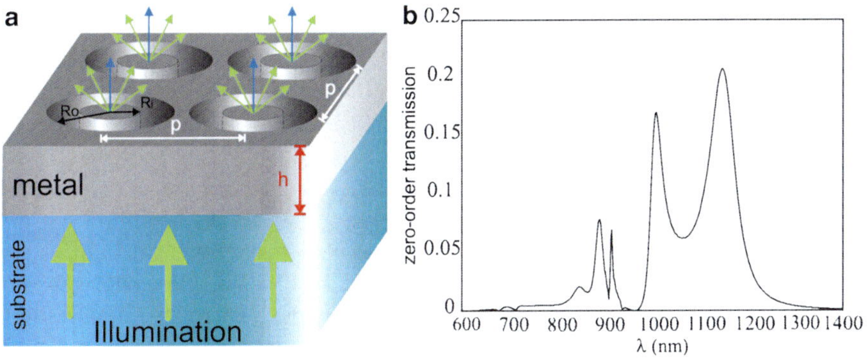

Fig. 9.2 (**a**) Schematic of the AAA structure and (**b**) the corresponding transmission spectrum. As in Fig. 9.1, the metal is gold and the geometrical parameters of the structure are: $p = 600$ nm, $h = 250$ nm, $R_0 = 100$ nm, $R_i = 50$ nm, and both substrate and superstrate are dielectrics ($\varepsilon = 2.34$)

of the EOT phenomenon and will, maybe, reconcile the differences that exist between those who affectionate guided modes and those who prefer the surface waves. More recently, Gordon et al. published a general review paper [22] on EOT through hole-arrays where a lot of applications are referenced and discussed.

In the context of EOT, the challenge is to conceive periodic structures of small apertures made in metallic opaque plates that allow light transmission for wavelength much larger than the array period. A simple idea is then emerged: it is to obtain and improve the EOT based on a guided mode that could be transmitted through the openings independently. The periodicity of openings plays only the role of phase matching allowing the control of the energy of the quantified diffracted orders. Thus, the concept of waveguide becomes essential for the design of these structures. Therefore, the first structure that was proposed [23, 24] is the annular aperture arrays (AAAs) because a coaxial waveguide supports a guided mode without cutoff, namely, the TEM mode even if this latter can only be excited under specific conditions [25]. Nevertheless, the AAA has become the first structure allowing an EOT five times higher than that obtained with cylindrical holes [24, 26–30] even if the comparison is questionable. Note that AAA structures made in perfectly electric conductors (PEC) have been studied since 1988 by Roberts and Mc Pherdran [31] but outside the context of the EOT. Experimental demonstration of EOT through AAA structures was first achieved by W. J. Fan since 2005 [32] in the infrared domain followed by other several studies [33–52] performed over a large spectral range from visible to microwave through Terahertz domain.

The first numerical demonstration of EOT through AAA structure in comparison with cylindrical aperture array was published in [24] and is given in Fig. 9.2. The considered annular apertures are simply obtained by adding small metallic cylinders inside the cylindrical apertures of Fig. 9.1. So, decreasing the hollow area leads to an increased transmission for larger values of the wavelength. Consequently, this transmission deserves to be qualified by "Super Enhanced Optical Transmission."

2 Why Annular Aperture Arrays?

To understand the principle of the EOT through AAA structure, let us consider the Fig. 9.3 where the cutoff wavelengths of some specific waveguides are given as a function of their geometrical parameters.

For a square or rectangular waveguide (widths a and b with $a > b$) in a PEC, the fundamental mode with the largest cutoff wavelength is the TE_{10} with $\lambda_c = 2a$.

Replacing a rectangular cross-section by a circular one with the same width (diameter $a = 2R$), a smaller cutoff is obtained for the fundamental mode: $\lambda_c \approx 1.7a$ for the TE_{11} mode. If we consider a coaxial waveguide, a very interesting result is obtained. The cutoff wavelength of all the modes, except two, depends on the difference between the outer and inner radii, i.e., the wavelength cutoff are very small. But the TEM_0 mode has no cutoff $\lambda_c \rightarrow \infty$ and the TE_{m1} mode has a cutoff proportional to the sum of the radii: $\lambda_c^{TE_{11}} \approx \pi(R_o + R_i)$, where R_o and R_i are the outer and inner radii, respectively. The maximum that can be reached is $\lambda_c^{TE_{11}} \approx 2\pi R_o$ which is obtained for a very small gap between the two radii.

Consequently, only the TE_{11} mode of the coaxial waveguide can allow a large value of the cutoff wavelength that is proportional to the sum of the coax radii. Nevertheless, noble metals such as gold, silver, or aluminum are not perfect conductors in the visible domain and their dispersion should be taken into account if we

Structures	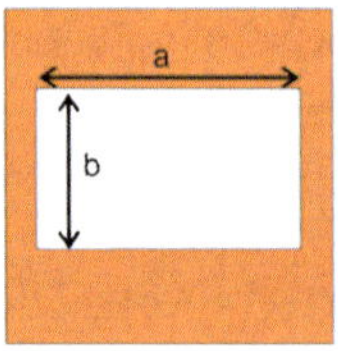	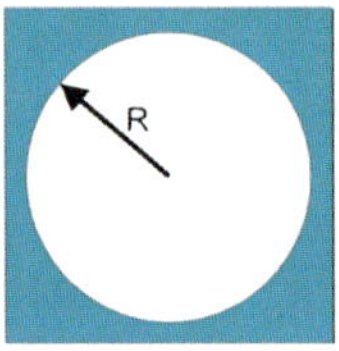	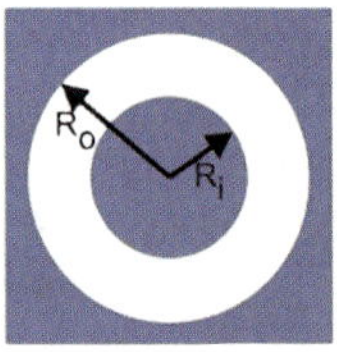
First modes	$\lambda^c_{TM_{11}} = \frac{2ab}{\sqrt{a^2+b^2}}$ $\lambda^c_{TE_{10}} = 2a$	$\lambda^c_{TE_{11}} = \frac{2\pi R}{1.841}$ $\lambda^c_{TM_{01}} = \frac{2\pi R}{2.405}$	$\lambda^c_{TE_{m1}} \cong \frac{(R_o+R_i)\pi}{m}$ $\lambda^c_{TE_{mn}} \cong \frac{2(R_o - R_i)}{n-1}$ $\lambda^c_{TM_{mn}} \cong \frac{2(R_o - R_i)}{n}$
Examples	$a = b = 250nm$ $\lambda^c_{TE_{10}} = 500nm$	$R = 125nm$ $\lambda^c_{TE_{11}} = 426.6nm$	$R_o = 125nm$, $R_i = 75nm$ $\lambda^c_{TE_{11}} = 628.3nm$

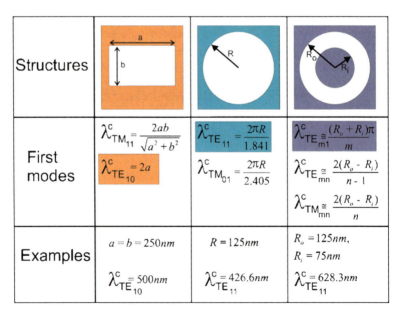

Fig. 9.3 Cutoff wavelengths of the first guided modes for three different kinds of waveguide made in perfectly electric conductor (PEC)

attempt to guide light through them. Therefore, metallic waveguides are prohibited in the optical domain because of the losses preventing efficient propagation over large distances. Moreover, in this case, *the guided modes become hybrid rather than pure TE or TM ones. In fact, longitudinal components of the electromagnetic field appear but are small compared to the transverse ones. In spite of this, we will keep the same notation (TE, TM, or TEM) in order to have a simplified nomenclature.* Consequently, it is important to determine the effect of the metal nature on the guided modes.

2.1 Properties of a Metallic Subwavelength Coaxial Waveguide in the Visible Range

Because of the metal dispersion, the guided mode's dispersion relations that express the propagation constants of the guided modes versus the geometrical and physical parameters of the waveguide are no more analytical. Recently, Catarysse and Fan established an analytical parametrical dispersion relation that need to be numerically solved allowing the determination of the modal propagation constants together with the propagation losses in the case of nano-coaxial waveguides made in real metal [53]. In our case, an N-Order BOR-FDTD (body-of-revolution-finite difference time domain) code was used to determine the dispersion diagrams of these modes. The BOR-FDTD algorithm used by this code is based on the discretization of the Maxwell equations after expressing them in cylindrical coordinates [54–56]. The *N*-order FDTD method [57, 58] is then adapted to the cylindrical symmetry and the final algorithm allows the determination of the dispersion curves for an azimuthal family of guided modes. Indeed, each mode is characterized by two numbers (m, n) where m is the azimuthal quantum number and n is the radial one. Consequently, the TE_{11} mode is characterized by $m=1$ and $n=1$ and its dispersion relation is obtained by running the N-Order BOR-FDTD code with $m=1$.

To confirm the results of Fig. 9.3 in the case of a real metal, we determine in Fig. 9.4 the cutoff wavelengths of the first guided modes for both cylindrical and coaxial waveguides made in silver and compare them to the PEC case.

As expected, the cutoff wavelength of the TE_{11} mode of the silver waveguides exhibits a red shift with regard to the PEC case. In addition, the "gain" of the coaxial structure compared to the cylindrical one is accentuated by the real nature of the metal. In fact, the difference between the two cutoffs is of only 130 nm in the case of PEC while it reaches 200 nm in the silver case.

In order to justify our interest about the TE_{11} mode, we present in Fig. 9.5 the intensity light distributions of the first three modes of a coaxial waveguide made in silver. It is clear from this figure that only the TE_{11} mode exhibits a nonzero overlap with a linearly polarized plane wave because both of them have an $m=1$ azimuthal dependence.

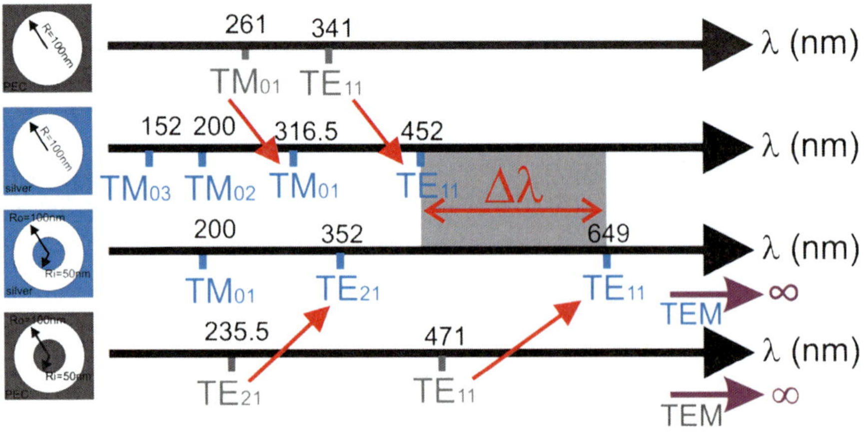

Fig. 9.4 Nomenclature of the guided modes cutoff wavelengths in the case of cylindrical and coaxial nano-waveguides made in perfect conductor (*gray*) and in silver (*blue*). The radius of the cylindrical section is set to $R=100$ nm and corresponds to the outer radius of the coaxial waveguide while the inner radius of this latter is set to $R_i=50$ nm. $\Delta\lambda$ is the difference between the two cutoff positions of the two fundamental modes. This value, estimated here as $2R$, can be used to quantify the gain due to the coaxial structure

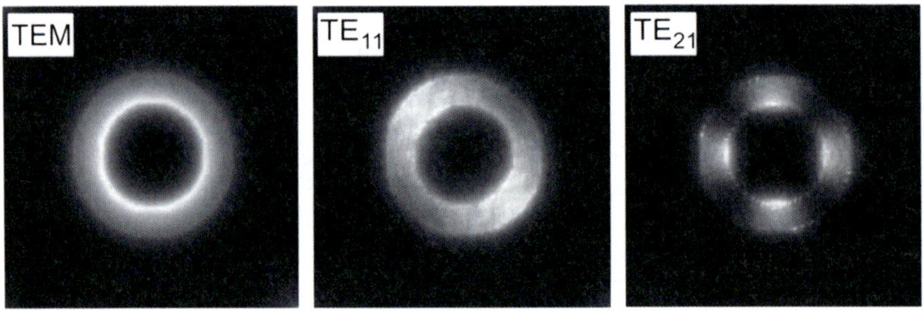

Fig. 9.5 Intensity light distribution of the first three-guided modes of a silver coaxial waveguide ($R_o=75$ nm and $R_i=50$ nm)

2.2 Modal Study

Let us now take a closer look at the effect of the geometrical parameters on the cutoff wavelength of the TE_{11} mode. Figure 9.6 shows the variations of this latter mode as a function of the outer radius for two coaxial waveguides having the same radii but made in PEC and in silver, respectively. In both cases, the inner radius is fixed to $R_i=50$ nm and the outer one R_o varies from 50 to 400 nm. As suspected, a red shift (called α in Fig. 9.6) appears in the case of real metal. This can be basically explained by the penetration of the electromagnetic field inside the metal and then corresponding to a virtual larger waveguide.

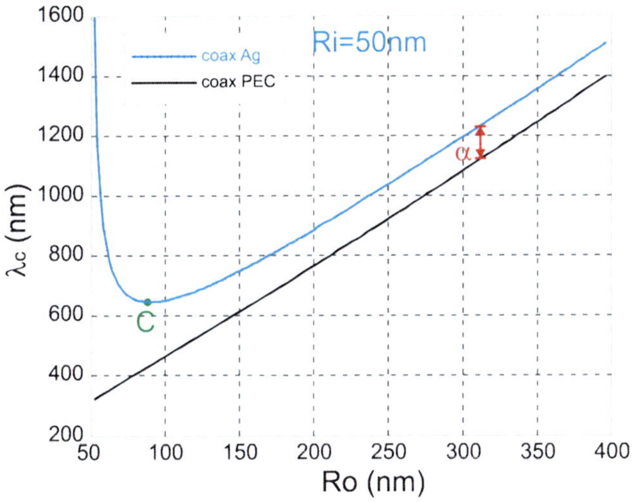

Fig. 9.6 Cutoff wavelengths of the TE_{11} mode of both PEC and silver coaxial waveguides. α denotes the red shift introduced by the metal nature in the case of a relatively large dielectric gap

Nevertheless, the most important result of Fig. 9.6 is that, in case of silver, the variation of the cutoff wavelength of the TE_{11} mode exhibits an unusual behavior when the external radius decreases. For a large external radius, the cutoff wavelength first decreases linearly, but, below a limiting value (point *C* on the Fig. 9.2) $R_{omin} = 80$ nm, the cutoff wavelength increases when R_o decreases. This demonstrates a peculiar behavior of the TE_{11} mode for very small gaps between the two metallic parts of the waveguides and, maybe indicates the presence of a coupling between these two interfaces. This coupling can only be possible if surface modes exist on both these two interfaces.

In order to explain this unusual behavior, we study the dispersion relation of the TE_{11} guided mode for coaxial waveguides having different geometrical parameters. Figure 9.7 shows the obtained results for four values of R_o around the point *C* of Fig. 9.6. The higher modes are propagative ones and, as the conventional modes of a PEC waveguides, they tend asymptotically to the light line for large values of the propagation constant (kz). For small values of R_o, these modes are shifted toward high frequencies; they do not appear in Fig. 9.6 for $R_o = 80$ nm. The two interesting modes are of course the lowest ones and more precisely the first one that was identified as the TE_{11} mode. Actually, they cut the light line and tend asymptotically to $\lambda_c = \omega_p/\sqrt{2}$ which is the surface plasmon frequency on a flat metal vacuum interface. When R_o decreases, the two curves repels each other and a large gap is obtained for $R_o = 80$ nm. *These two modes are plasmonic ones* [59, 60]. Note that for cylindrical waveguides, only one plasmonic mode exists [21].

It is clear that when $R_o \rightarrow R_i$ the cutoff frequency of the TE_{11} mode tends to zero while it tends to ω_p for the second plasmonic mode. This can be schematically understood by the fact that, locally, the curvature of the guide tends to infinity (very large radius compared to the gap). This means that the modes tend to those of a slot.

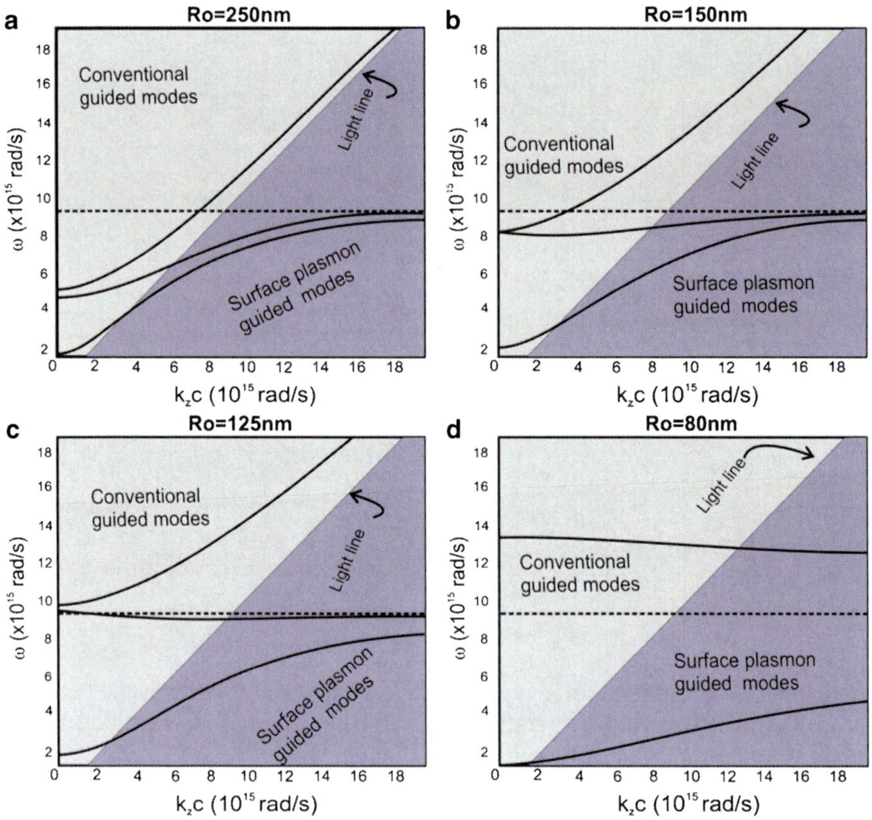

Fig. 9.7 Dispersion diagrams of the first guided modes of silver nano-waveguides for different values of the outer radius. In all cases, the inner radius is fixed to $R_i = 75$ nm

Therefore, the TE_{11} mode becomes the TE_{01} one (also named TEM mode) of an infinitely long slit that does not have cutoff wavelength.

According to these results, the TE_{11} plasmonic mode of a nano-coaxial waveguide made in real metal is clearly involved in the EOT obtained with AAA structures. In fact, it was early established [58] that the peak of the largest wavelength appearing in the transmission spectrum of an AAA structure occurs exactly at the cutoff wavelength of the TE_{11} mode of the infinitely long waveguide.

2.3 *Transmission Study*

Figure 9.8 shows transmission spectra corresponding to different self-standing (in air) AAA structures. The *FP0* peak located around $\lambda = 700$ nm corresponds to the cutoff (kz=0). In this case, the effective index of the guided mode vanishes so it

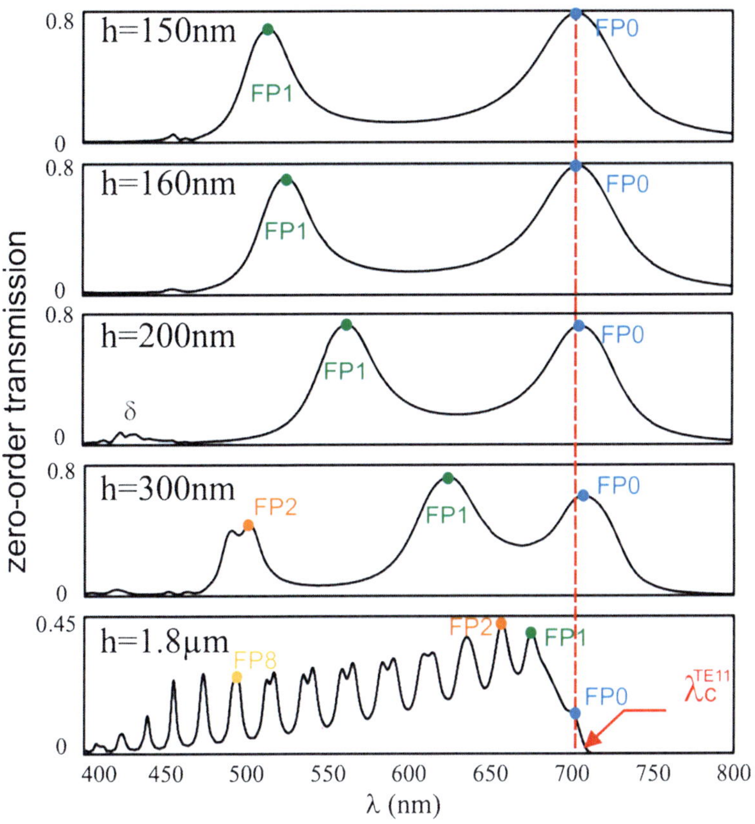

Fig. 9.8 Transmission spectra of different thickness silver AAA structures showing the apparition of additional transmission peaks. The geometrical parameters of the annular apertures are $R_o = 75$ nm, $R_i = 50$ nm, and the period of the AAA is fixed to $p = 300$ nm

does not really propagate. In addition, its position is independent of metal thickness, so, as demonstrated in Fig. 9.8, it appears at almost the same position whatever the thickness value is. Other peaks (*FPi*, $i>0$) appear in the transmission spectrum when the thickness increases [23, 61]. They are directly related to a Fabry–Perot phase matching between signals traveling along the annular apertures in the metal thickness (in the upward and downward directions). According to this peak's nomenclature, the FP0 can be interpreted also as the zero harmonic of a Fabry–Perot. In fact, this harmonic does not exist in Fabry–Perot conventional devices because the propagation constant can never vanish contrarily to the case of a guided mode at its cutoff.

Another way to confirm this result is presented in Fig. 9.9 where the light distributions inside the annular apertures are presented in a longitudinal plane (plane along the z direction and the electric field of the incident wave) at wavelengths corresponding to the three peaks of the transmission spectrum.

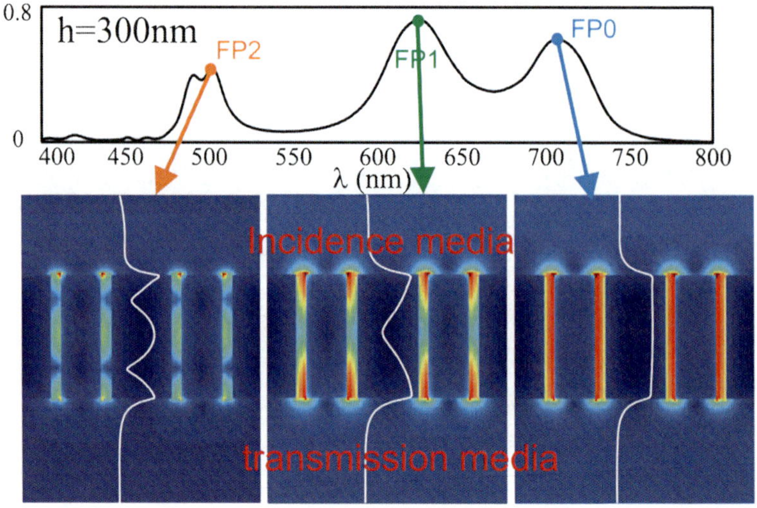

Fig. 9.9 Electric field intensity distributions along a longitudinal plane corresponding to the three transmission peaks of one AAA structure with $h = 300$ nm. Note that these three light distributions are separately normalized

Figure 9.9 clearly demonstrates the origin of the EOT for each peak. For FP0 peak, an almost uniform tint is obtained confirming the zero value of the effective guided mode index. At this peculiar wavelength, the phase velocity tends to infinity and, consequently, a very small value of the group velocity is expected. Therefore, *the FP0 is assisted by a slow light vertical mode and thus a light confinement can be obtained and exploited to greatly enhance nonlinear effects* [39, 62].

2.4 Single Annular Aperture

The following results are original and demonstrate, once again, the origin of the EOT transmission of AAA structures. To this end, let us consider the structure presented in Fig. 9.10a.

The considered annular aperture has the same geometrical parameters as the one considered in Figs. 9.8 and 9.9. So, the cutoff wavelength of the TE_{11} mode is around $\lambda = 700$ nm. The transmission spectrum presented in Fig. 9.10b is calculated as the ratio of two quantities P_t and P_0. P_t is the Poynting vector flux through the cylindrical surface drawn in red on Fig. 9.10a and P_0 is the same flux calculated over the blue surface located in the incidence side below the aperture. This latter corresponds then to the incident energy on the opening area of the aperture while P_t presents the transmitted total energy in the far-field only if the red cylinder radius is very large. In fact this condition is necessary to avoid the consideration of energy carried by an

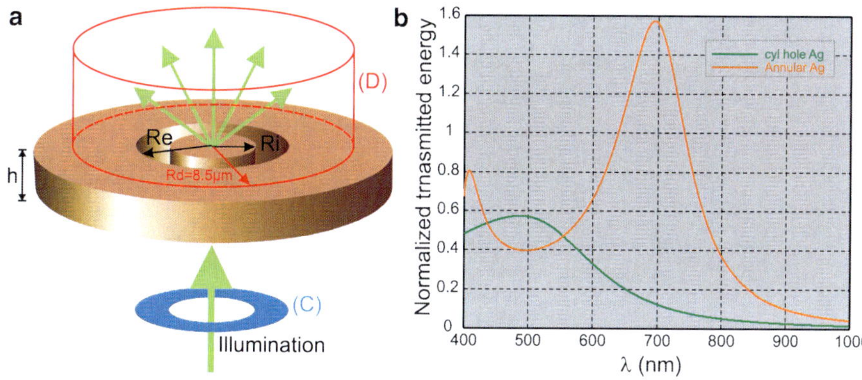

Fig. 9.10 (**a**) Schema of the single annular aperture and the surfaces used for the determination of the normalized transmitted energy: the incident one corresponds to the Poynting flux through the blue surface while the transmitted energy is determined using the red cylinder. (**b**) Transmission spectrum of the single AA (*red curve*) compared to the case of a single cylindrical one (*green curve*)

eventual surface plasmon wave (or simply any surface wave) that can be excited by the light scattered by the aperture.

$$\text{Normalized transmitted energy } = \frac{P_t}{P_0} = \frac{\iint_D \vec{P}(r,\varphi,z)\mathrm{d}S}{\iint_C \vec{P}(r,\varphi,z)\mathrm{d}S}.$$

As seen in Fig. 9.10b, the transmission of the annular aperture exhibits a transmission peak at the same position of the FP0 peak of the periodic structure. So, this peak corresponds again to the excitation of the TE_{11} mode at its cutoff wavelength. Note that the normalized transmission at this peak directly corresponds to its cross-section that is greater than one.

Using the same developed BOR-FDTD code, we also validated the results on the unusual behavior of the cutoff of the TE_{11} mode as a function of distance $R_o - R_i$. Figure 9.11a presents N-Order BOR-FDTD results showing the trace of the cutoff wavelengths of two coaxial waveguides (in PEC and silver) in comparison with those of two cylindrical waveguides versus their outer radius or their radius, respectively. Once again, we note the unusual behavior of the TE_{11} mode obtained for a different value of the inner radius ($R_i = 75$ nm here instead of $R_i = 50$ nm in Fig. 9.6) while the silver cylindrical waveguide seems to not present this phenomenon.

BOR-FDTD simulations are then performed to determine the transmission spectra (Fig. 9.11b) of the diffracted zero-order of a 100-nm-thick silver layer perforated by only one aperture. Four apertures were studied corresponding to points A, B, C, and D of Fig. 9.11a. Two phenomena are clearly shown in Fig. 9.11b: first, the use of silver instead of PEC leads to a shift in the transmission peak toward a larger value of wavelength and, second, this shift can be amplified by decreasing the outer radius.

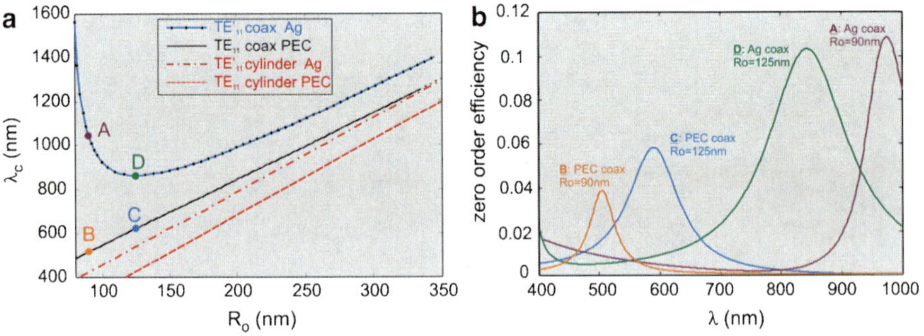

Fig. 9.11 (**a**) Traces of the cutoff wavelength of the TE11 mode as a function of the outer radius for different waveguides. The inner one is fixed to $R_i = 75$ nm. Only the real metal (silver, for instance) coaxial waveguide exhibits an unusual behavior corresponding to a minimum value of λc (point D). (**b**) Transmission spectra of single AA engraved into 100-nm-thick PEC (*orange* and *blue*) or silver (*green* and *purple*) layer corresponding to points A, B, C, and D of (**a**). As expected, the peak position greatly depends on the outer radius value, but so does the efficiency

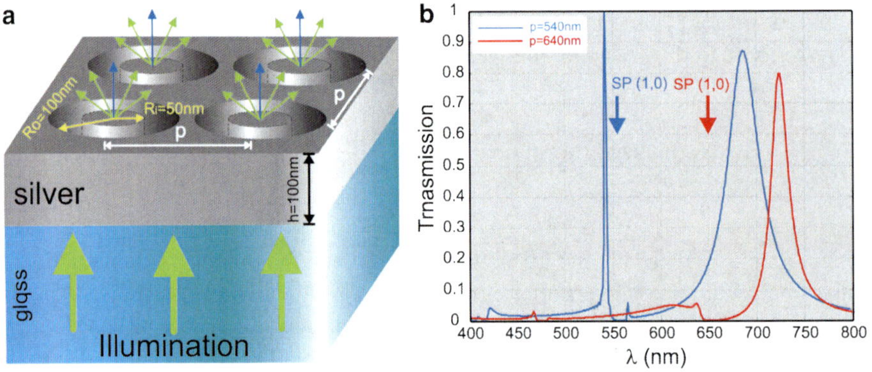

Fig. 9.12 (**a**) Schematic of a silver AAA deposited on a glass substrate. The geometrical parameters are indicated. (**b**) Two transmission spectra corresponding to two different values of the periods. The *vertical arrows* indicate the surface plasmon resonance positions corresponding to a zero transmission. Nevertheless, the presence of these resonances allows large red shift of the TE_{11} transmission peak especially in the case of the *red curve*

According to the previous results, it is irrevocable that the AAA EOT is directly linked to the guided mode inside the apertures and that it is uncorrelated to any surface wave that can be excited by tuning the periodicity of the whole structure. In fact, for all the AAA structures considered before, the period of the arrays was chosen to be very small compared to the cutoff wavelength of the TE_{11} mode. This ensures the obtaining of a transmission peak well beyond the one due to the plasmon resonance accentuating, by this way, the phenomenon of EOT. It should be noted that the study of coupling between the two modes remains an open question because of its complexity. Indeed, as shown in Fig. 9.12, this coupling can be both positive and negative.

Before presenting the experimental results confirming some of these simulations, we will finish this part by showing other specific theoretical properties of the AAA structure, thus opening the way for practical applications in different fields of physics (detection radar, radome, photovoltaic, etc.), in addition to those already planned and announced by researcher's community in the field of EOT.

2.5 Additional Properties of the AAA Structure

This section concerns the robustness of the AAA structure according to the characteristics of the incident beam [63, 64]. Indeed, we demonstrate that, because of the good overlap between a linearly polarized plane wave and the TE_{11} mode, the transmission spectrum through an AAA structure is almost independent of the angle of incidence, the direction of the plane of incidence, and the polarization. Results of 3D-FDTD simulations presented in Fig. 9.13 clearly show these properties.

The structure depicted in Fig. 9.13 is designed to work in the Gigahertz domain. It consists of a 2D array of annular apertures perforated into an $h=3.6$ mm thick perfect metal. The values of the inner and the outer radii are chosen to get enhanced transmission at a desired frequency (here $f=13$ GHz). Thus, we set $R_i=2.7$ mm, $R_o=3.6$ mm, and we fix the array period to $p=12.6$ mm in order to avoid the Rayleigh anomaly at normal incidence. In fact, this last appears when the first diffracted order lies on the grating surface, i.e., when the tangential component of the wave vector associated with this first diffracted order is equal to ω/v with v being the light velocity in the incident or in the transmission media. The fine details of the structure are well described through a nonuniform meshing that was integrated into the 3D-FDTD algorithm for the study of very small samples such as resonant particles [56] or very thin films [65].

Transmission spectra for four different values of the azimuthal angle ϕ, in the cases of TE and TM polarized incident plane waves, are presented in Fig. 9.13a, b respectively. For the two cases, the angle of incidence is fixed to $\theta=20°$. One notices that, except the Rayleigh anomalies, the transmission response of the structure remains almost the same when both polarization and azimuthal angle vary. In addition, staircase artifacts due to the Cartesian discretization of the rotational symmetrical apertures remain present and lead to small alterations on the transmission responses when the illumination parameters vary (especially ϕ). The robustness of the AAA structure is also verified when the angle of incidence θ is varied. Figure 9.13c, d presents the transmission spectra for the TE and TM polarizations, respectively, when the angle of incidence is varied from 0 up to 40°. One can note that in the TM case, the influence of the Rayleigh anomalies appears as discontinuities and alters the transmission spectra. To bypass this problem it is possible to optimize the geometrical parameters probably by decreasing the period value.

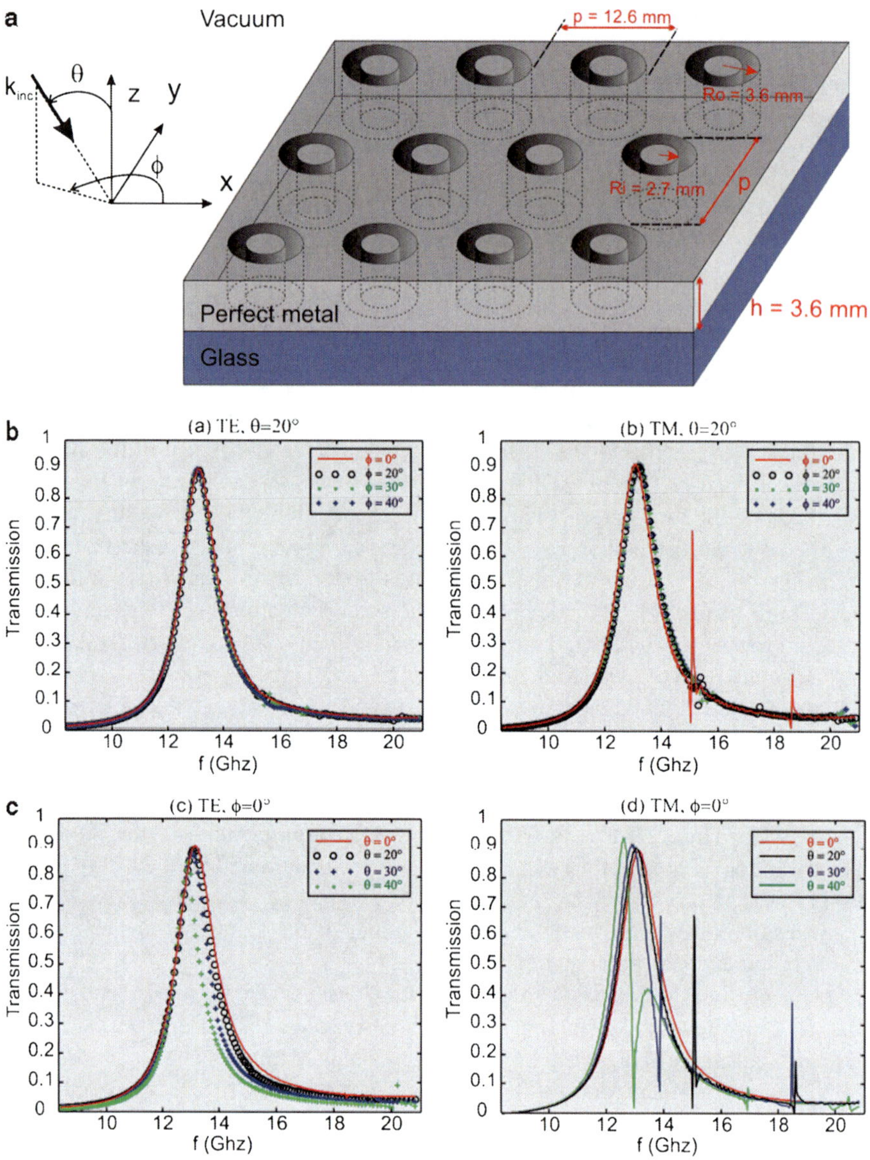

Fig. 9.13 (**a**) Schematic of an AAA structure designed for application in the Thz domain. (**b**) Transmission spectra for an arbitrary value of the angle of incidence versus the azimuthal angle for both TE (*left*) and TM (*right*) polarizations. (**c**) Transmission spectra at $f = 0°$ for different values of the angle of incidence in TE (*left*) and TM (*right*) polarizations

3 Experimental Studies

This part deals with experiments which validate the theoretical development of the previous part. After having briefly exposed the fabrication process of the AAA, we present scanning near-field experiments validating the existence of a TE_{11} mode inside the annular cavities. Experimental spectral responses exhibit transmission peaks with maximum efficiency up to 90%: both fundamental mode and first harmonic of the TE_{11} mode have been highlighted.

3.1 Fabrication of the Nanostructures

The design of the fabricated structure (thickness, diameters, and metal nature) was already determined by theoretical calculations in order to fulfill a large transmission in the visible range. Different AAAs have been fabricated in gold or silver layers deposited by evaporation. The first set (outer diameter 330 nm, inner diameter 250 nm, and period 600 nm) in a 150-nm-thick gold layer has been generated by e-beam lithography and gold lift-off by Perentes et al. from Ecole Polytechnique Fédérale in Lausanne (EPFL, Switzerland): one can refer to [36] and [66] for more details. The second set was milled by focused ion beam (FIB) in a 100-nm-thick silver film at FEMTO-ST Institute. The FIB operates at 30 keV and the intensity of the beam is 25 pA which leads to a beam diameter around 40 nm. SEM and AFM images in Fig. 9.14 show the good quality of our samples. The geometrical parameters of the AAA are: inner diameter 100 nm, outer diameter 200 nm, and period 350 nm.

A first qualitative optical characterization of the AAA can be made, thanks to a conventional microscope working in transmission mode, using a thermal white source. Figure 9.15 shows three different matrices in a silver layer in real colors and their theoretical spectral responses calculated by FDTD. In order to compare the AAAs responses, samples 1 and 3 have the same diameters but a different period and samples 2 and 3 have the same period but different diameters.

The three arrays appear in different color (yellow, purple, and orange) according to the theoretical transmission peak in the visible range. As an example, high peaks of transmission exist for the array number 1 around 550 nm (green) and 700 nm (red) which obviously lead to a yellow color of the matrix. The same reasoning can be applied for the two other structures. More qualitative results concerning the electric field distribution above the nanostructures and the EOT spectra are introduced in the following chapters.

3.2 Near-Field Study

As already explained before, the high transmission peaks exhibited by the AAA are due to a guided mode. In order to experimentally validate the existence of this TE_{11}-like mode inside the annular cavity, scanning near-field optical microscope (SNOM) is a suitable tool because it allows reaching the fine structure of the electromagnetic field

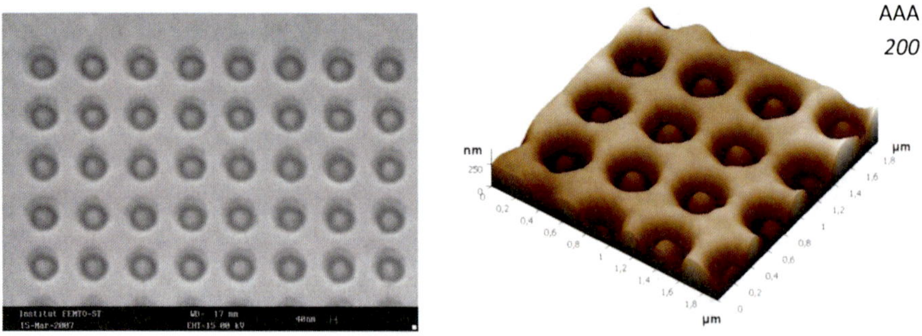

Fig. 9.14 *Left*: SEM images of fabricated structures on a 100-nm-thick silver film (FIB milling). The coaxes diameters are 200/100 nm, the period is 350 nm. *Right*: AFM image of another sample on a gold film

above/in the nano-apertures below the diffraction limit. If the optical near-field distribution inside the cavities has been computed many times [24, 26, 58], it has not been much experimentally studied because AAAs present significant challenges to near-field optical characterization. Nevertheless, Orbons et al. [67] used a metal coated tip to characterize the near-field behavior in annular apertures. However, the coupling between the metal coating of the near-field microscope probe and the metallic sample leads to significant distortions of the electromagnetic field from the computed field which ignores the probe influence. To avoid these interactions and to get round the important losses occurring at the subwavelength aperture of the tip, the authors of [46] suggest using a metal-free probe based on a fractal fiber probe and they showed an enhanced collection capability.

We report here near-field experiment using a simple dielectric tip working in reflection mode [68]. The SNOM works in reflection mode (see [69]: Figure 9.16 shows the principle of the Reflection-SNOM). The sample is an AAA in a 100-nm-thick gold layer milled by FIB. The period is 500 nm and the inner and outer diameters are 200 and 280 nm, respectively. The theoretical responses (3D-FDTD) of this sample are plotted in Fig. 9.17 (squared modulus of the diffracted field in the close vicinity of the sample and transmission spectrum). The mode inside the annular cavity is excited at 633 nm thanks to a dielectric tip which also picks up the backward signal. A coupler separates incident and reflected signals; the latter is detected by a photomultiplier. We chose an excitation wavelength (633 nm) outside the theoretical transmission window of this AAA (see Fig. 9.17) so that the cavity mode cannot propagate inside the annular aperture and it is completely reflected: it allows a better detection of the field above the structure. Both optical and shear force images are depicted in Fig. 9.18. The optical images are blurred by fringes which are very common in this kind of experiments (various and uncertain origins: interference between the incident and the diffracted fields, feedback defects, surface plasmons, oscillations of the probe, etc.) and consequently the expected field (Fig. 9.17) is not clearly resolved. Moreover, the incident polarization cannot be fully monitored at the apex of the probe. However, a simple Fourier Transform filtering can suppress these parasitical interferences from the optical image. The near-field structure is then resolved (images at the bottom of Fig. 9.18) and one can observe a "coffee-bean" like structure which corresponds to

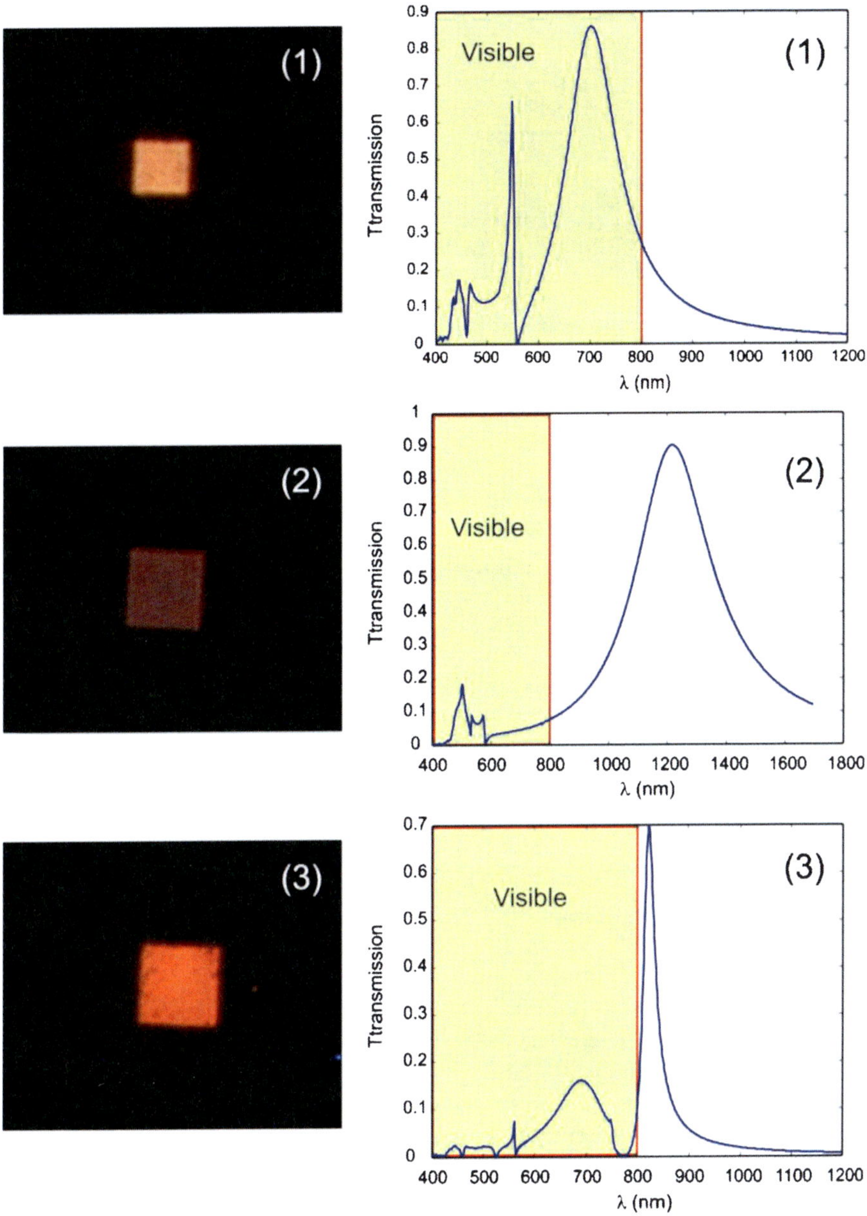

Fig. 9.15 Far field image (real color) obtained with a classical microscope working in transmission mode of three different AAAs in a silver film (100 nm thick). The period, the inner, and outer diameters are, respectively: for sample 1 (7×7 μm^2): 350, 100, 200; for sample 2 (10×10 μm^2): 500, 230, 330 nm, and for sample 3 (10×10 μm^2): 500, 100, 200 nm

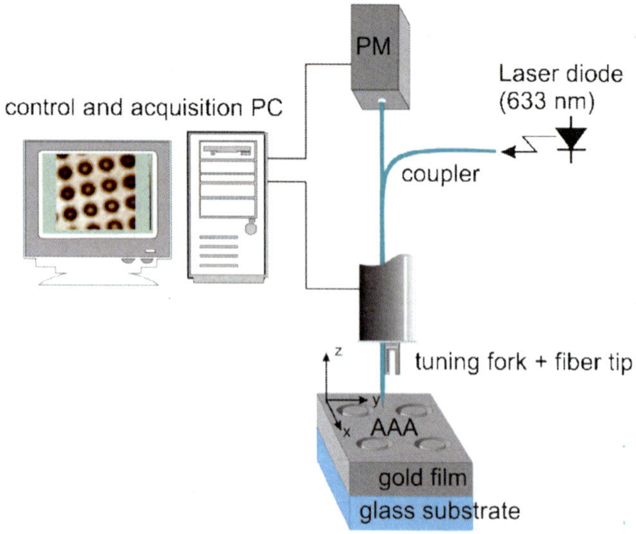

Fig. 9.16 Near-field characterization: experimental set-up of the reflection scanning near-field optical microscopy (RSNOM)

the theoretical calculations (Fig. 9.17). The two-lobes structure is characteristic of the field distribution of the TE_{11} mode excited in the cavity (the orientation of the two-lobes depends on the polarization direction). As a conclusion, we here experimentally demonstrated the excitation of the TE_{11} mode inside the subwavelength annular apertures.

3.3 *EOT Spectra*

In order to record the EOT spectra, the sample is illuminated at normal incidence thanks to a supercontinuum. The collimated incident light is linearly polarized and the zero-order transmission (intensity *It*) is then propagated to a spectrometer by a cleaved-end multimode fiber (core diameter 62.5 μm) placed above the structure (see Fig. 9.19 for the experimental set-up). The reference intensity *Ii* and its spectrum are measured the same way through a bare squared hole in the metallic film presenting the same area than the AAA. We can then define the absolute transmission of the structure by the ratio between *It* and *Ii*.

The transmissions have been measured through three different AAAs: one is embedded in a 150-nm-thick gold film (inner diameter 200 nm, outer diameter 280 nm, and period 500 nm), the two others are embedded in a 100-nm-thick silver film, present the same geometrical parameters (inner diameter 100 nm, outer diameter 200 nm, and period 350 nm) but a different whole area (30 × 30 periods, i.e., ~100 μm and 40 × 40 periods, i.e., ~190 μm^2).

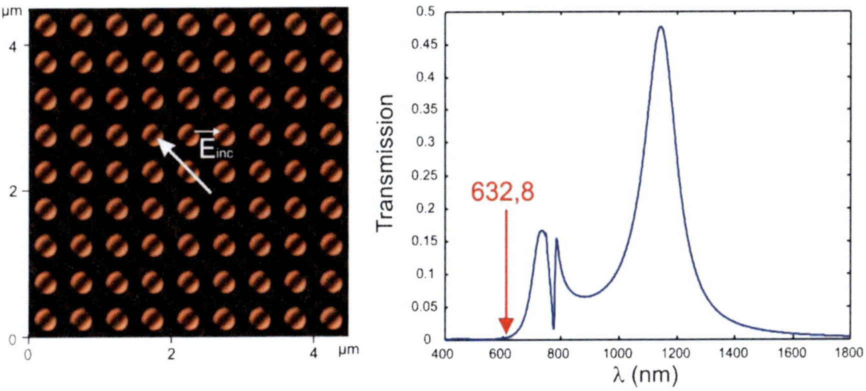

Fig. 9.17 Theoretical responses (3D-FDTD) of an AAA embedded in a 100-nm-thick gold film (diameters 200–280 nm and period 500 nm). *Left*: squared modulus of the tangential components of the electric field at a distance of 30 nm above the sample. The tip is supposed to act as a dipole and consequently it is not sensitive to the vertical component of the field [70, 71]. The *white arrow* marks the incident plane wave polarization (45°). *Right*: theoretical transmission spectrum and incident wavelength used in the near-field experiment

The results are depicted in Fig. 9.20 for the AAA in a gold layer and in Fig. 9.21 for the AAAs in a silver layer. The computed spectra are defined as the energy associated with the diffracted zero-order transmitted through the infinite AAA structure divided by the same quantity calculated through the substrate without the metal layer. First, one can note that the experimental curves are in good agreement with the theoretical predictions for the three samples. The small discrepancy between the experimental and the theoretical curves observed in both cases can be related to different factors. First, we do not take into account in our calculation the thin adhesion layer (titanium or chromium). Second, the real parameters (dimensions, permittivity, and slope of the sides) slightly differ from the theoretical ones which can induce a shift of the transmission peak. As an example, a 5 nm difference on the inner radii drastically shifts the position of the TE_{11} mode (see Fig. 9.22) and this small difference is under the resolution of our FIB. Finally, the finite size of the real AAA may induce a blue shift of the peak transmission compared to the theoretical peak calculated with an infinite object [72].

Other comments can be deduced from the transmission spectra in Figs. 9.20 and 9.21. First of all, the experimental spectral responses are reproducible: both experimental curves in Fig. 9.21 are obtained for two different samples and there are similar in spite of the fact that the two AAAs do not have the same area (arrays with 30×30 and 40×40 periods). Then, the EOT in both cases are more important than those assisted by surface plasmon with simple holes [1, 9] or even with holes surrounded by corrugations [73, 74]. Second, the transmission is larger for silver than for gold as expected. The losses due to the metal absorption are not the only explanation. Indeed, the transmission peak in the case of gold corresponds to the first harmonic of the TE_{11} mode: the fundamental occurs around 1,330 nm and reaches

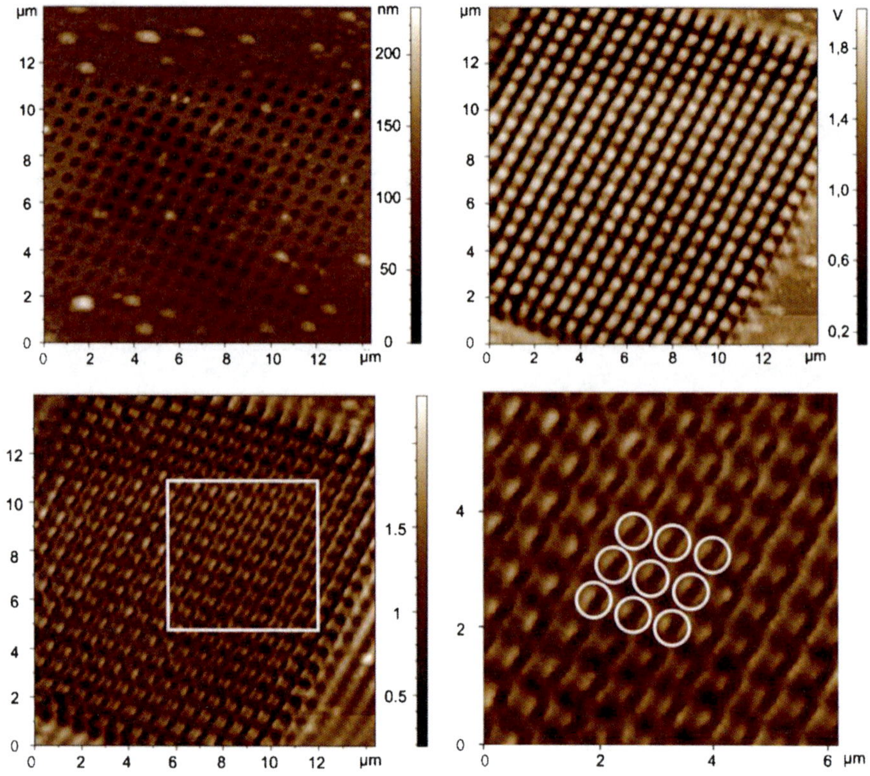

Fig. 9.18 Reflection scanning near-field optical microscope (RSNOM) images at 633 nm of an AAA in a 100-nm-thick gold film. The inner diameter is 200 nm, the outer diameter 280 nm, and the period 500 nm. *Above*: shear force (*left*) and raw optical images (*right*). *Bottom*: optical image filtered by Fourier Transform; the *white square* in the *bottom left* picture shows the location of picture in the *bottom right* (numerical zoom-in, 6×6 μm²). The *white circles* mark the apertures position deduced from shear force image. This last image is to compare with the theoretical one in Fig. 9.17

almost 100% transmission (see right part of Fig. 9.20). Using silver instead gold shifts the fundamental of the TE_{11} mode in the visible range [58]: it corresponds to the wide transmission peak measured at 675 nm in Fig. 9.21 (theoretically, at 690 nm). This transmission peak reaches 90% and we emphasize here the fact that it is to our knowledge the most important EOT measured in the visible range.

The last experiments we present deals with the influence of the incident polarization (Fig. 9.23). For a perfect object, this influence is supposed to be quasi null due to the symmetry of the structure. However, the noncylindrical symmetry of the real apertures induced a substantial modification of the measured transmission versus the polarization direction (see bottom curves in Fig. 9.23). In order to theoretically take into account the lack of symmetry and to fit the experimental condition as much as possible, we introduced in our calculation an infinite object deduced from a SEM image. One coaxial cavity has been chosen by chance in SEM image in Fig. 9.14 and digitalized (see image on the right of Fig. 9.23). We defined this way

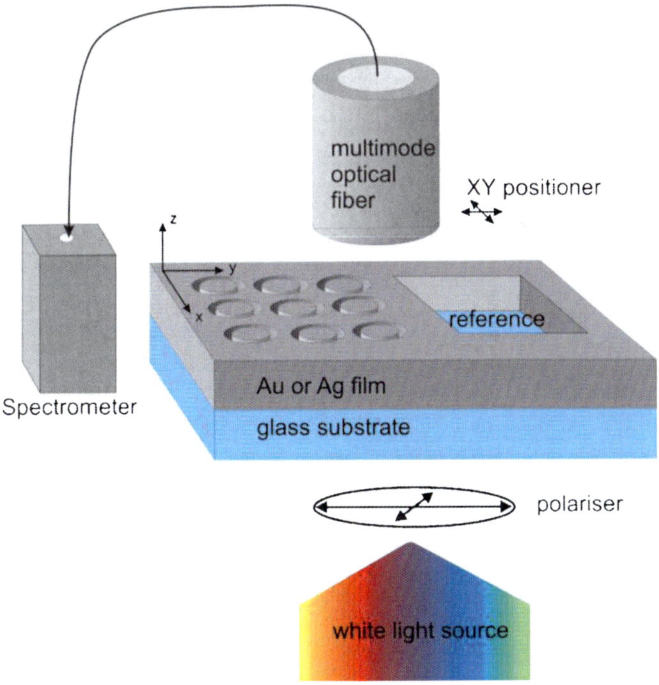

Fig. 9.19 Experimental set-up used to record the transmission spectra through different AAAs embedded in a metallic film. The transmission is defined as the ratio between the spectra through the AAA and through a reference. This reference consists of a bare squared hole having the same size than the AAA

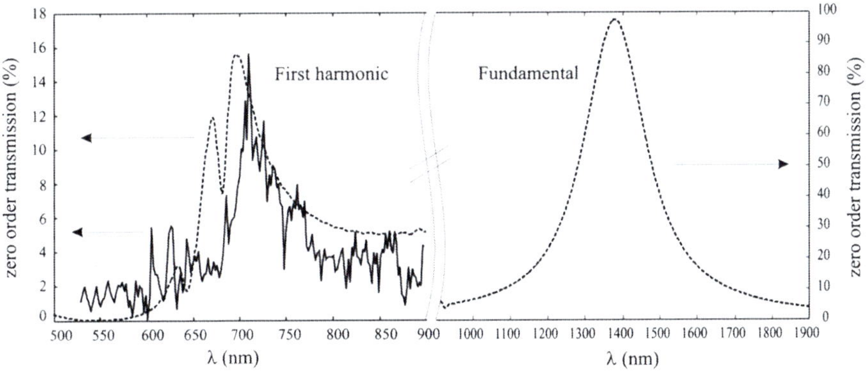

Fig. 9.20 Theoretical (*dashed line*) and experimental transmission through an AAA in a gold film (period 600 nm, inner/outer diameters 250/330 nm) (*left part*: first harmonic in the visible range, *right part*: fundamental mode in the infrared region)

the periodic pattern of the infinite theoretical object and the computed spectra are then in good agreement with the experimental ones, showing a diminution of the maximum transmission with an increase of the angle of polarization with respect to

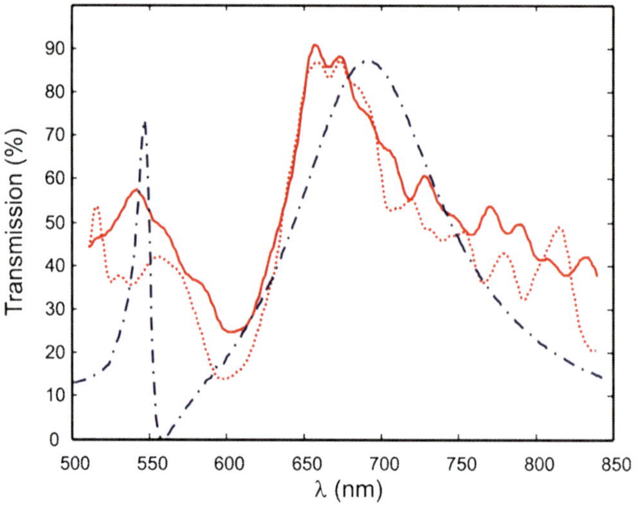

Fig. 9.21 Theoretical and experimental transmission spectra due to the fundamental of the TE_{11} mode through two AAAs having the same period and inner and outer diameters but different areas. *Dotted curve*: matrix with 30×30 periods (~100 μm²); *solid curve*: matrix with 40×40 periods (~190 μm²); *dotted–dashed curve*: theoretical curve (FDTD) obtained with an infinite periodic structure

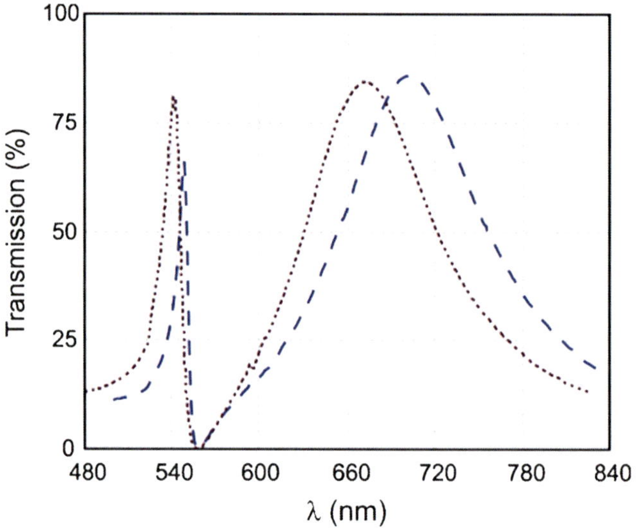

Fig. 9.22 Theoretical transmission spectra through an AAA calculated for a period of 350 nm, an outer diameter of 200 nm, and for two inner diameters of 90 nm (*dotted line*) and 100 nm (*dashed line*) which is supposed to correspond to the experimental object

one of the axes of periodicity. It seems that the position of the transmission maxima is directly connected to the inner and the outer radii measured along the direction of the incident electric field (Figs. 9.22–9.23).

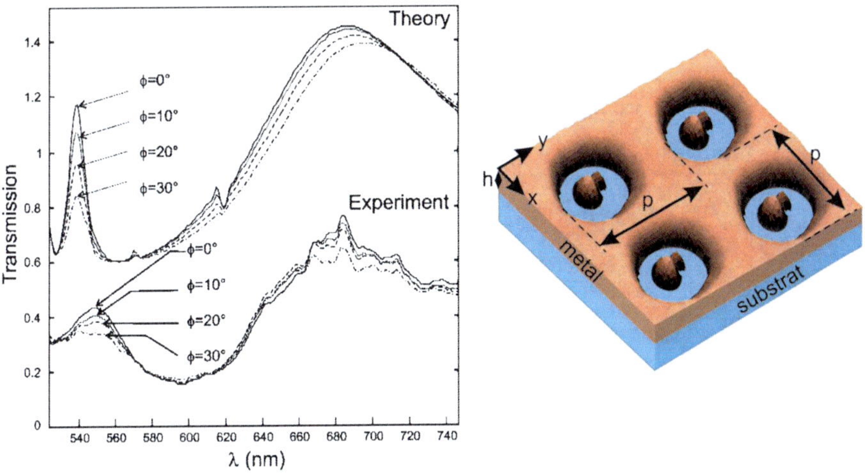

Fig. 9.23 *Left*: transmission spectra for different directions of polarization of the incident field: 0° (*solid curve*), 10° (*dotted*), 20° (*dashed*), and 30° (*dashed–dotted*) with respect to the *y* axis of the image on the right. *Bottom curves*: experimental data, upper curves: theoretical curves vertically shifted by 0.6 for the sake of clarity. *Right*: 2×2 periods of the theoretical object used for the calculation. The infinite periodic structure is deduced from one annular hole of the SEM image in Fig. 9.14

4 EOT by Excitation of the TEM Mode

As mentioned before, experimental EOT with AAA was only experimentally obtained thanks to the excitation of the TE_{11} mode. So, it is rightful to ask about the excitation of the TEM of such a structure. More precisely, is it possible to get EOT through AAA structures by means of this TEM mode? The answer is given in [25]. The demonstration is performed by expressing the electromagnetic (EM) field of the incident plane wave through its angular spectrum. For a perfect metallic coaxial waveguide, the TEM mode is completely determined by only two components of the EM, i.e., E_r and H_ϕ. In fact, it is well-known that the electric field associated with the TEM has only radial component while its magnetic field is azimuthal. In addition, there is a perfect cylindrical symmetry so these two components do not depend on the azimuthal angle. The basic way to excite this mode is to illuminate the structure with an EM field having simultaneously nonzero values for E_r and H_ϕ components.

By projecting a linearly polarized plane wave obliquely propagating at θ degrees from the z axis of the annular aperture and presenting an arbitrary polarization direction (both TE and TM components) over the Fourier angular component ($m=0$), it is demonstrated in [25] that two conditions should be fulfilled according to the following obtained values of E_r and H_ϕ:

$$E_r^0(r,z,t) = C\cos\theta E_{\mathrm{TM}} J_1(kr\sin\theta),$$

$$H_\phi^0(r,z,t) = \frac{-Ck}{\mu\omega} E_{\mathrm{TM}} \left\{ \left(\frac{3}{4}\cos 2\theta + \frac{1}{4} \right) J_1(kr\sin\theta) + \frac{\sin^2\theta}{2} J_3(kr\sin\theta) \right\}.$$

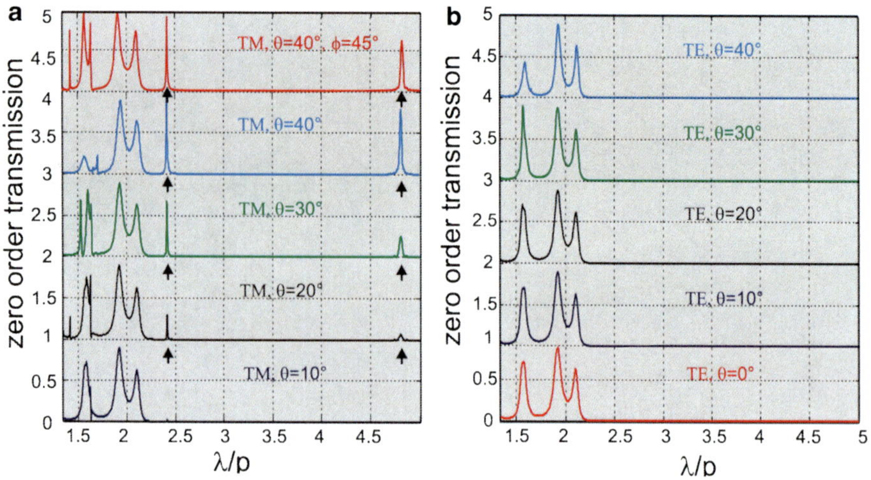

Fig. 9.24 Transmission spectra for different angles of incidence in both TM (**a**) and TE (**b**) polarizations obtained for an AAA made in perfect conductor. The *black vertical arrows* indicate the position of the additional peaks appearing due to the excitation, the propagation, and the transmission of the TEM mode that only occur in the TM case

According to these last equations, the incident plane wave must strike the structure at oblique incidence and must have a nonzero axial component (along the axis of symmetry of the apertures, here Oz) of the electric field. Strictly speaking, a TM polarized plane wave under off-normal incidence can allow the excitation of the TEM mode inside the annular apertures and thus its transmission through the nano-structured metallic film.

A numerical demonstration of the validity of these two conditions is shown in Fig. 9.24 where transmission spectra of an AAA structure made in PEC are presented for different illuminations. The considered structure has $h=2p$, $R_o=p/3$, and $R_i=p/4$ where p is the period of the AAA. When comparing the TM case (Fig. 9.24a) to the TE one (Fig. 9.24b), additional transmission peaks appear beyond the cutoff wavelength of the TE_{11} mode located at $1.83p$. The amplitude of these last peaks increases with θ and their positions remain independent of the angle of incidence. In fact, the position of the TEM peaks only depends on the length of each annular cavity, i.e., on the metal thickness because they correspond to a Fabry-Perot harmonics of each annular aperture exactly as the peaks FPi, $i>0$ of Fig. 9.8. Consequently the position of the TEM peaks is directly connected to the metal thickness and to the effective index of the TEM-guided mode traveling inside the apertures. One can thus easily understand that, in the visible range, the metal dispersion will play a key role for the localization of the TEM peaks in the transmission

9.25a shows the theoretical transmission spectra calculated for a free lver AAA with $R_o=100$ nm, $R_i=50$ nm, $h=220$ nm, and $p=300$ nm. The ; clearly appear only in the TM polarization with an efficient transmission

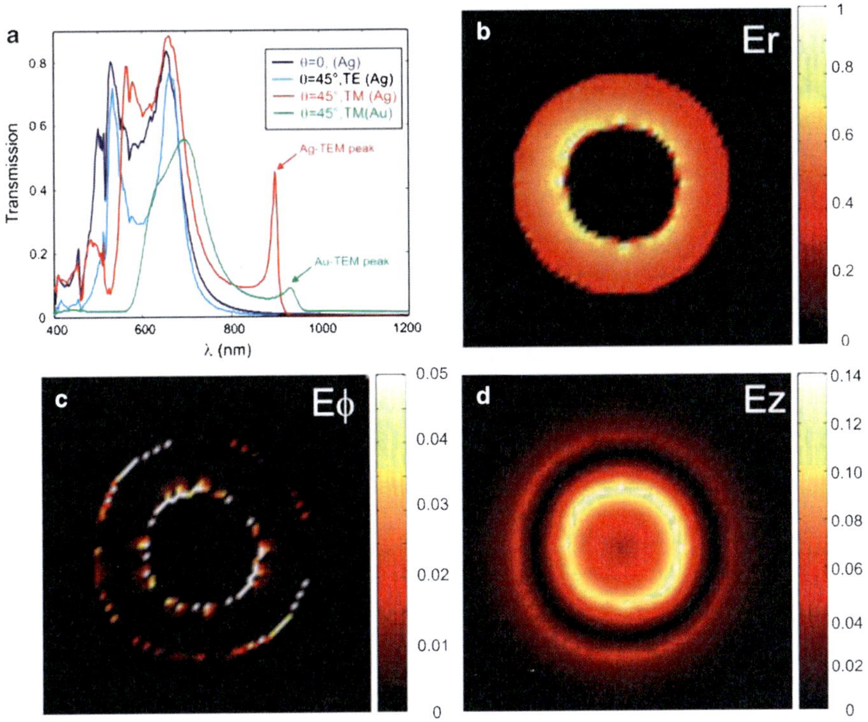

Fig. 9.25 (**a**) Transmission spectra of silver and gold AAA ($R_o = 100$ nm, $R_i = 50$ nm, $p = 300$ nm, and $h = 220$ nm) demonstrated the occurrence of a TEM-like at oblique incidence responsible for additional peaks. The position of these latter greatly depends on the metal nature. Gold seems to allow more red shift but also more absorption manifested by less efficiency than the silver case. Spatial amplitude distribution of the radial (**b**), azimuthal (**c**), and axial (**d**) components of the electric field inside silver annular aperture (at the mid-thickness) at $\lambda = \lambda_{TEM} = 900$ nm normalized with respect to the radial component

coefficient (40%) for the silver metal. The nature of the guided mode inside the aperture at the peak wavelength is demonstrated in Fig. 9.25b–d where the amplitude distributions of the three cylindrical components of the electric field are plotted. It is clear that the real nature of the metal leads to a hybrid mode but with small longitudinal components (see Fig. 9.25c, d in comparison with b).

Before concluding, let us only mention that the TEM mode of a real metal nanocoaxial waveguide is a pure plasmonic mode. This is theoretically demonstrated on Fig. 9.26 where the dispersion diagram of this mode is presented in the case of silver. The dispersion curve presents an almost linear variation versus the propagation constant, but is still located under the line of light. So, its effective index is larger than 1 in the case of a coax filled by air. This property is of great importance because it allows finding a compromise between the propagation losses and the large value of the thickness required to push the TEM peak beyond the TE_{11} one, in order to benefit from it.

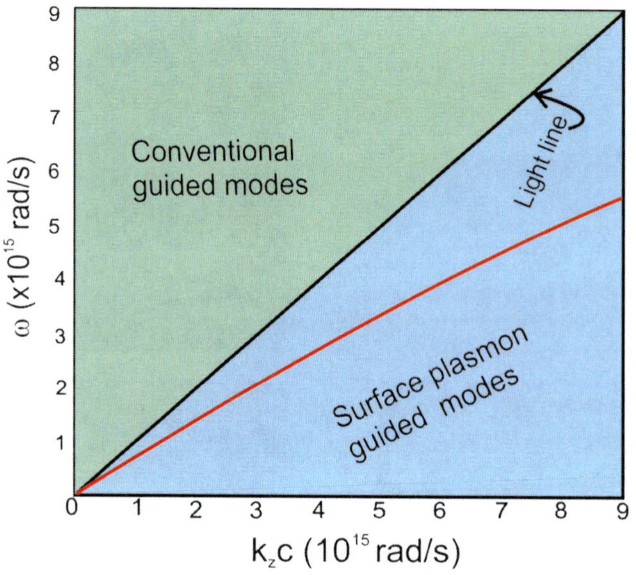

Fig. 9.26 Dispersion relation of the TEM mode of an infinitely long nano-coaxial waveguide made in silver. The outer and inner diameters are 100 nm and 50 nm, respectively as in Fig. 9.25

Experimental demonstration of light transmission, thanks to the excitation of this mode, is not already achieved even if it is claimed by some authors [75–77] for the simple reason that the AAAs were never experimentally illuminated under oblique incidence. In our case, we try to progress in this direction to solve the technological problems related to the fabrication of these structures, i.e., the high aspect factor of the desired structure due to the need of thick metallic screens. FIB technique is currently tested in parallel to RIE or DRIE ones.

In conclusion, this main disadvantage of this mode is the fact that it has not a cutoff and thus it cannot be exploited to exalt nonlinear effects because it does not allow small group velocity for the light propagation.

5 Conclusion

This chapter has been dedicated to the presentation of the EOT phenomenon obtained using AAA structures. These have a major interest for the enhancement of light transmission compared with cylindrical apertures. The major property of AAA is linked to the TE_{11} mode which has a wavelength cutoff that can be much greater than the outer diameter. This allows designing periodic structures with periodicity smaller than this cutoff. Therefore, one can easily understand the occurrence of transmission peak beyond the Wood anomaly corresponding to EOT with cylindrical apertures. The real nature of the metal allows a red shift of the wavelength cutoff but also causes losses due to absorption by the metal. Although the plasmonic character of this TE_{11} mode, it is clear that losses are minimized by the fact that, at the cutoff, the electromagnetic field is rather localized in the interconductor zone. In fact, at the cutoff, the dispersion curve is necessarily above the line light making it a conventional guided mode.

We think that the mechanism of excitation of the TEM mode will certainly continue to stimulate scientific researches based on the AAA structure. This has already begun in the form of theoretical [78] and experimental [79, 80] studies in order to exploit this method for the design of new functionalities metamaterials [67, 81–83].

References

1. T.W. Ebbesen, H.J. Lezec, H.F. Ghaemi, T. Thio, and P.A. Wolff. Extraordinary optical transmission through sub-wavelength hole arrays. 391:667–669, February 1998.
2. T.J. Kim, T. Thio, T.W. Ebbesen, D.E. Grupp, and H.J. Lezec. Control of optical transmission through metals perforated with subwavelength hole arrays. *Opt. Lett.*, 24:256–258, 1999.
3. T. Thio, H.F. Ghaemi, H.J. Lezec, P.A. Wolf, and T.W. Ebessen. Surface-plamon-enhanced transmission through hole arrays in Cr films. *J. Opt. Soc. Am. B*, 16(10):1743–1748, October 1999.
4. D.E. Grupp, H.J. Lezec, T.W. Ebbesen, K.M. Pellerin, and Tineke Thioa. Crucial role of metal surface in enhanced transmission through subwavelength apertures. *Appl. Phys. Lett.*, 77(11):1569–1571, 2000.
5. Tineke Thio, H.J. Lezec, and T.W. Ebbesen. Strongly enhanced optical transmission through subwavelength holes in metal films. *Physica B*, 279:90–93, 2000.
6. U. Schröter and D. Heitmann. Surface-plasmon enhanced transmission through metallic gratings. *Phys. Rev. B*, 58:15419–15421, 1998.
7. H.F. Ghaemi, T.Thio, D.E. Grupp, T.W. Ebbesen, and H.J. Lezec. Surface plasmons enhance optical transmission through sub-wavelength holes. *Phys. Rev. B*, 58:6779–6782, 1998.
8. J.A. Porto, F.T. Garcia-Vidal, and J.B. Pendry. Transmission resonances on metallic gratings with narrow slits. *Phys. Rev. Lett.*, 83:2845–2848, 1999.
9. A. Krishnan, T. Thio, T.J. Kim, H.J. Lezec, T.W. Ebbesen, P.A. Wolff, J. Pendry, L. Martin-Moreno, and F.J. Garcia-Vidal. Evanescently coupled resonance in surface plasmon enhanced transmission. *Optics Commun.*, 200:1–7, 2001.
10. M. Sarrazin, J.P. Vigneron, and J.M. Vigoureux. Role of wood anomalies in optical properties of thin metallic films with a bidimensional array of subwavelength holes. *Phys. Rev. B*, 67:085415, 2003.
11. T. Lopez-Rioz, D. Mendoza, F.J. Garcia-Vidal, J. Sanchez-Dehesa, and B. Pannetier. Surface shape resonances in lamellar metallic gratings. *Phys. Rev. Lett.*, 81:665–668, 1998.
12. Ph. Lalanne, J.P. Hugonin, S. Astilean, M. Palamaru, and K.D. Möller. One-mode model and airy-like formulae for one-dimensional metallic gratings. *J. Opt. A: Pure Appl. Opt.*, 2:48–51, 2000.
13. S. Astilean, Ph. Lalanne, and M. Palamaru. Light transmission through metallic channels much smaller than the wavelength. *Optics Commun.*, 175:265–273, 2000.
14. E. Popov, M. Nevière, S. Enoch, and R. Reinisch. Theory of light transmission through subwavelength periodic hole arrays. *Phys. Rev. B*, 62:16100–16108, 2000.
15. L. Martin-Moreno, F.J. Garcia-Vidal, H.J. Lezec, K.M. Pellerin, T. Thio, J.B. Pendry, and T.W. Ebbesen. Theory of extraordinary optical transmission through subwavelength hole arrays. *Phys. Rev. Lett.*, 86(6):1114–1116, February 2001.
16. Laurent Salomon, Frédéric Grillot, Anatoly V. Zayats, and Frédérique de Fornel. Near-field distribution of optical transmission of periodic subwavelength holes in a metal film. *Phys. Rev. Lett.*, 86(6):1110–1113, 2001.

17. J.M. Vigoureux. Analysis of the Ebbesen experiments in the light of evanescent short range diffraction. *Optics Commun.*, 198:257–263, 2001.
18. T.-K. Wu and S.-W. Lee. Multiple frequency selective surface with multiring patch elements. 42:1484–1490, 1994.
19. C. Winnewisser, F. Lewen, J. Weinzierl, and H. Helm. Transmission features of frequency-selective components in the far field determined by terahertz time–domain spectroscopy. *Appl. Opt.*, 38(18):3961–3967, 1999.
20. Garcia de Abajo. Light scattering by particles and hole arrays. *Reviews of Modern Physics*, 79:1267, 2007.
21. Peter B. Catrysse and Shanhui Fan. Propagating plasmonic mode in nanoscale apertures and its implications for extraordinary transmission. *Journal of Nanophotonics*, 2:021790, 2008.
22. R. Gordon, A.G. Brolo, D. Sinton, and K.L. Kavanagh. Resonant optical transmission through hole-arrays in metal films: physics and applications. *Laser & Photon. Rev.*, 4:311, 2010.
23. F.I. Baida and D. Van Labeke. Light transmission by subwavelength annular aperture arrays in metallic films. *Optics Commun.*, 209:17–22, August 2002.
24. F.I. Baida and D. Van Labeke. Three-dimensional structures for enhanced transmission trough a metallic film: Annular aperture arrays. *Phys. Rev. B*, 67(155314):1–7, 2003.
25. F. I. Baida. Enhanced transmission through subwavelength metallic coaxial apertures by excitation of the tem mode. *Applied Phys. B*, 89(2–3):145–149, 2007. Rapid Communication.
26. A. Moreau, G. Granet, F. I. Baida, and D. Van Labeke. Light transmission by subwavelength square coaxial aperture arrays in metallic films. *Opt. Express*, 11(10):1131–1136, May 2003.
27. F.I. Baida, Y. Poujet, B. Guizal, and D. Van Labeke. New design for enhanced transmission and polarization control through near-field optical microscopy probes. *Optics Commun.*, 256: 190–195, 2005.
28. Michael I. Haftel, Carl Schlockermann, and Girsh Blumberg. Role of cylindrical surface plasmons in enhanced transmission. *Appl. Phys. Lett.*, 88:193104, 2006.
29. S. M. Orbons, M. I. Haftel, C. Schlockermann, D. Freeman, M. Milicevic, T. J. Davis, B. Luther-Davies, D. N. Jamieson, and A. Roberts. Dual resonance mechanisms facilitating enhanced optical transmission in coaxial waveguide arrays. *Opt. Lett.*, 33:821–823, 2008.
30. F. I. Baida, Y. Poujet, J. Salvi, D. Van Labeke, and B. Guizal. Extraordinary transmission beyond the cut-off through sub-λ annular aperture arrays. *Optics Commun.*, 282:14631466, 2009.
31. A. Roberts and R. C. McPhedran. Bandpass grids with annular apertures. *IEEE Trans. Antennas Propag.*, 36:607–611, 1988.
32. W. J. Fan, S. Zhang, B. Minhas, K.J. Malloy, and S.R.J. Brueck. Enhanced infrared transmission through subwavelength coaxial metallic arrays. *Phys. Rev. Lett.*, 94(33902):1–4, January 2005.
33. W. Fan, S. Zhang, K. J. Malloy, and S. R. J. Brueck. Enhanced mid-infrared transmission through nanoscale metallic coaxial-aperture arrays. *Opt. Eng.*, 13(12):4406–4413, June 2005.
34. Matthew J. Lockyear, Alastair P. Hibbins, and J. Roy Sambles. Microwave transmission through a single subwavelength annular aperture in a metal plate. *Phys. Rev. Lett.*, 94:193902, 2006.
35. H. Caglayan, I. Bulu, and E. Ozbay. Extraordinary grating-coupled microwave transmission through a subwavelength annular aperture. *Opt. Eng.*, 13:1666, 2005.
36. J. Salvi, M. Roussey, F. I. Baida, M.-P. Bernal, A. Mussot, T. Sylvestre, H. Maillotte, D. Van Labeke, A. Perentes, I. Utke, C. Sandu, P. Hoffmann, and B. Dwir. Annular aperture arrays: Study in the visible region of the electromagnetic spectrum. *Opt. Lett.*, 30(13):1611–1613, July 2005.
37. W. Fan, S. Zhang, K. J. Malloy, and S. R. J. Brueck. Large-area, infrared nanophotonic materials fabricated using interferometric lithography. *J. Vac. Sci. Technol. B*, 23:2700, 2005.
38. H. Caglayan, I. Bulu, and E. Ozbay. Plasmonic structures with extraordinary transmission and highly directional beaming properties. *Microwave and Optical Technology Letters*, 48:2491, 2006.
39. Wenjun Fan, Shuang Zhang, N.-C. Panoiu, A. Abdenour, S. Krishna, R.M. Osgood, K.J. Malloy, and S. R. J. Brueck. Second harmonic generation from a nanopatterned isotropic nonlinear material. 6(5):1027–1030, 2006.

40. A.J. Gallant, J.A. Levitt, M. Kaliteevski, D. Wood, M.C. Petty, R.A. Abram, S. Brand, and J.M. Chamberlain. Enhanced thz transmission through micromachined sub-wavelength annular apertures. *IET Seminar on MEMS Sensors and Actuators*, 2006(11367):169–175, 2006.
41. Yannick Poujet, Jérôme Salvi, and Fadi Issam Baida. 90% extraordinary optical transmission in the visible range through annular aperture metallic arrays. *Opt. Lett.*, 32(20):2942–2944, 2007.
42. Hongfeng Gai, Jia Wang, Qian Tian, Wei Xia, and Xiangang Xu. Experimental investigation of the performance of an annular aperture and a circular aperture on the same very-small-aperture laser facet. *Appl. Opt.*, 46(25):6449–6453, 2007.
43. C.K. Chang, D.Z. Lin, Y.C. Chang, M.W. Lin, J.T. Yeh, J.M. Liu, C.S. Yeh, and C.K. Lee. Enhancing intensity of emitted light from a ring by incorporating a circular groove. *Opt. Express*, 15(23):15029–15034, 2007.
44. A.-A. Yanik X. Wang, S. Erramilli, M.-K. Hong, and H. Altug. Extraordinary midinfrared transmission of rectangular coaxial nanoaperture arrays. *Appl. Phys. Lett.*, 93:081104, 2008.
45. M.J. Kofke, D.H. Waldeck, Z. Fakhraai, S. Ip, and G.C. Walker. The effect of periodicity on the extraordinary optical transmission of annular aperture arrays. *Appl. Phys. Lett.*, 94:023104, 2009.
46. C.M. Rollinson, S.M. Orbons, S.T. Huntington, B.C. Gibson, J. Canning, J.D. Love, A. Roberts, and D.N. Jamieson. Metal-free scanning optical microscopy with a fractal fiber probe. *Opt. Eng.*, 17(3):1772–1780, 2009.
47. E. Verhagen, L. Kuipers, and A. Polman. Field enhancement in metallic subwavelength aperture arrays probed by erbium upconversion luminescence. *Opt. Eng.*, 17:14586, 2009.
48. Yuh-Yan Yu, Ding-Zheng Lin, Long-Sun Huang, and Chih-Kung Lee. Effect of subwavelength annular aperture diameter on the nondiffracting region of generated bessel beams. *Opt. Eng.*, 17(4):2707–2713, 2009.
49. Tsung-Dar Cheng, Ding-Zheng Lin, Jyi-Tyan Yeh, Jonq-Min Liu, Chau-Shioung Yeh, and Chih-Kung Lee. Propagation characteristics of silver and tungsten subwavelength annular aperture generated sub-micron non-diffraction beams. *Opt. Express*, 17(7):5330–5339, 2009.
50. Ahmet Ali Yanik, Ronen Adato, Shyamsunder Erramilli, and Hatice Altug. Hybridized nanocavities as single-polarized?plasmonic antennas. *Opt. Express*, 17(23):20900–20910, 2009.
51. Feng Wang, Min Xiao, Kai Sun, and Qi-Huo Wei. Generation of radially and azimuthally polarized light by optical transmission through concentric circular nanoslits in ag films. *Opt. Express*, 18(1):63–71, 2010.
52. C.R. Williams, M. Misra, S.R. Andrews, S.A. Maier, S. Carretero-Palacios, S.G. Rodrigo, F.J. Garcia-Vidal, and L. Martin-Moreno. Dual band terahertz waveguiding on a planar metal surface patterned with annular holes. *Appl. Phys. Lett.*, 96:011101, 2010.
53. P.B. Catrysse and Shanhui Fan. Understanding the dispersion of coaxial plasmonic structures through a connection with the planar metal-insulator-metal geometry. *Appl. Phys. Lett.*, 94:231111, 2009.
54. D.B. Davidson and R.W. Ziolkowski. Body–of–revolution finite–difference time–domain modeling of space–time focusing by a three–dimensional lens. *J. Opt. Soc. Am. A*, 11(4):1471–1490, April 1994.
55. A. Taflove and S.C. Hagness. *Computational Electrodynamics, the Finite-Difference Time–Domain Method*. Artech House, Norwood, MA, second edition, 2005.
56. F.I. Baida, D. Van Labeke, and Y. Pagani. Body-of-revolution FDTD simulations of improved tip performance for scanning near-field optical microscopes. *Optics Communications*, 255:241–252, 2003.
57. C.T. Chan, Q.L. Yu, and K.M. Ho. Order-n spectral method for electromagnetic waves. *Phys. Rev. B*, 51(23):16635–16642, June 1995.
58. F.I. Baida, D. Van Labeke, G. Granet, A. Moreau, and A. Belkhir. Origin of the super-enhanced light transmission through a 2-D metallic annular aperture array: a study of photonic bands. 79:1–8, 2004.
59. F.I. Baida, A. Belkhir, D. Van Labeke, and O. Lamrous. Subwavelength metallic coaxial waveguides in the optical range: Role of the plasmonic modes. *Phys. Rev. B*, 74:205419, 2006.

60. Michael I. Haftel, Carl Schlockermann, , and G. Blumberg. Enhanced transmission with coaxial nanoapertures: Role of cylindrical surface plasmons. *Phys. Rev. B*, 74:235405, 2006.
61. R. de Waele, S.P. Burgos, A. Polman, and H.A. Atwater. Plasmon dispersion in coaxial waveguides from single-cavity optical transmission measurements. *Nano Lett.*, 9:2832–2837, 2009.
62. E.H. Barakat, M.-P. Bernal, and F. I. Baida. Second harmonic generation enhancement by use of annular aperture arrays embedded into silver and filled by lithium niobate. *Optics Express*, 18:6530, 2010.
63. D. Van Labeke, D. Gérard, B. Guizal, F. I. Baida, and L. Li. An angle-independent frequency selective surface in the optical range. *Opt. Express*, 14(25):11945–11951, 2006.
64. A. Belkhir and F. I. Baida. Three-dimensional finite-difference time-domain algorithm for oblique incidence with adaptation of perfectly matched layers and nonuniform meshing: Application to the study of a radar dome. *Phys. Rev. E*, 77:056701, 2008.
65. J. Seidel, F. I. Baida, L. Bischoff, B. Guizal, S. Grafstrom, D. Van Labeke, and L.M. Eng. Coupling between surface plasmon modes on metal films. *Phys. Rev. B*, 69:121405, 2004.
66. A. Perentes, I. Utke, B. Dwir, M. Leutenegger, T. Lasser, P. Hoffmann, F. Baida, M.-P. Bernal, M. Roussey, J. Salvi, and D. Van Labeke. Fabrication of arrays of sub-wavelength nanoapertures in an optically thick gold layer on glass slides for optical studies. *Nanotechnology*, 16:S273–S277, 2005.
67. S.M. Orbons, D. Freeman, B. Luther-Davies, B.C. Gibsonc, S.T. Huntingtonc, D.N. Jamiesona, and A. Roberts. Optical properties of silver composite metamaterials. *physica B*, 394:176–179, 2007.
68. Y. Poujet, M. Roussey, J. Salvi, F.I. Baida, D. Van Labeke, A. Perentes, C. Santschi, and P. Hoffmann. Super-transmission of light through subwavelength annular aperture arrays in metallic films: Spectral analysis and near-field optical images in the visible range. *Photon. Nanostruct.*, 4:47–53, 2006.
69. D. Courjon, J.-M. Vigoureux, M. Spajer, K. Sarayeddine, and S. Leblanc. External and internal reflection near field microscopy: experiment and results. *Appl. Opt.*, 29:3734–3740, 1990.
70. D. Van Labeke and D. Barchiesi. Probes for scanning tunneling optical microscopy: A theorical comparison. *J. Opt. Soc. Am. A*, 10(10):2193–2201, October 1993.
71. T. Grosjean and D. Courjon. Polarization filtering induced by imaging systems: Effect on image structure. *Phys. Rev. E*, 67(4):046611, Apr 2003.
72. J. Bravo-Abad, F.J. García-Vidal, and L. Martín-Moreno. Resonant transmission of light through finite chains of subwavelength holes in a metallic film. *Phys. Rev. Lett.*, 93:227401, 2004.
73. C. Genet and T. W. Ebbesen. Light in tiny holes. *Nature*, 445:39–46, 2007.
74. T. Thio, K.M. Pellerin, R.A. Linke, H.J. Lezec, and T.W. Ebbesen. Enhanced light transmission through a single subwavelength aperture. *Opt. Lett.*, 26(24):1972–1974, December 2001.
75. Y. Wang, K. Kempa, B. Kimball, J. B. Carlson, G. Benham, W. Z. Li, T. Kempa, J. Rybczynski, A. Herczynski, and Z. F. Ren. Receiving and transmitting light-like radio waves: Antenna effect in arrays of aligned carbon nanotubes. *Appl. Phys. Lett.*, 85:2607, 2004.
76. J. Rybczynski, K. Kempa, A. Herczynski, Y. Wang, M. J. Neughton, Z. F. Ren, Z. P. Huang, D. Cai, and M. Giersig. Subwavelength waveguide for visible light. *Appl. Phys. Lett.*, 90:021104, 2007.
77. T. Thio. Coaxing light into small spaces. *Nature Nanotechnology*, 2:136–138, 2007.
78. A. Roberts. Beam transmission through hole arrays. *Opt. Express*, 18(3):2528–2533, 2010.
79. X. Wang Y. Peng and K. Kempa. Tem-like optical mode of a coaxial nanowaveguide. *Opt. Eng.*, 16(3):1758–1763, 2008.
80. K. Kempa, X. Wang, Z. F. Ren, and M. J. Naughton. Discretely guided electromagnetic effective medium. *Appl. Phys. Lett.*, 92:043114, 2008.
81. Jun Wang, Wei Zhou, and Er-Ping Li. Enhancing the light transmission of plasmonic metamaterials through polygonal aperture arrays. *Opt. Express*, 17(22):20349–20354, 2009.
82. F. J. Rodríguez-Fortuño, C. García-Meca, R. Ortuño, J. Martí, and A. Martínez. Coaxial plasmonic waveguide array as a negative-index metamaterial. *Opt. Lett.*, 34(21):3325–3327, 2009.
83. Zhu Wei-Ren, Zhao Xiao-Peng, Bao Shi, and Zhang Yan-Ping. Highly symmetric planar metamaterial absorbers based on annular and circular patches. *Chinese Physics Letters*, 27(1):014204, 2010.

Chapter 10
Melting Transitions of DNA-Capped Gold Nanoparticle Assemblies

Sithara S. Wijeratne, Jay M. Patel, and Ching-Hwa Kiang

1 Introduction

DNA is responsible for the storage and realization of genetic information [1]. The double-helical structure of DNA has been widely used due to its chemical, biological, and physical properties [2, 3]. Composed of four nitrogenous bases, DNA base pairing (AT and CG) through hydrogen bonding influences the chemical properties of DNA. The base-pair combination potential allows DNA to store tremendous amounts of information, significant for biological processes. Their recognition capabilities, notable in power and accuracy, have allowed DNA to show great promise as a basis for various self-assembled nanostructures [4, 5]. DNA's capacity for self-assembly is rooted in the simple, solid nature of its complimentary base pair interactions.

The self-assembly of DNA-capped nanoparticles involves interactions of surface-bound DNA, usually used as the probe, and the complementary, free DNA, which links the nanoparticle together to form aggregates. Probe DNA with known sequences is used to detect nucleotides with sequences complementary to the probe DNA. For example, DNA microarrays use a lattice of distinct DNA sequences fixed to a solid support to monitor gene expression levels or gene mutation levels [6]. For nanostructure self-assembly, probe DNA can be fixed to surfaces or to nanoparticles, as shown in Fig. 10.1 [7, 8].

DNA melting is the process of separating the double helical DNA structure into two single strands [9–14]. The mechanism of melting depends on the length of the strands involved; when melted DNA has fewer than 12–14 base pairs, a two-state model provides an accurate picture of the process [15, 16], with a broad transition curve [17], as shown in Fig. 10.2. However, when the DNA is bound to surfaces, the melting curves exhibit a much sharper transition, as shown in Fig. 10.3. The transition temperature

S.S. Wijeratne • J.M. Patel • C.-H. Kiang (✉)
Department of Physics and Astronomy, Rice University, Houston, TX 77005, USA

C.D. Geddes (ed.), *Reviews in Plasmonics 2010*, Reviews in Plasmonics,
DOI 10.1007/978-1-4614-0884-0_10,

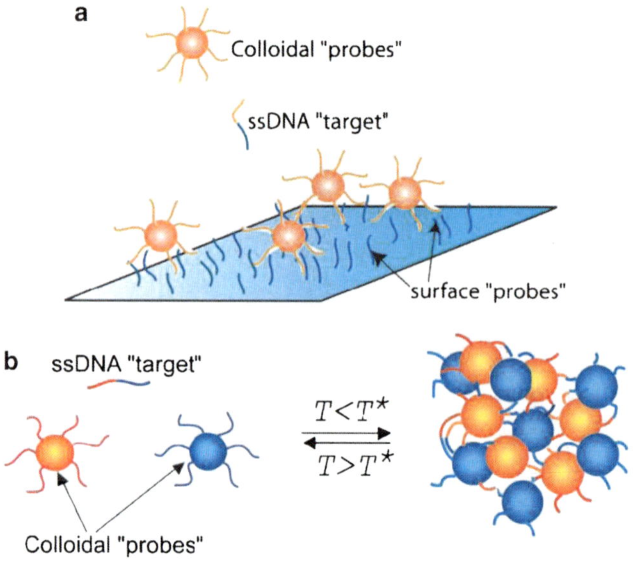

Fig. 10.1 Schematic representation of DNA-linked colloids (**a**) in the surface and (**b**) bulk formats. From [7]

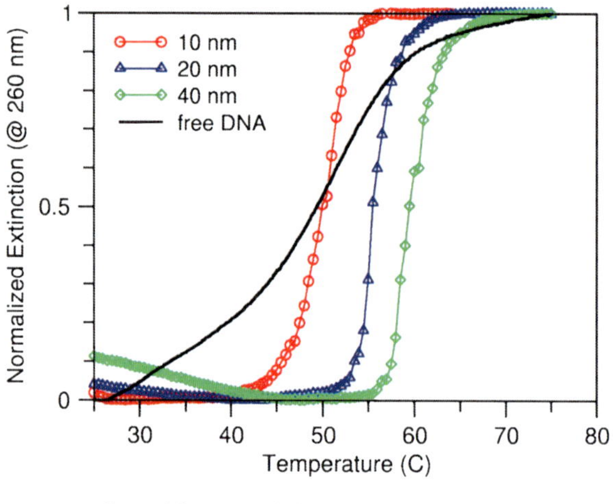

Fig. 10.2 Melting curves for gold nanoparticle systems monitored at 260 nm. The free DNA shows a broad melting transition. Adapted from [18]

depends on the particle size. Such results have been observed experimentally and supported by theory (see Fig. 10.3). The hybridization of DNA results in self-assembly of these gold nanoparticles (see Fig. 10.4), affecting the optical properties of the solution containing these assemblies. This change can provide a tool for DNA detection technology. In addition, such microscopic DNA behavior can be mapped onto macroscopic phase transitions which offers a unique control over colloidal aggregation [20].

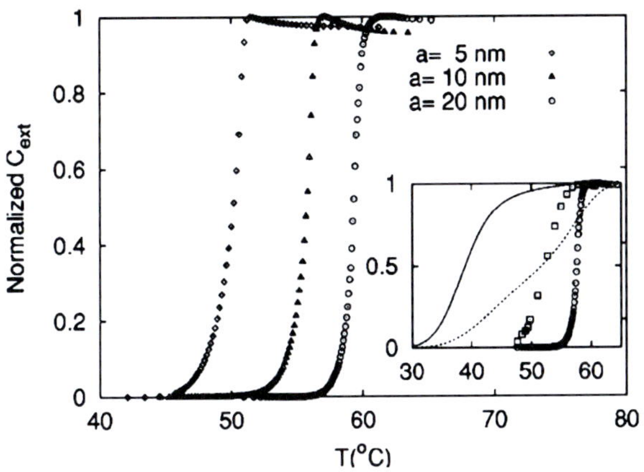

Fig. 10.3 Theoretical melting curve for several particle radii *a*, 5 nm, 10 nm, and 20 nm. From [19]

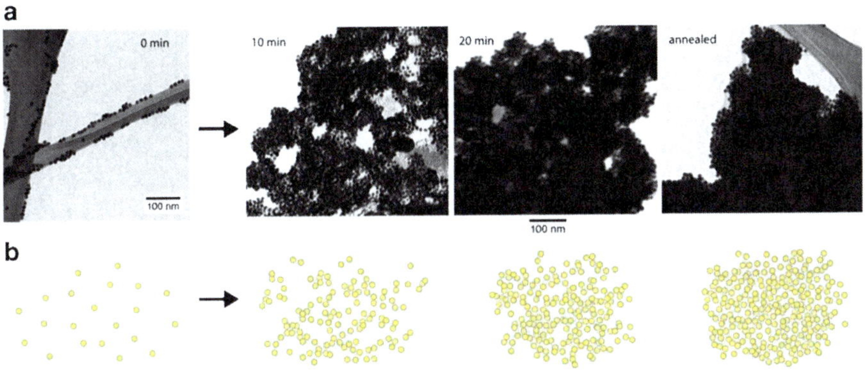

Fig. 10.4 (**a**) Transmission electron microscopy image of 10-nm gold colloids capped with thiol-modified DNA. (**b**) The growth mechanism of DNA-linked gold colloids. From [18]

Phase-transition experiments of surface-bound DNA have been shown to be unique, resembling a new class of complex fluids [18, 21–24].

Potential arrangements of these DNA nanostructures include two-dimensional arrays [6], two-dimensional patterns [25], and three-dimensional nano-objects [26]. Examples of synthesized DNA-based nanostructures include metallic nano-wires and nanoparticles [25–29], fullerene molecules [30], and carbon nanotubes [31, 32]. The use of these nanostructures requires a detailed understanding of the interactions between DNA strands. The conjugation of these nanoscale solids (dots, wires, nanotubes) with recognition-capable DNA allows for the creation and miniaturization of biological electronics and other bionanotechnology. Applications of DNA-conjugated nanostructures have shown improvement in not only size, but also performance. For example, noncomputational tiling arrays have been used as molecular scale circuit components [33, 34], either by the chemistry between DNA and molecular electrodes [34], or by the use of gold beads [35]. Nanomechanical devices have also been created that undergo conformational change due to environmental change [36–38], strand displacement [39], such as the nanowalker [40, 41], or enzymatic activity [42]. This is only one type of many novel biosensors developed using this technology [43–46]. Lastly, the optical properties brought about by aggregation and network formation can be used as a tool in DNA-detection. Some examples include the diagnosis of genetic diseases, RNA profiling, biodefense [47–49], gene chips [50], detection of UV damage [51], and single-molecule sequencing [52], including the use of nanopores [53–55].

2 Experimental Procedures

Gold nanoparticles (10–40 nm, <10% polydispersity) were conjugated with either 3′ or 5′ thiol-modified single-stranded DNA (target DNA) [21]. These gold nanoparticles were fully covered with DNA and do not form aggregate without complementary DNA linkage. When noncomplementary target DNA was used, the aggregation begins only when complementary linker DNA was added. The kinetic and equilibrium properties of particle aggregation are governed by DNA hybridization. The DNA network can be formed either with linker DNA or without linker DNA if the sequences of single-stranded DNA on gold nanoparticles are complementary.

The aggregation process and phase transitions can be monitored by optical spectroscopy, taking advantage of DNA bases' strong absorption in the ultraviolet region and optical extinction from the gold surface plasmon at the visible region. DNA base pairs have strong optical absorption at approximately 260 nm [16] and the intensity decreases when the hydrogen bonds between base pairs are broken, i.e., the nonpaired DNA bases have less absorption than the paired bases due to hypochromism. On the other hand, the surface plasmon of gold nanoparticles exhibits strong extinction at approximately 520 nm [19, 56, 57], with the size of the aggregate determining the specific extinction coefficient. This makes optical extinction spectroscopy a vital analysis tool for studying the kinetics of aggregation.

Extinction spectra of DNA–gold nanoparticles were taken using a PerkinElmer Lambda 45 spectrophotometer. The experimentally observed (Fig. 10.5a) and theoretically predicted (Fig. 10.5b) spectra for this process show that the initial aggregation occurs alongside first a volume fraction increase, followed by a network size increase [18, 21]. Equally important, spectroscopy techniques can also be used to study the thermal melting behavior of these DNA networks. Unlike melting of free DNA, DNA-bound to gold nanoparticles have a sharper melting transition, with a mechanism related to colloidal phase transition, as shown in Fig. 10.2.

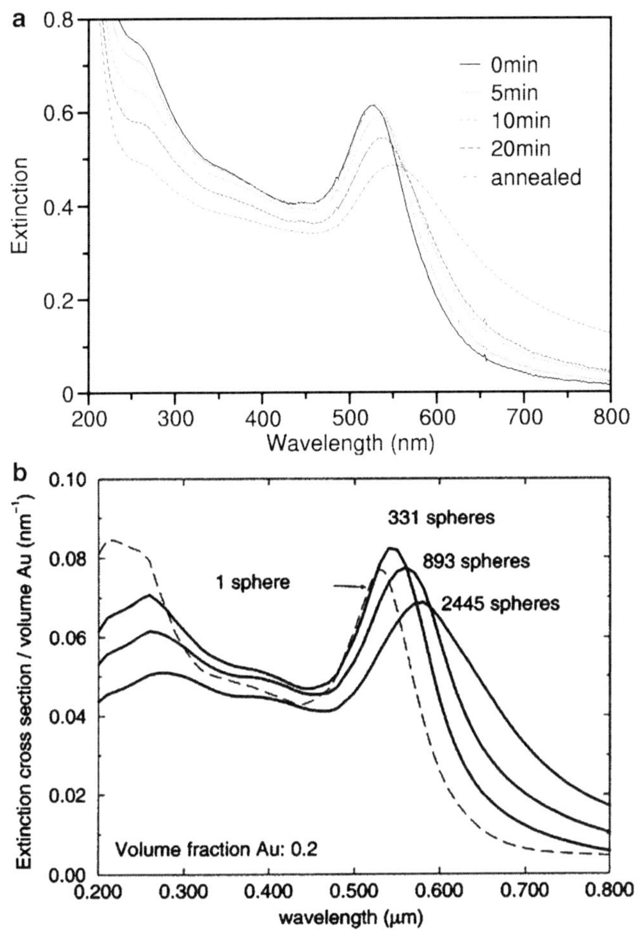

Fig. 10.5 (**a**) Optical absorption spectra during aggregation. From [21]. (**b**) Calculated extinction spectra for nanoparticle assemblies of increasing size. From [56]

3 Results and Discussion

3.1 Particle Size

Figure 10.2 demonstrates the effect of particle size on the melting transition. The melting transition width (FWHM) is 5°C, compared to the 12°C as demonstrated by melted free DNA [43]. The transition width and melting temperature, T_m, of DNA have been modified by the binding to gold particles [18]. The curves for 10, 20, and 40 nm gold particles with linker DNA show that the melting properties depend on the

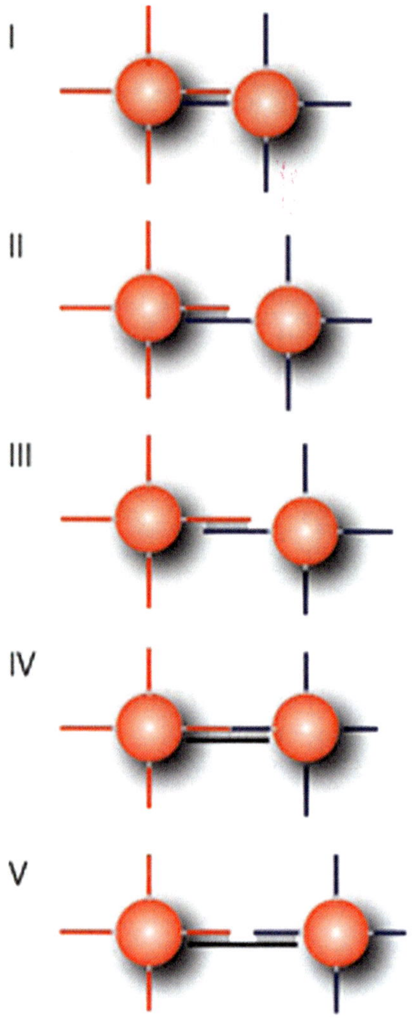

Fig. 10.6 DNA-capped gold nanoparticle samples with different lengths of spacers. Samples I, II, and III are directly linked nanoparticles without a linker (target) DNA. Samples IV and V utilize DNA linkers to facilitate aggregation. Inclusion of spacer DNA affects the melting temperatures, and the effects are different for systems linked with and without linker DNA

size of the particles. Large gold nanoparticles have a higher melting temperature, perhaps due to the multiple DNA connections between gold nanoparticles. Results from theoretical calculations provide an explanation of this observation as shown in Fig. 10.3 [7].

3.2 Direct Linking

The melting transitions of gold nanoparticles with and without linker DNA have distinct features. As an example, we looked at a system containing samples prepared with different spacers as shown in Fig. 10.6. Samples I–III, all have 12 complementary bases, and the only difference is the length of noncomplementary DNA,

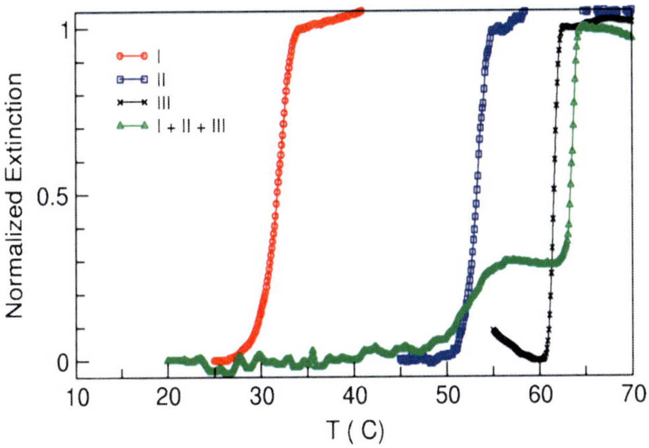

Fig. 10.7 Melting curves of directly linked (no linker DNA) nanoparticle assemblies. Three melting curves are from three samples with varying lengths of spacers. When mixing particles with DNA of different spacer length, the curve exhibits a multistep melting transition. Adapted from [58]

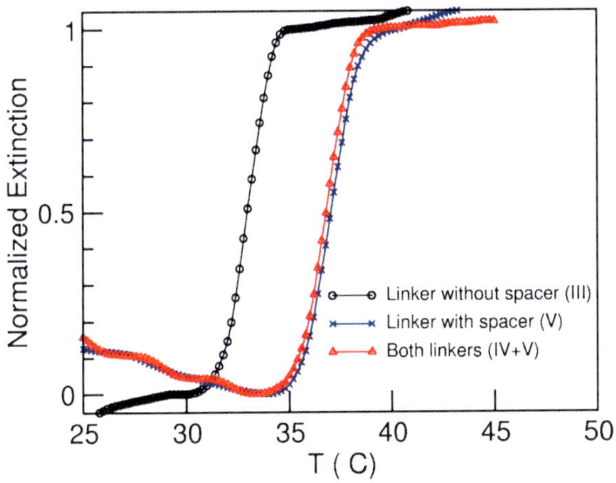

Fig. 10.8 Melting curves of DNA nanoparticle assembly. With the same length and sequence for the paired DNA, the inclusion of spacers raises the melting temperatures. Adapted from [58]

i.e., spacers. Sample I does not have spacer DNA, sample II has 6-base spacer DNA on one particle, and sample III, has 6-base spacer DNA on both of the gold particles. The distance between nanoparticles are, therefore, 12, 18, and 24 bases for samples I, II, and III, respectively. Since the distance between nanoparticles affects the melting temperature, with larger interparticle distance melting at a higher temperature, we have three distinct melting curves for the three different samples, which have melting temperatures ranging from 32°C to 62°C (see Fig. 10.7).

The kinetics of nanoparticle aggregation for systems with linkers, however, is different from systems that are directly linked. Figure 10.8 shows the melting curves of systems with linkers with and without spacers (see samples IV and V in Fig. 10.8).

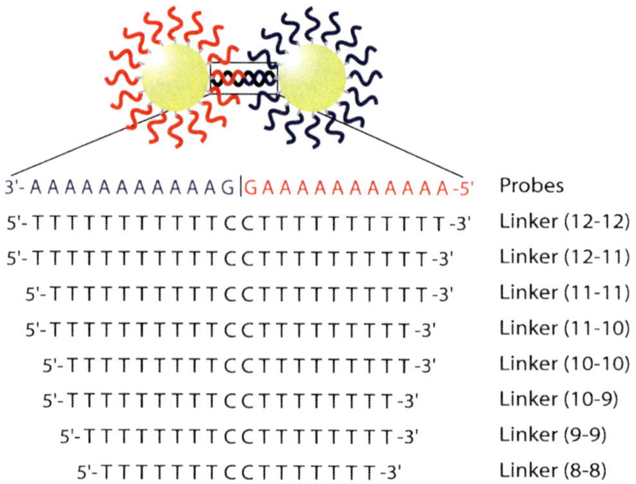

Fig. 10.9 DNA linkers of varying lengths used to form aggregation. Adapted from [59]

Systems with linkers having spacer DNA melt at higher temperatures, compared to those without spacers. This is expected since a change in interparticle spacing affects melting temperatures. However, when mixing linkers of different lengths (with and without added spacers), competition between linkers results in the system settling down at a more stable state, i.e., those with higher melting temperatures. This indicates that most of the linkers that contribute to the linking of the nanoparticles in the aggregates are the ones with spacers, which form more stable assemblies. This is distinct from the system connected without linkers, with aggregates likely to cluster with particles that have similar DNA lengths. This may be due to the difficulty for the systems without linkers to bind and unbind in the process of searching for the most stable cluster, and the system stays in the kinetic trap that has lower melting temperatures.

3.3 *Disorder*

To study the dependence on the length of the linker DNA, simple DNA sequences with uniform base composition were used to avoid complications resulting from sequences-dependent effects. A schematic of the basic building block of DNA along with the linker sequences is shown in Fig. 10.9. DNA linker length ranging from 26 to 16 bases, composed of mostly T bases, was used in the study. An unusual melting temperature dependence on DNA linker length has been observed. This characteristic is not seen in free DNA, whose melting temperature rises when the number of bases is higher (see Fig. 10.10). The melting temperature of gold–DNA systems oscillates, i.e., the increase in melting temperature is non-monotonic with linker length there are an even.

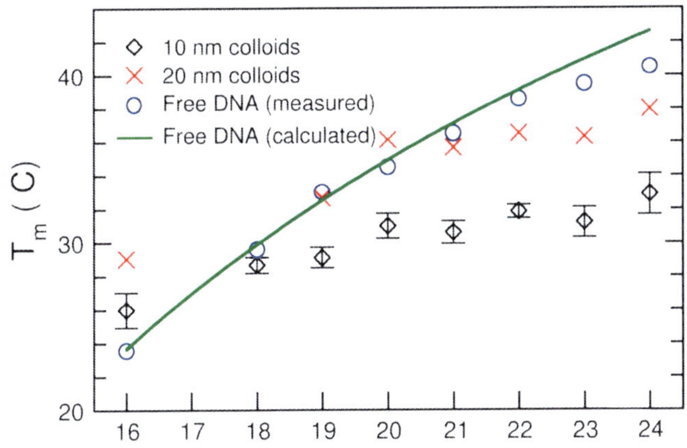

Fig. 10.10 Melting temperature as a function of linker length for free DNA, and 10- and 20-nm gold-attached DNA. The *line* represents the calculated melting temperatures for free DNA. Adapted from [59]

In DNA-linked nanoparticle systems, using duplexes with different lengths introduces binding energy disorder into the system, lowering its stability. Because this is not observed in free DNA duplexes, we believe that the lower temperature stems from an entropic effect specific from nanoparticle assembly. More importantly, the interaction energy is tunable by changing the DNA composition [59], making this system suitable for studying various aspects of colloidal phase transitions.

3.4 *Defects*

Melting temperatures for linkers with either complementary DNA sequences and with various defects such as mismatches and deletions were compared. Defects include mismatched base pairs and deletions at various locations of DNA, as shown in Fig. 10.11. Unlike the free DNA, where defects in DNA base-pairing almost always results in lower melting temperatures, the inclusion of defects in the DNA linkers sometimes raises the melting temperature. The melting temperatures for free and surface-bound DNA with various sequence mismatch and deletions are shown in Fig. 10.12. The melting temperature of free DNA always decreases when noncomplementary DNA is introduced. However, certain mismatches and deletions of surface-bound DNA increase the melting temperature compared to fully complementary DNA sequences. Some single base mismatches and deletions near the surface of the probe particles increase the melting temperature of the nanoparticle assembly.

Deviation in DNA hybridization thermodynamics for surface-bound DNA may be a result of Coulomb blockage [61]. The increase in melting temperature due to deletions at the particle surface can be explained by a decrease in electrostatic repulsion between the particle surface and the nearby paired base. Nonspecific binding

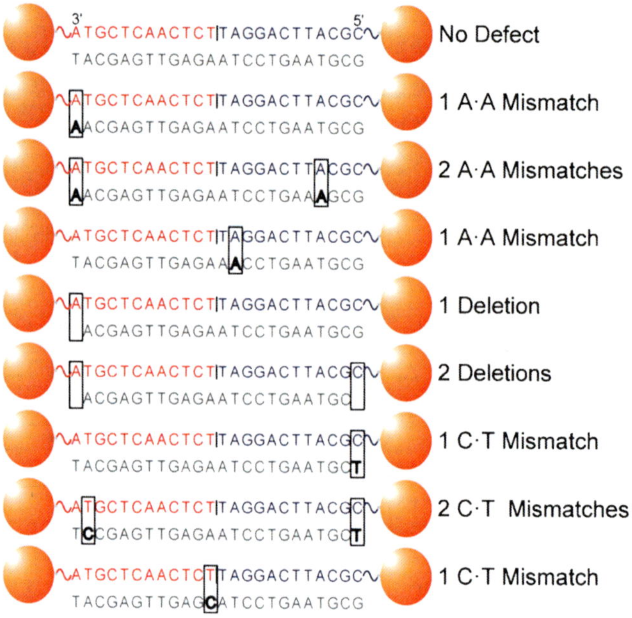

Fig. 10.11 Linker and probe DNA sequences used in the study. The *boxed area* indicates base-pairing defects. From [60]

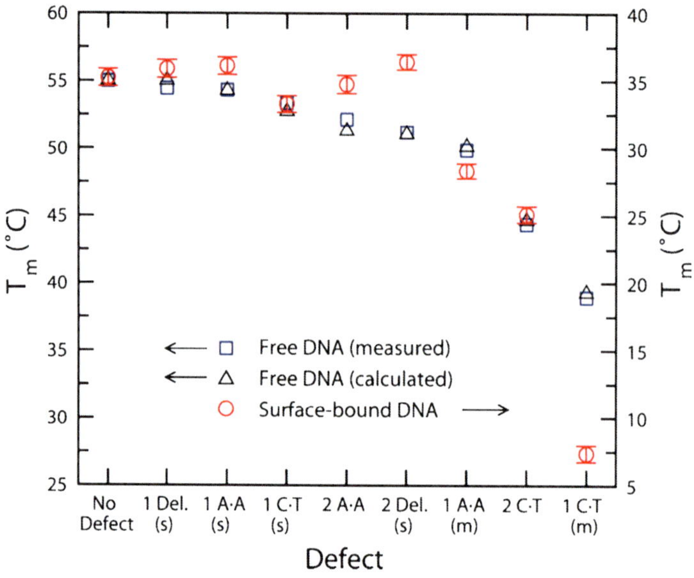

Fig 10.12 Melting temperatures with respect to DNA defect for free, unattached DNA and surface-bound DNA. From [59]

between the surface and the unpaired, dangling base may also contribute to the increased stability of systems with mismatches and deletions. This contrasts with free DNA, for which the inclusion of defects consistently lowers the melting point. A mismatched base pair would give the dangling base the flexibility necessary for nonspecific binding to the particle surface, which may result in higher melting temperature. Furthermore, the nature and degree of the change in melting temperature are determined by the sequence and location of the responsible defect, as well as the specific base pair involved. These results show that mismatches and deletions have different effects on the melting temperature for free and surface-bound DNA. Therefore, for quantitative interpretation of the DNA hybridization to surface-bound DNA, the system needs to be well characterized, particularly if one intends to use them for the detection of single-nucleotide polymorphism (SNP). Such a strong surface effect should be taken into account when quantifying the data from DNA detection systems such as nanoparticle aggregates and gene chips. The increased stability may even be used to improve the sensitivity of the detection limit.

4 Conclusion

The melting properties of DNA–gold nanoparticles were investigated using optical extinction spectroscopy. Various parameters such as particle size, DNA linker length, interparticle distance, and degree disorder were varied to get a better understanding of the DNA–gold nanoparticle system. The change in optical property due to DNA–gold nanoparticle self-assembly has the potential to be used for DNA detection. The network formation in DNA–gold particle assemblies exhibits unique phase behavior, which is different from that of free DNA. The observed melting transition was also sharper for the nanoparticle assemblies than for free DNA. The phase behavior is sensitive to defects in DNA base pairing and disorder. These results have implications for the design of DNA detection technology and fundamental understanding of DNA melting behavior.

Acknowledgments We thank the National Science Foundation DMR-0907676 and the Welch Foundation C-1632 for financial support.

References

1. Frank-Kamenetskii MD (1997) Biophysics of the DNA molecule: Phys Rep 288:13–60
2. Watson JD, Crick FHC (1953) Molecular structure of nucleic acids. Nature 171:737–738
3. Zheng J, Birktoft JJ, Chen Y, Wang T, Sha R, Constantinou PE, Ginell SL, Mao C, Seeman NC (2009) From molecular to macroscopic via the rational design of a self-assembled 3D DNA crystal. Nature 461:74–77
4. Yao H, Changqing Y, Tzang C-H, Zhu J, Yang M (2006) DNA-directed self-assembly of gold nanoparticles into binary and ternary nanostructures. Nanotechnology 18: 015102-1-7

5. Maiti PK, Pascal TA, Vaidehi N, Goddard III WA (2007) Understanding DNA based nanostructures. J Nanosci and Nanotech 7:1712–1720
6. Tomiuk S, Hofmann K (2001) Microarray probe selection strategies. Brief Bioinform 2:329–340
7. Lukatsky DB, Frenkel D (2005) Surface and bulk dissolution properties, and selectivity of DNA-linked nanoparticle assemblies. Journal of Chem Phys 122:214904
8. Sun Y, Harris NC, Kiang C-H (2007) Phase transition and optical properties of DNA-gold nanoparticle assemblies. Plasmonics 2:193–99
9. Wartell RM, Benight AS (1985) Thermal denaturation of DNA molecules: a comparison of theory with experiment. Phys Rep 126:67–107
10. Cule D, Hwa T (1997) Denaturation of heterogenous DNA. Phys Rev Lett 79:2375–2378
11. Gelfand CA, Plum GE, Mielewczyk S, Remeta DP, Breslauer KJ (1999) A quantitative method for evaluating the stabilities of nucleic acid complexes. Proc Natl Acad Sci 96:6113–6118
12. Lubensky DK, Nelson DR (2000) Pulling pinned polymers and unzipping DNA. Phys Rev Lett 85:1572–1575
13. Danilowicz C, Kafri Y, Conroy RS, Coljee V, Weks J, Prentiss M (2004) Measurement of the phase diagram of DNA unzipping in the temperature-force plane. Phys Rev Lett 93: 078101-1-17
14. Erie DA, Jones RA, Olson WK, Sinha NK, Breslauer KL (1989) Melting behavior of a covalently closed, single-stranded, circular DNA. Biochemistry 28(1):268–273
15. Bloomfield VA, Crothers DM, Tinoco Jr I (2000) Nucleic acids. University Science Books, California
16. Cantor CR, Schimmel PR (1980) Chemistry, Part II: Techniques for the Study of Biological Structure and Function. WH Freeman and Company, New York
17. Naef F, Lim DA, Patil N, Magnasco M (2002) DNA hybridization to mismatched templates: a chip study. Phys Rev E 65: 040902-1-4
18. Sun Y, Harris NC, Kiang C-H (2005) The reversible phase transition of DNA-linked colloidal gold assemblies. Physica A 354:1–9
19. Park SY, Stroud D (2003) Theory of melting and the optical properties of gold/DNA nanocomposites. Phys Rev B 67:212202
20. Anderson VJ, Lekkerkerker HNW (2002) Insights into phase transition kinetics from colloid science. Nature 416:811–815
21. Kiang C-H (2003) Phase transition of DNA-linked gold nanoparticles. Physica A 321: 164–169
22. Lukatsky DB, Frenkel D (2004) Phase behavior and selectivity of DNA-linked nanoparticle assemblies. Phys Rev Lett 92:068302-1-4
23. Drukker K, Wu G, Schatz GC (2001) Model simulations of DNA denaturation dynamics. J Chem Phys 114:579–590
24. Kim J-Y, Lee J-S (2009) Synthesis and thermally reversible assembly of DNA-gold nanoparticle cluster conjugates. Nano Lett 9:4564–4569
25. Maeda Y, Tabata H, Kawai T (2001) Two-dimensional assembly of gold nanoparticles with a DNA network template. Appl Phys Lett 79:1181–1183
26. Becerril HA, Stoltenberg RM, Wheeler DR, Davis RC, Harb JN, Woolley AT (2005) DNA-templated three-branched nanostructures for nanoelectronic devices. J Am Chem Soc 127:2828–2829
27. Mirkin CA (1996) A DNA-based method for rationally assembling nanoparticles in macroscopic materials. Nature 382:607–609
28. Braun E, Eichen Y, Sivan U, Ben-Yoseph G (1998) DNA-templated assembly and electrode attachment of a conducting silver wire. Nature 391: 775–778
29. Liu D, Park SH, Reif JH, LaBean TH (2005) DNA nanotubes self-assembled from triple-crossover tiles as templates for conductive nanowires. P Natl Acad Sci 101: 717–722
30. Cassell AM, Scriven WA, Tour JM (1998) Assembly of DNA/fullerene hybrid materials. Angew Chem Int Ed 37:1528–1531

31. Dwyer C, Guthold, M, Falvo M, Washburn S, Superfine R, Erie D (2002) DNA-functionalized single-walled carbon nanotubes. Nanotechnology 13:601–604
32. Daniel S, Rao TP, Rao KS, Rani SU, Naidu GRK, Lee H-Y, Kawai T (2007) A review of DNA functionalized/grafted carbon nanotubes and their characterization. Sens Act B 122:672–682
33. Reif, JH, LaBean TH, Seeman NC (2001) Challenges and applications for self-assembled DNA nanostructures. Lecture Notes in Computer Science 2054:173–198
34. He J, Lin L, Liu H, Zhang P, Lee M, Sankey OF, Lindsay SM (2009) A hydrogen-bonded electron-tunneling circuit reads the base composition of unmodified DNA. Nanotechnology 20:075102-1-8
35. Martin BR, Dermody DJ, Reiss BD, Fang M, Lyon LA, Natan MJ, Mallouk TE (1999) Orthogonal self-assembly on colloidal gold-platinum nanorods. Adv Mater 11:1021–1025
36. Mao C, Sun W, Shen Z, Seeman NC (1999) A nanomechanical device based on the B-Z transition of DNA. Nature 397:144–146
37. Chen Y, Lee SH, Mao C (2004) A DNA nanomachine based on a duplex-triplex transition. Angew Chem Int Ed 43: 5335–5338
38. Liu H, Xu Y, Li F, Yang Y, Wang W, Song Y, Liu D (2007) Light-driven conformational switch of i-motif DNA. Angew Chem Int Ed 46:2515–2517
39. Yan H, Zhang X, Shen Z, Seeman NC (2002) A robust DNA mechanical device controlled by hybridization topology. Nature 415:62–65
40. Sherman WB, Seeman NC (2004) A precisely controlled DNA biped walking device. Nano Lett 4:1203–1207
41. Shin JS, Pierce NA (2004) A synthetic DNA walker for molecular transport. J Am Chem Soc 126:10834–10835
42. Yin P, Yan H, Daniell XG, Turberfield AJ, Reif JH (2004) A unidirectional DNA walker that move autonomously along a track. Angew Chem Int Ed 43:4906–4911
43. Elghanian R, Storhoff JJ, Mucic RC, Letsinger RL, Mirkin CA (1997) Selective colorimetric detection of polynucleotides based on the distance dependent optical properties of gold nanoparticles. Science 277:1078–1081
44. Storhoff JJ, Elghanian R, Mucic RC, Mirkin CA, Letsinger RL (1998) One-pot colorimetric differentiation of polynucleotides with single base imperfections using gold nanoparticle probes. J Am Chem Soc 120:1959–1964
45. Chan WCW, Nie S (1998) Quantum dot bioconjugates for ultrasensitive nonisotopic detection. Science 281:2016–2018
46. Park SJ, Taton TA, Mirkin CA (2002) Array-based electrical detection of DNA with nanoparticle probes. Science 295:1503–1506
47. Kushon SA, Bradford K, Marin V, Suhrada C, Armitage BA, McBranch D, Whitten D (2003) Detection of single nucleotide mismatches via fluorescent polymer superquenching. Langmuir 19:6456–6464
48. Lockhart DJ, Winzeler EA (2000) Genomics, gene expression and DNA arrays. Nature 405: 827–836
49. Hill AA, Hunter CP, Tsung BT, Tucker-Kellogg G, Brown EL (2000) Genomic analysis of gene expression in C. elegans. Science 290:809–812
50. Lipshutz RJ, Fodor SP, Gingeras TR, Lockhart DJ (1999) High density synthetic oligonucleotide arrays. Nat Genet 21:20–24
51. Jiang Y, Ke C, Mieczkowski PA, Marszalek PE (2007) Detecting UV damage in single DNA molecules by atomic force microscopy. Biophys Journal 93:1758–1767
52. Austin RH, Brody JP, Cox EC, Duke T, Volkmuth W (1997) Stretch genes. Phys Today 50:32–38
53. Branton et al (2008) The potential and challenges of nanopore sequencing. Nature Biotechnology 26:1146–1153
54. Storm AJ, Chen JH, Ling XS, Zandbergen HW, Dekker C (2003) Fabrication of solid-state nanopores with single-nanometre precision. Natural Materials 2:537–540
55. Gerland U, Bundschuh R, Hwa T (2004) Translocation of structured polynucleotides through nanopores. Phys Biol 1:19–27

56. Lazarides AA, Schatz GC (2000) DNA-linked metal nanosphere materials: structural basis for the optical properties. J Phys Chem B 104:460–467
57. Link S, El-Sayed MA (1999) Spectral properties and relaxation dynamics of surface plasmon electronic oscillations in gold and silver nanodots and nanorods. J Phys Chem B 103:8410–8426
58. Sun Y, Harris NC, Kiang C-H (2005) Melting transition of directly linked gold nanoparticle DNA assembly. Physica A 350:89–94
59. Harris NC, Kiang C-H (2005) Disorder in DNA-linked gold nanoparticle assemblies. Phys Rev Lett 95: 046101-1-4
60. Harris NC, Kiang C-H (2006) Defects can increase the melting temperature of DNA-nanoparticle assemblies. J Phys Chem B 110:16393–16396
61. Vainrub A, Pettitt BM (2002) Coulomb blockage of hybridization in two-dimensional DNA arrays. Phys Rev E 66:041905-1-4

Chapter 11
Plasmonic Gold and Silver Films: Selective Enhancement of Chromophore Raman Scattering or Plasmon-Assisted Fluorescence

Natalia Strekal and Sergey Maskevich

1 Introduction

Progress achieved in nanosized materials technology has renewed interest to surface-enhanced phenomena having history from 1974, when Fleischmann et al. [1] observed an unusual experimental result with the Raman scattering of pyridine molecules on roughened silver electrodes. Of particular interest are the nanoscale noble metals, which have important applications in surface-enhanced Raman scattering (SERS) [2, 3], single-molecule spectroscopy [4, 5], surface-enhanced fluorescence (SEF) [6–8], radiative decay engineering [9], plasmon-assisted fluorescence, chemical and biological sensoring, and optoelectronic nanodevises [10]. Intensive trends in all these surface phenomena and in the near-field optical microscopy lead to the origin of the emerging field of science, which is now called plasmonics.

In the hierarchy of the nanoplasmonic objects one may select three following types – plasmonic crystals [11] (PC), resonance plasmonic particles [12] (RPP) and plasmonic films [13]. Plasmonic film presents the disordered nanotextured metal surface in which surface plasmon resonance (SPP) and/or localized plasmons (LP) could be excited under resonance conditions. Such a substrate has old, *pre*plasmonic terminology – the island vacuum deposited film, just as plasmon resonance particles earlier were named as metal colloids.

Silver island vacuum-evaporated films were likely the most widely used substrates for the enhancement of both Raman and fluorescence [14–16]. The SEF effect, however, has been described as the "weak cousin of the SERS effect" [2], owing to its rather tiny enhancement factors. Moreover, fluorescence quenching was considered as the main advantage of surface-enhanced resonance Raman scattering (SERRS)

N. Strekal (✉)
Grodno State University, University of Maryland Grodno, Belarus Baltimore, MD, USA
e-mail: nat@grsu.by

S. Maskevich
Ministry of Educaton of Republic of Belarus, Minsk, Belarus Baltimore, MD, USA

C.D. Geddes (ed.), *Reviews in Plasmonics 2010*, Reviews in Plasmonics,
DOI 10.1007/978-1-4614-0884-0_11, © Springer Science+Business Media, LLC 2012

[17]. Nevertheless, SEF in application to analyte molecules with low fluorescence quantum yield gives the excellent result because it was shown [18] that fluorescence enhancements increase as the quantum yield of a free molecule is decreased.

In spite of nonperiodic, nonuniform, and primitive surface morphology, plasmonic films possess some unique properties. The tailoring of its spectral properties allows to selectively excite the SERRS or the enhanced fluorescence of the same biomolecule without changing of light source [7]. This chapter is aimed at comprehensive characterization of these plasmonic substrates, including AFM study of surface morphology, fractal properties, near-field optical scanning microscopy of analyte distribution, distance and polarization dependence of surface-enhanced secondary emission.

Working with vacuum deposition technology, we undertake systematical and comparative analysis of plasmonic properties of silver and gold, because the growth parameters of metal islands on dielectric substrate have common features.

2 Vacuum Deposition and Postdeposition Treatment of Plasmonic Films

Plasmonic gold and silver films (PGFs and PSFs) were fabricated with a VUP-5 vapor deposition system. The pressures in the diffusion-pumped chamber were $<10^{-5}$ Torr. The deposition source was a tungsten boat filled with a Au or Ag powder (2–5 mg). To measure the resistivity of the films, two silver electrodes were deposited at the edges of the film surface. A custom-built temperature-controlled aluminum block served as a substrate holder for thermal experiments. The substrate temperatures were measured with a calibrated thermocouple placed in direct contact with the substrate.

Glass microscope and quartz slides were obtained from Fisher Scientific Inc. The cleaning procedure involved washing the slides in the chromium salt solution, followed by successive sonications in deionized water, ethanol, and acetone. The slides were dried at 120°C for 20 min and cooled to room temperature before introduction into the vapor deposition system. The slides were placed in the deposition unit 12 cm above a tungsten boat that served as the heating source. The controllable parameters of vacuum deposition are the deposition rate and the mass thickness of the film. These parameters were controlled by the current passed through a tungsten boat and quartz microbalance.

Postdepositon annealing of plasmonic films on glass or quartz substrates was performed in the muffle heating system at several stages for 6 min each: 160, 240 and 350°C. After each stage, the films were cooled to room temperature, and their extinction spectra were recorded. AFM images were taken in air using a Nanotechnology P4-SPM AFM/STM microscope. The images are collected in the constant force mode. Commercial Si_3N_4 cantilever tips (Park Scientific Instruments) with spring constants of ~0.12 nm^{-1} were used. More detailed report concerning the film deposition and characterization are reported in [19] and [7].

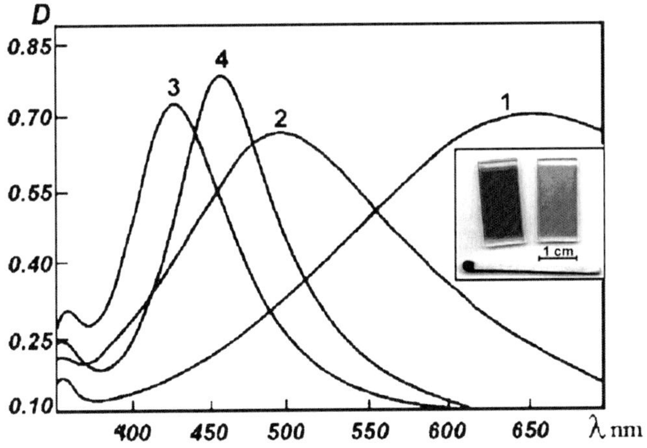

Fig. 11.1 Extinction spectra of as-deposited plasmonic silver films (1) and plasmonic silver films annealed at 160 (2), 240 (3), and 340°C (4). *Inset*: photo image of as-deposited (*left*) and annealed at 340°C (*right*) plasmon silver films

Figure 11.1 shows the extinction spectra of plasmonic silver films obtained at deposition rate 0.4 Å/s and annealed at different temperatures in air atmosphere.

It is well known that optical properties of silver island films depend on their mass thickness [20], the rate of silver deposition [8, 21] the temperature of a solid support [21] and the procedure of postdeposition annealing of the film [20, 21]. Figure 11.1 shows nonpolarized extinction spectra of the plasmon silver films at normal incidence of light for each of stages of the annealing procedure. The as-deposited TSFs (Fig. 11.1, spectrum 1) possess a broad extinction band peaking at 650 nm. This maximum corresponds to the LP resonance excitation in metal particles on the film surface. The annealing induces dramatic changes in both the position and width of LP band. The maxima of LP band for plasmonic films annealed at 160–240°C undergo a dramatic blue shift (from ~650 to 420 nm) and significant narrowing versus those of the as-deposited film (Fig. 11.1, spectrum 1). The next stage of the annealing (up to 340°C) leads to a further narrowing of the LP band, its slight hyperchromicity and red shift back to 450 nm (Fig. 11.1, curve 4). All the annealed films exhibit a loss of absorption in the deep red region and change the color (Fig. 11.1, inset).

A new extinction band at 350 nm (Fig. 11.1, spectra 1, 3 and 4) also becomes prominent after annealing. This band is assigned to excitation of the transverse plasmon resonance within regular uniform surface features [22], whereas long wavelength band – to longitudinal one. The blue-shift of extinction maxima were interpreted [23, 24] as thermally induced breakup of silver particles, so that new nucleation sites appear, resulting in formation of particles with increased height (Fig. 11.2). Increasing of particle size accompanied by the red-shift of extinction maxima was interpreted as being consistent with the electromagnetic theory.

Figure 11.2 shows the AFM images of plasmonic silver films, annealed at different temperatures. The distributions of lateral dimensions of silver particles (after

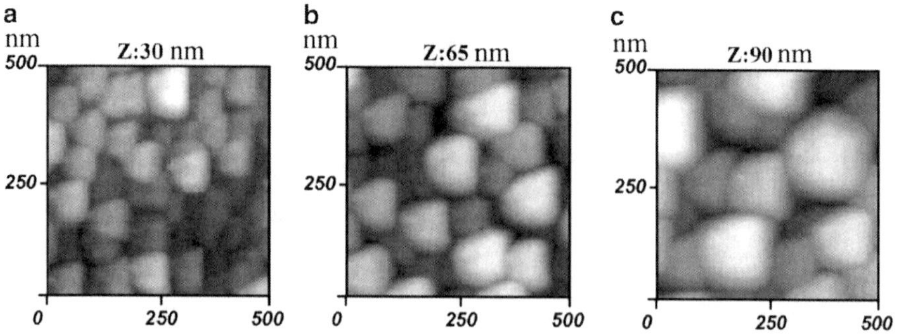

Fig. 11.2 AFM images of plasmonic silver films annealed at 160°C (**a**), 240°C (**b**) and 340°C (**c**)

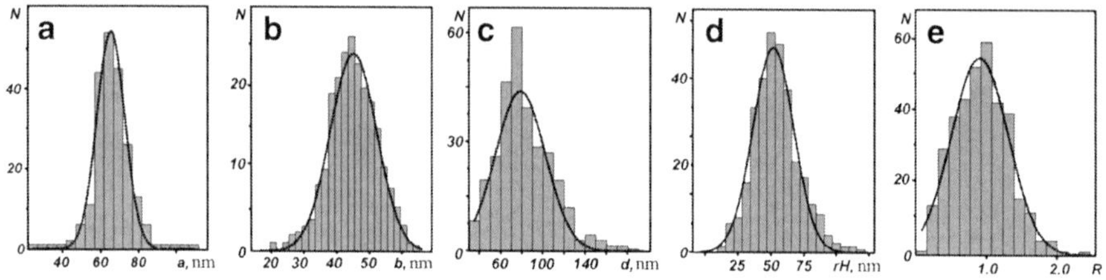

Fig. 11.3 The distributions of lateral dimensions of silver particles along major (**a**) and minor (**b**) axis, minimal distance between neighboring particles (**c**), visible height (**d**), and form factor (**e**) for the plasmonic silver particles annealed at 340°C

correction for the tip-induced broadening) along major and minor axis, minimal distance between neighboring particles, visible height and form factor of the silver particles are presented in the form of histograms in the Fig. 11.3. Visible height of each of the particles was determined as difference between maximal and minimal height, form factor – as ratio of visible height to major axis. Average values and standard deviations were calculated for the parameters listed above.

All the distributions are characterized by good fitting to Gauss function. The analysis of presented distributions shows the mean lateral sizes and height of particles. The most probable value of distance between silver particles is comparable with its sizes (Fig. 11.3a–d). The mean form factor of silver particles changes in the range of 0.8–1.3 (Fig. 11.3e). It was revealed that form factor increases with annealing temperature. It means that growth of particles is carried out in lateral directions more slowly than in high up.

Let us consider the results, concerning the vacuum deposition of gold onto quartz slide and effects of postdeposition annealing on spectral and morphological properties of obtained plasmonic gold films. Whereas the LP bands of plasmonic silver films, annealed at different temperatures may overlap all optical range (Fig. 11.1), the plasmonic gold films are optically confined in the wavelength range above 520 nm (Fig. 11.4).

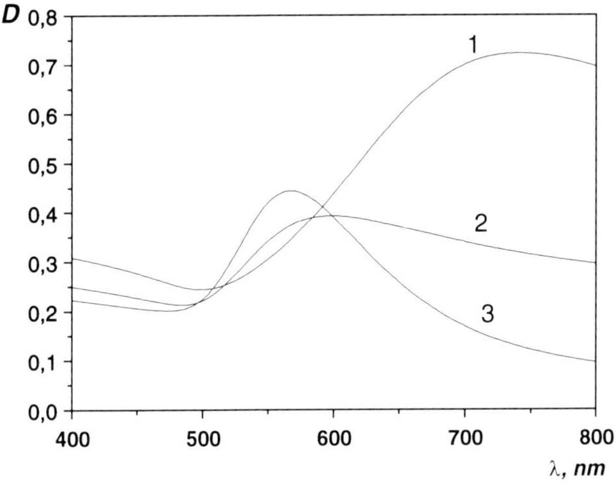

Fig. 11.4 Extinction spectra of as-deposited plasmonic gold films (1) and plasmonic gold films annealed at 240 (2) and 340°C (3)

Extinction spectrum of as-deposited plasmonic gold films presents the broad LP band with the maximum at 720 nm (Fig. 11.4, spectrum 1). The LP band of plasmonic gold films annealed at 240°C undergoes low-wavelength shift to 600 nm with significant hypochromism. The same effect was observed upon annealing of as-deposited Ag films and was explained by breaking of a continuous as-deposited thick metal film and self-assembling of superficial clusters and grains in islets of the ellipsoidal form. An annealing of gold films at 340°C results in the further low-wavelength shift of the LP resonance excitation band (Fig. 11.4, spectrum 3). This shift is induced by the growing of the self-assembled metal clusters detected by the AFM (Fig. 11.5).

The surface of as-deposited TGF has a microscopically granular structure with surface undulations of up to 3 nm. Annealing of TGF at 240°C results in breaking up of the continuous film and self-assembling of superficial clusters and grains of gold in islets of a roughly conical shape and variable size, typically around 60 nm base diameter and 6 nm height (Fig. 11.5a, c, e). The annealing of the as-deposited TGF at a temperature of 340°C results in growth of the gold islets both laterally and in height. Morphological analysis of the corresponding AFM images (Fig. 11.5b, d, f) demonstrates that the gold islets slightly increase in lateral size and became two times bigger in height. The distributions of lateral diameter *d*, visible height *h*, and radius of curvature *R* are shown in Fig. 11.5 for plasmonic gold films, annealed at 240 (left) and 340°C (right).

At the same time, the mean radius of curvature decreases upon decreasing the annealing temperature (Fig. 11.5g, h). It may be realized if the distance between neighboring gold particles increase and its spatial overlap decrease. It is important that decrease of radius and increase of distance between gold particles correlate with low-wavelength shift and narrowing of LP band of plasmonic film annealed at

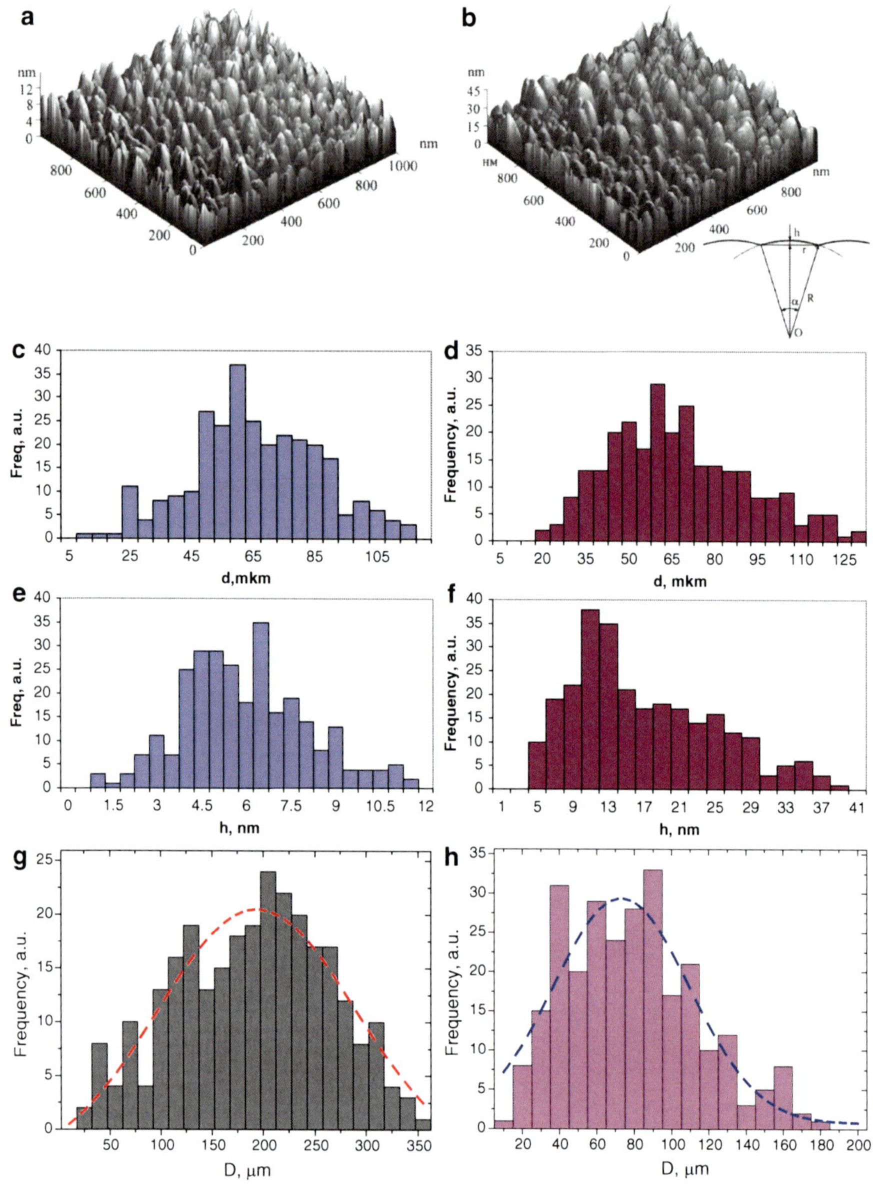

Fig. 11.5 AFM images (**a** and **b**), diameter d of gold particles in lateral plane (**c** and **d**), visible height h of gold particles (**e** and **f**), and double radius of curvature $2R = D$ (**g** and **h**) for the plasmonic gold films annealed at 240 (**a**, **c**, **e**, **g**) and 340°C (**b**, **d**, **f**, **h**)

340°C in comparison with film annealed at 240°C (Fig. 11.4, spectra 3 and 2). The alf width of LP band 2 on Fig. 11.4 is equal to ~300 nm and this value of LP band on Fig. 11.4 is ~65 nm. That is the half width of LP bands differs approximately in 3 times, although the excess of sizes statistical distribution for these films differs no more than 2 times. In other words, the narrowing of LP band cannot be explained by narrowing in particle size distribution. It is possible that the electromagnetic dipole–dipole interaction between gold particles is of lesser importance in plasmonic gold films annealed at higher temperature due to increase of particle–particle distance.

3 Surface-Enhanced Secondary Emission of Mitoxantrone Remoted from Plasmonic Silver and Gold Films in the Near Field

The plasmonic silver films, prepared with vacuum deposition and postdeposition annealing were reported earlier [19] as extremely time- and organic solvent-stable versus as-deposited films. These PSFs were found to be nondisturbing SERS-active substrates in the application to studies of complexation of crown ether styryl dyes with metal ions. A pronounced SERS signal of the analyte rhodamine 6G was observed with PSFs, even when the analyte was separated from the silver surface with five Langmuir–Blodgett monolayers of stearic acid. At the same time, depositing only a monolayer of stearic acid on the as-deposited film completely suppressed the SERS signal of the analyte. Finally, the self-assembling of Ag clusters on the surface of the aTSF, stimulated by the high-temperature annealing, results in the creation of a time- and organic solvent-stable SERS substrate with nanometer scale quasi-periodical roughness, and this substrate exhibits an increased contribution of the electromagnetic component to the overall Raman enhancement.

Plasmonic gold films were reported [7] afterwards as selective substrates that may be specifically annealed to serve as a substrate for SEF or SERS spectroscopy of the same molecule. It was the first experimental evidence of possible mode separation in surface-enhanced secondary emission. High-resolved SERRS spectra of mitoxantrone were recorded on the PGFs annealed at 340°C, whereas no Raman enhancement but an increase of mitoxontrone fluorescence signal were detected on the TGF annealed at 240°C (Fig. 11.6). The possible practical application of plasmonic films in histology was reported earlier [25], but this chapter we aimed mainly on enhancement mechanism and selectivity of PGFs to enhance of SERS or fluorescence. From our point of view, it is the more interesting aspect of our results, taking into account a lot of publications and patents devoted to practical use of SERS and SEF substrates in the world.

Figure 11.6 presents the surface-enhanced secondary emission of mitoxantrone, adsorbed on the surface of as-deposited PGF (Fig. 11.6, spectrum 1) and PGFs annealed at 240°C (Fig. 11.6, spectrum 2) and 340°C (Fig. 11.6, spectrum 3).

As it is clearly seen from Fig. 11.6, weak SERRS signal of mitoxantrone (spectrum 1) is replaced by enhanced fluorescence (spectrum 2) and later is replaced by high-resolved SERRS (spectrum 3) if we change only the PGFs. Extinction and fluorescence spectra of mitoxantrone aqueous solution (10^{-6} M) are presented in

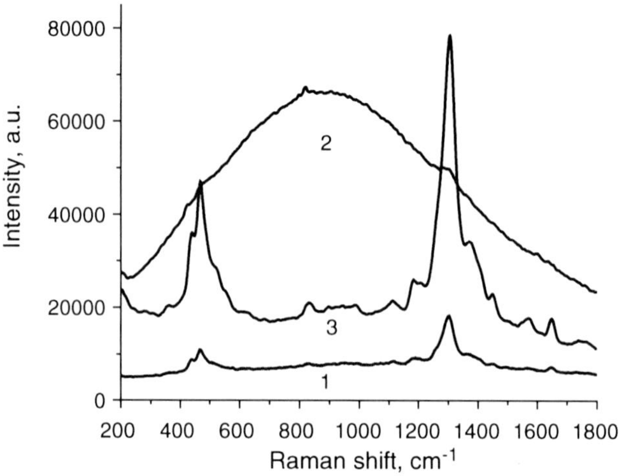

Fig. 11.6 Surface-enhanced secondary emission of mitoxantrone, adsorbed on as-deposited PGF (1) and PGFs, annealed at 240 (2) and 340°C (3)

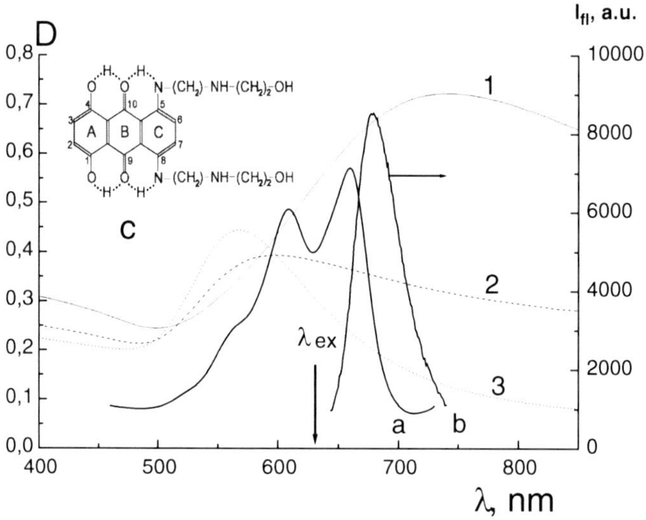

Fig. 11.7 Extinction (**a**) and fluorescence (**b**) spectra of mitoxantrone (**c**) aqueous solution (10^{-6} M) and extinction spectra of PGFs transferred from Fig. 11.4 (*dash* and *dot lines*) for the spectral overlap visualization. Fluorescence excitation wavelength 633 nm is marked by *arrow*

Fig. 11.7, where the extinction spectra of PGFs also transferred from Fig. 11.4 (dot and dashed lines) for the visualization of spectral overlap between the chromophore and substrates. The temperature used for annealing of fluorescently-active PGFs is varied in the range of intermediate temperatures 240–290°C. For

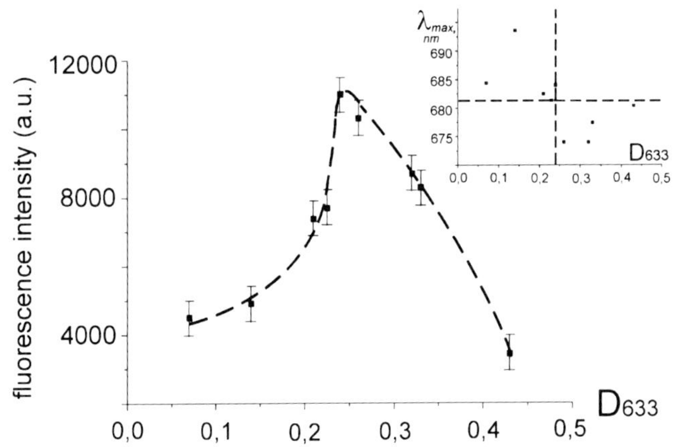

Fig. 11.8 Dependence of mitoxantrone fluorescence intensity and band position (λ_{max}, *inset*) on optical density D_{633} of fluorescence-active PGFs at excitation wavelength 633 nm

the tuning of PGF to activity in SERS, it is necessary to anneal it at terminus temperature in the range of 340–350°C.

The surface enhancement of secondary emission of the molecules deposited on the metal films depends on the overlapping of the molecular electronic transitions and the band of the localized plasmon excitation. An example of rodamine 6G [26] shows that its secondary emission is maximal if the position of LP-band of a metal film "optimally" overlaps with the bands of electronic transitions of the molecule. One may propose that the selectivity of enhancement of the Raman or fluorescence component of the secondary emission is determined by the way in which the electronic spectra of the molecule and SES-substrate overlap under the conditions of resonance excitation.

We select several fluorescently-active PGFs with gliding of LP band position to vary the spectral overlap. Figure 11.8 shows that dependence of the mitoxantrone fluorescence intensity on PGF optical density at excitation wavelength (D_{633}) is non-monotonic [27]. Horizontal dashed line in inset to Fig. 11.8 shows the wavelength of mitoxantrone fluorescence band maximum in aqueous solution; vertical dashed line in inset to Fig. 11.8 shows the D_{633} value for PGF with which the maximal enhancement effect is reached. Two conclusions follow from the analysis of Fig. 11.8. Firstly, there is some spectral range of tuning the fluorescently active plasmonic gold films through annealing temperature regime. This tuning shifts the LP band to prescribed spectral overlapping range. Secondly, the shift of mitoxantrone fluorescence maxima in dependence on PGF parameter (inset to Fig. 11.8) shows that it is not chromophore fluorescence but in more degree the plasmon-assisted chromophore fluorescence.

It is important that PGF's selectivity was revealed for separate chromophore (mitoxantrone in our case) and this selectivity strongly depends on spectral overlapping between chromophore and LP band. It is interesting to note that both kinds of secondary emission – SERRS and enhanced fluorescence – are detected only for resonance excitation 633 nm. In spite of the fact that mitoxantrone fluorescence successfully detected with laser excitation line 514 nm [28] in aqueous solution, the surface-enhanced secondary emission of any types does not detect with use of PGFs at this excitation wavelength. It may be explained by the well-known fact about gold absorbance. It is impossible to excite plasmon oscillation below 520 nm due to interband d→sp transition in gold. That is why the extinction spectra of gold films present only one LP band (Fig. 11.4) corresponding to longitudinal plasmon, whereas in extinction spectra of silver films, the second short wavelength band appears at 350 nm (Fig. 11.1) due to transverse plasmon oscillation. The impossibility to excite chromophore SERS or SEF below the threshold wavelength of LP excitation explicitly indicates on plasmon-assisted nature of both phenomena.

With use of as-deposited and annealed-at-160°C plasmon silver films (Fig. 11.1, spectra 1 and 2), one can obtain SERRS spectra of mitoxantrone under both excitation – 514 and 633 nm. But it is impossible to obtain the enhanced fluorescence signal in pure form (without vibrational structure) or SERRS without fluorescence background with use of silver films. Enhanced Raman scattering decrease and fluorescence background increase if mitoxantrone is separated from PSFs by monomolecular layers as it was predicted [26] and usually observed by other authors for other molecules [3, 29]. The distance dependence for selective SERRS and plasmon-assisted fluorescence of mitoxantrone on the surface of annealed PGFs have another character and are reported in the next section.

In this section, concerning the "first layer" effect, let us consider the simple assumption about gold particles with particular sizes as responsible for selective reemission of chromophore resonance Raman or structureless fluorescence. This assumption was indirectly confirmed in our experiments with chemically modified PGFs.

Figure 11.9a presents the electron microscopy image of PGF exposed to 1 mM solution of sodium mercaptoethylsulfonate (SMES) in ethanol for 60 min. Figure 11.9b shows extinction spectra of PGF chemically modified by SMES. From Fig. 11.9, it is clear that the chemical modification results in strong broadening of LP band (Fig. 11.9, upper panel). The result of this band deconvolution is also presented in Fig. 11.9b (bottom panel). It is obvious that the LP band of chemically modified PGF may be shown as overlapping doublet. It consists of two main bands with maxima at 572 and 681 nm (Fig. 11.9b bottom panel). Two LP bands in extinction spectra of chemically modified PGF (Fig. 11.9b) are assigned to gold particles with two typical sizes, presented on image (Fig. 11.9a). The bigger particles have the mean diameter in the range of 100–75 nm and the smaller one – 20–40 nm. We suppose that gold particles with LP band at 570 nm promote the fluorescence emission and particles with LP band at 680 nm – resonance Raman scattering of mitoxantrone. As it was mentioned above, the PGFs annealed at intermediate and terminus temperatures have different distribution of curvature radius (Fig. 11.5g, h) too and develop different activity in SERRS or fluorescenece.

It should be mentioned that mitoxantrone secondary emission, detectable from chemically modified PGF with dichroic PL band (Fig. 11.9a), present the

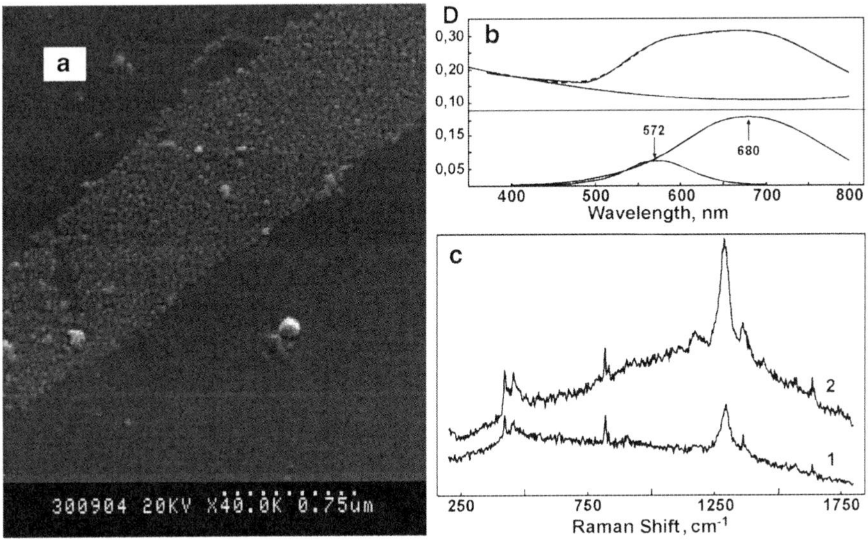

Fig. 11.9 Electron microscopy image of PGF modified with SMES (**a**), extinction spectra of modified PGF (**b**, *upper panel*) and result of its deconvolution (**b**, *bottom panel*), SERRS spectra of mitoxantrone, deposited on modified PGF by SMES and covered by one (**c**, spectrum 1) and three (**c**, spectrum 2) layers of polyelectrolyts (one layer consists two oppositely charged polymers) [43, 44] polydiallyldimethylammonium chloride (PDADMAC) and sodium polystyrene sulfonate (PSS)

superposition of SERRS and enhanced fluorescence. Thus chemical modification with SMES allows to partially select all the particles on two size groups on the PGF, which become simultaneously Raman- and fluorescence-active. The distance dependence (Fig. 11.9c) shows that both signal increase with distance as it is observed (Sect. 4) for mitoxantrone secondary emission on Raman- and fluorescence-active PGFs.

4 Surface-Enhanced Secondary Emission of Analyte Molecules in the Near Field of Plasmonic Silver and Gold Films

There are several elegant technologies, to separate the chromophore and plasmonic substrate at the desired distance. Among them the self-assembling monolayers of tiols [30] and Langmuir–Blodgett [31] (LB) techniques were widely used in surface-enhanced spectroscopy last decades. Layer-by-layer deposition of polyelectrolytes [30] (PE) with opposite charge are now widely used in Forster resonance energy transfer (FRET) experiments and in SERS spectroscopy. We apply all these techniques for systematic study of light interaction with plasmonic films and chromophores.

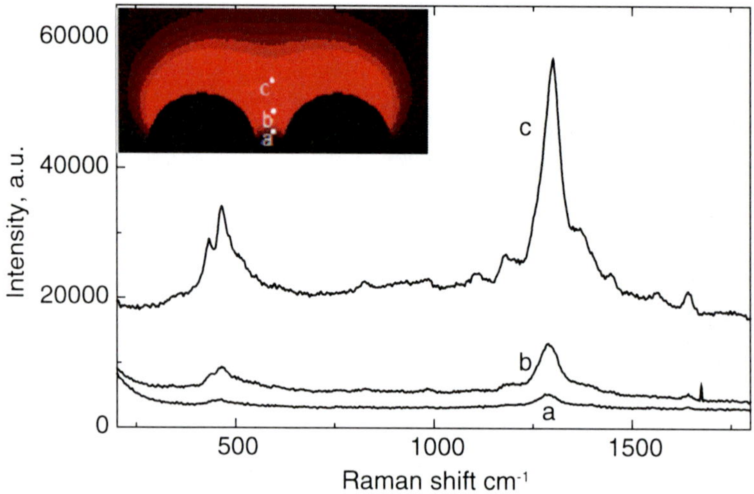

Fig. 11.10 SERRS spectra of mitoxantrone deposited on Raman-active PGF, covered by 1 (**a**), 3 (**b**) and 5 (**c**) Langmuir layers of cadmium behenat. *Inset*: electric field intensity near the two charged particles in quasistatic approach and possible chromophore location on the axis between particles for the cases of 1 (**a**), 3 (**b**), and 5 (**c**) LB layers covering

It was revealed that mitoxantrone fluorescence was further enhanced by deposition of monolayers of pentathiol or poly-L-lysine on the surface of annealed fluorescence-active PGF. The maximal fluorescence enhancement factor per mitoxantrone molecule of approximately 50 was obtained for the fluorescence-active PGF covered with poly-L-lysine [7]. Nonmonotonic distance dependence for SEF was discussed earlier [6, 7] and also revealed for CdSe/ZnS quantum dots (QDs) near the gold colloidal films [32] as a result of at least two competing processes – long-range field enhancement and short-range fluorescence quenching. For the first time, we presented [33] the nondecreasing distance dependence (in close ptoximity to the surface) for SERRS of mitoxantrone separated by LB of behenat cadmium from Raman-active PGF.

Figure 11.10 shows SERRS spectra of mitoxantrone, deposited on Raman-active PGFs, covered by different number of behenat cadmium LB monolayers. The sharp increase of mitoxantrone SERS signal is observed as number of the LB layers increased. It is 10 times higher for 5 layers of separated spacer than for 1 layer (Fig. 11.10a, c). Each behenat cadmium molecule consists of 20 CH_2 groups; for five Langmuir layers, this number is equal 100. Most crude estimates give the thickness of 1 and 5 layered BC spacer as 3 and 15 nm, correspondingly. Note that no fluorescence background is detected with the use of Raman-active films even if thickness of spacer is ca 15 nm, what is enough to exclude quenching due to FRET. It is possible that any selection rule forbid the radiative transition in this system.

Our results, concerning the increase of SERS with distance from plasmonic Raman-active gold surface, are in good agreement with predictions, obtained by

O. J. F. Martin and coworkers [34] for the interacting plasmon resonant nanoparticles. The possible geometry of molecular probe disposition in some points indicated as *a*, *b*, and *c* between two gold particles on PGF is presented in inset to Fig. 11.10. Point *a* corresponds to 1 LB layer and to spectrum a on Fig. 11.10, points *b* and *c* – to 3 and 5 LB layers, and spectra b and c on Fig. 11.10, correspondingly. It is possible that these points generate so-called "hot spots" [35] in assembles of more than two particles on self-aggregated films after annealing. These "hot spots" can develop at the surface not only with respect to field amplitudes but with respect to photon density of states (DOS) and scattering rates as well [36]. Spontaneous emission and scattering of light is not an intrinsic property of chromophores but is essentially a result of their interaction with zero electromagnetic fields (electromagnetic vacuum). For the first time, S. V. Gaponenko reported that similar to spontaneous emission, spontaneous Raman scattering should experience modification if photon DOS redistribution over frequency and solid angle occurs [37].

On the other hand, a very careful control of the quality of mono-and multilayer deposited atop the nanometer scale rough surfaces is required particularly when surface enhanced optical techniques are used to exclude possible artifacts. To control of analyte distribution on rough plasmon films, we combine the LB technique with near-field scanning optical microscopy [38] (NSOM) and applied NSOM to study the spatial distribution of fluorescently labeled phospholipids monolayer deposited on PGF. The location of the lipid molecules can be determined with spatial resolution ~50 nm with analysis of the near-field fluorescence signal from the fluorescently labeled lipids. Simultaneously, the near-field transmission signal from the plasmonic gold film can map the position of the gold particles and interparticle areas on the metal surface. NSOM was chosen for our purpose as it is more chemically specific than AFM topography as to the location of lipid molecules on rough surfaces.

It was showed that near field (NF) fluorescence of the dye-labeled monolayer deposited on the plasmon gold films is spatially heterogeneous. Maximum fluorescence intensity in NF fluorescence images was observed to spatially coincide with minimum transmission in NF transmission images suggesting preferential location of the labeled molecules between the particles of metal films. A similar NF fuorescence experiment was performed with a single lipid monolayer of Dipalmitoylphosphocholine (DPPC) labeled with BODIPY 581/591-PC and transferred by the LB technique on the plasmon gold film. Overlay of the bright spots in NF fluorescence images and the dark areas in the NF transmission images was observed. Thus the observed effect of the plasmonic gold film topography on the phospholipid monolayer transfer is most probably a general phenomenon independent on the layer composition. Taking these data into consideration, the distance dependence of mitoxantrone SERRS on Raman-active PGF, covered by LB layers may be really explained by location of chromophore in interparticle "hot" areas (Fig. 11.10, inset).

More homogeneous surface coverage provides the layer-by-layer PE deposition. A layer-by-layer deposited polyelectrolyte spacer was, for example, used to attach CdSe QDs [34], dye molecules and fluorescein-labeled bovine serum albumin [39]

(BSA-FITC) to flat glass and plasmonic silver films in order to study the effect of spacer thickness on homogeneity and surface concentration of fluorophore coverage. Three different methods of fluorophore deposition atop the polyelectrolyte spacer are examined using steady-state spectroscopy, fluorescent microscopy and statistical analysis. The best homogeneous covering with CdSe Qds at a controllable concentration was found for deposition from a solution of QDs which are electrostatically bound to polyelectrolyte macromolecules. This fluorophore deposition method allows one to achieve highly homogeneous distribution of fluorophores and to evaluate the fluorescence enhancement factor of BSA-FITC [40] adsorbed on plasmonic silver films. The maximum ninefold enhancement coefficient for the fluorescence of FITC corresponds to a thickness of the intermediate layer of ~4 nm, or three layers of the polyelectrolyte. In this case, we observed a significant decrease in the average photoluminescence decay time for the label near the PSF compared with a dielectric medium.

Turning back to the unique property of fluorescence-active PGF, we apply layer-by-layer PE deposition to compare the excitation efficiency of the secondary emission of mitoxantrone adsorbed both on plasmonic gold and silver films and quartz slide depending on spacer thickness between dye and metal, polarization, and angle of incidence of the exciting light.

5 Comparative Analysis of Plasmonic Silver and Gold Films

As it is follow from preceding data, plasmonic silver and gold films, prepared with the same technology, differs on capability to selectively enhance Raman scattering or fluorescence. Plasmonic films on the base of silver did not develop the separately Raman- and separately fluorescence-activity, as it is observed for gold plasmonic films. The aim of this section is to compare the fluorescence excitation efficiency as depending on spacer thickness between dye and metal, polarization, and angle of incidence of the exciting light.

Figure 11.11 shows the dependence of optical density on incidence angle for *s*- and *p*-polarized light for fluorescently-active PGF (Fig. 11.11a) and for PSF (Fig. 11.11b). Optical density (D) were measured at the wavelength corresponding to the maxima of LP bands. There are one LP band for PGF and two LP bands for PSF, as it were presented above on Figs. 11.4 and 11.1, correspondingly. If the incidence angle is equal to zero, the *s*- and *p*-polarized light are characterized by the same value of optical density. If the incidence angle increased up to 65°, the D value increased for *s* and decreased for *p*-polarization with the exception of D_{352} of transverse LP band on PSF (Fig. 11.11b). It indicates, firstly, localized nature of plasmons generated in PGF, since *s*-polarized light did not excite the propagating plasmons or SPP at all. Secondly, these observations correlate very well with assignment of these bands to longitudinal and transverse localized plasmon in gold and silver films. Laser light with *s*-polarization more effective excites lateral or longitudinal LP and with *p*-polarization – transverse one.

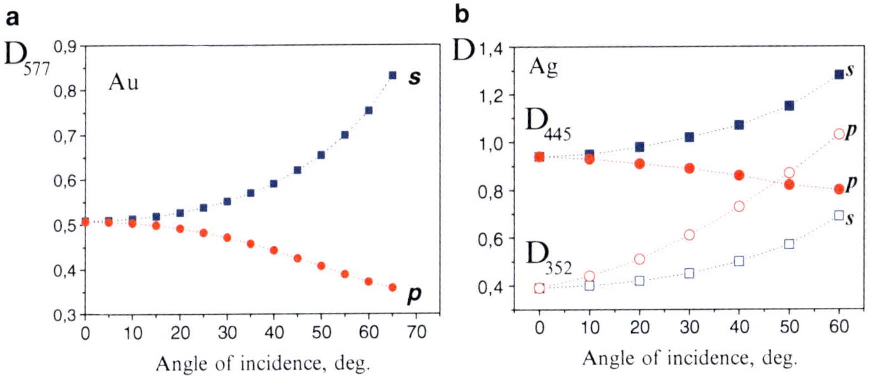

Fig. 11.11 The dependence of plasmonic gold (**a**) and silver (**b**) films optical density at wavelength, corresponding to LP bands maxima on incidence angle of *s*- (*squares*) and *p*-polarization (*circles*) light

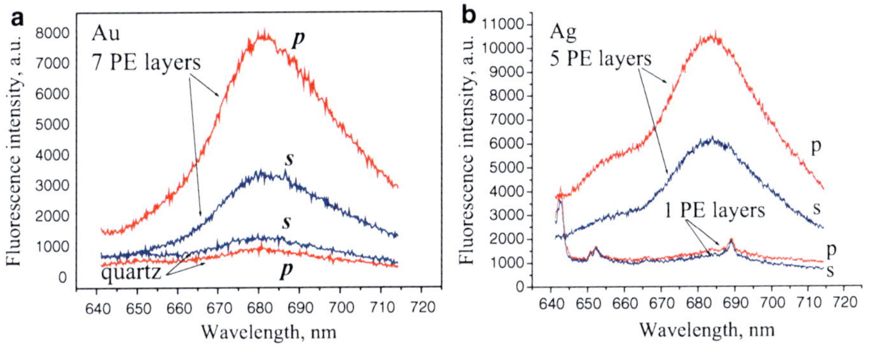

Fig. 11.12 Fluorescence spectra of mitox deposited on plasmonic gold (**a**), silver (**b**) films and reference quartz slide (**a**, two *bottom spectra*) covered with 7 layers of PE spacer (**a**) in the case of gold and with 1 and 5 PE layers in the case of silver (**b**). Excitation wavelength was 633 nm, laser power – 10 mW/mm^2, incident angle –70°, polarization is indicated near each spectrum

Optical density at fluorescence excitation wavelength D_{633} qualitatively changes with incidence angle as D_{577} and D_{445} for gold and silver film for *s*- an *p*-polarization.

Figure 11.12 demonstrates typical fluorescence spectra of mitoxantrone adsorbed on plasmonic gold (a) and silver (b) films and reference spectra on quartz slide, covered with the PE.

Spectra for spacer coating in thickness of 1 layer on PSF is presented (Fig. 11.12b) to show the Raman bands of mitoxantrone and practically absence of fluorescent background at these conditions. Different PE numbers (7 for PGF and 5 for PSF) are presented, because the mitoxantrone fluorescent level under *s*-polarization

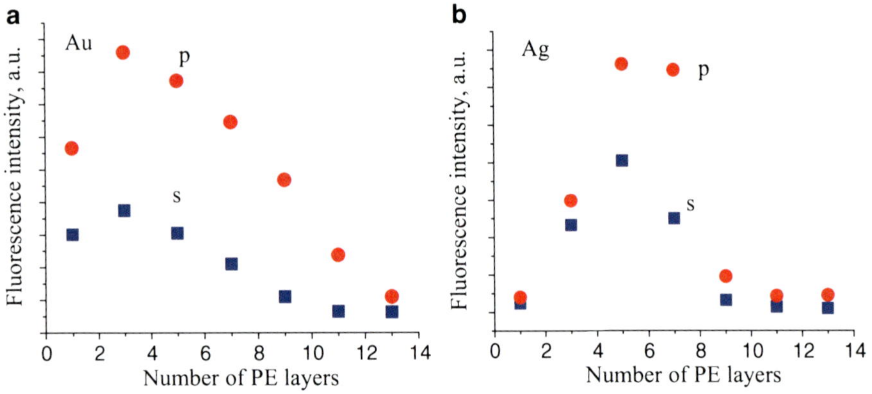

Fig. 11.13 Fluorescence intensity of mitoxantrone deposited on PGF (**a**) PSF (**b**) versus PE thickness (in layers number) for *p*-polarized (*circles*) and *s*-polarized (*squares*) excitation. Experimental parameters as for Fig. 11.12

practically coincides for these coatings on gold and on silver. Figure 11.13 shows full picture of mitoxantrone fluorescence distance dependence for gold and silver plasmonic films.

It is general regularity for all investigated plasmonic films that secondary emission enhancement of mitoxantrone is more effective for *p*-polarized incident light than for *s*-polarized one. Otherwise, the fluorescence intensity of mitoxantrone on quartz is higher for *s*-polarized excitation. Figure 11.13 demonstrates the more long-range effect of silver plasmonic films because the maximum of fluorescence signal in distance dependence accounts at bigger number of PE layers. To compare the values of fluorescence intensity (relatively to reference quartz slide) is not correct because the optical density at excitation wavelength D_{633} for PSF is two times smaller than for PGF.

In spite of the fact that Figs. 11.11–11.13 present qualitatively the similar data for the gold and silver plasmonic films with the exception of transverse LP band, which is absent for gold, the next illustration reveals essential distinctions (Fig. 11.14).

Firstly, in spite of more effective excitation of localized plasmons with *s*-polarization under various incidence angle (Fig. 11.11), the more effective excitation of plasmon-assisted fluorescence is observed under *p*-polarized excitation in the range of incidence angle from 55 to 80°. It is also true for both gold and silver plasmonic films. More effective excitation of mitoxantrone fluorescence under *p*-polarized light (Fig. 11.5) is in good agreement with well-known fact of complete *p*-polarizarion plasmon-coupled emission [41]. However, plasmon-controlled fluorescence decrease with incidence angle for both *p*- and *s*- polarization in the case of chromophore/silver plasmonic films (Fig. 11.14b). Decrease in plasmon-controlled fluorescence intensity with incidence angle may be caused by some LP damping channel, excitation of SPP, for example. It for one's turn may be if LP and SPP

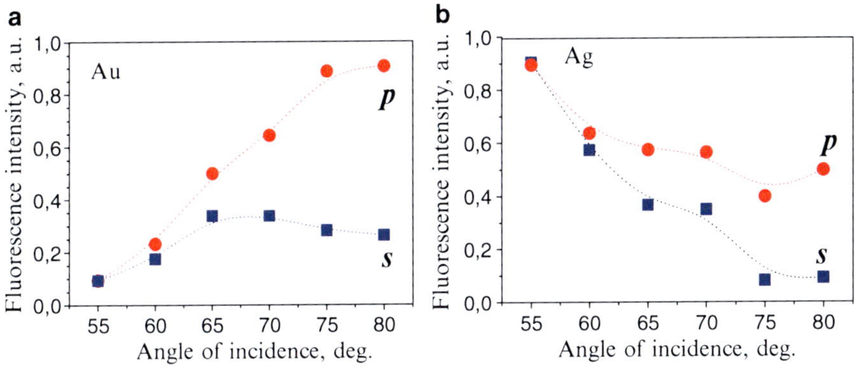

Fig. 11.14 Mitoxantrone fluorescence intensity versus the incidence angle of excitation light for *p*-polarization (*circles*) and *s*-polarization (*squares*) on gold (**a**) and silver (**b**) plasmonic films. Experimental parameters as for Fig. 11.12

frequencies and wave vectors coincide. This channel is especially actual for *p*-polarized light because SPP excitation is carried only under *p*-polarized light.

Nevertheless, photon DOS redistribution over frequency and solid angle may occur in mesoscopic structures [38, 39]. From this point of view, Raman-active PGFs demonstrate the system with high Raman photon DOS and fluorescently-active PGFs – the systems with low Raman photon DOS. Indeed SERRS is absent even mitoxantrone is deposited directly on its surface (Fig. 11.6, spectrum 2) and Raman-active PGFs develop exclusive nondecreasing distance dependence for the near-field location of chromophore (Fig. 11.10). In the case of fluorescently-active PGFs, the LP more effectively absorbs the *s*-polarized light in wide range of incidence angle (Fig. 11.11a), but fluorescence signal is more intense at these angles under *p*-polarization (Fig. 11.14a) as for plasmon-coupled emission.

On the contrary, plasmonic silver films may be presented as Raman active in this frequency and solid angle range and in the close proximity to its surface (Fig. 11.14b, 11.12, PE layer). Distance- and angle-dependence of fluorescence for the cromophore, which are located at spacer atop PSFs, may be explained as competition between long-range field enhancement and short-range fluorescence quenching as was mentioned above. We assume that fluorescence signal belongs to fluorophore in this case in more degree than to plasmons.

Thus the main difference between plasmonic gold and silver films is to promote plasmon-coupled emission or the enhanced fluorescence of mitoxantrone. It possibly may be explained by different plasmonic spectra of these substrates. Spectrum of LP and magnitude of field enhancement in the metal particles ensemble considerably depend on particle shape, size, and distances between them as well as on interaction among particles. SPP modes could not be realized if photon states density is zero. It may happen if frequencies have their values within forbidden states by analogy with photon crystals. However, on the edge of forbidden states band, SPP dispersion appears smooth and photon states density is high, which corresponds to field large enhancement near metal surface. Modes with large enhancement could be excited by incident light at different incident angles which makes substrate frequency-selective surface [42].

Acknowledgments This work was supported by Ministry of Education of Belarus, National Academy of Sciences of Belarus, BRFFI (Belarus) grants F07K-094 and F10R-232, "Nanotekh" program (Belarus) grant #6.18, "Crystal and molecular structures" program (Belarus) grant #KM-40. We are grateful to Dr. I. Sveklo for AFM, to V. Oskirko for SERS experiments, and to Dr O. Kulakovich for PE and analyte deposition. Ongoing discussions with Prof. S. Gaponenko and Dr. D. Guzatov are acknowledged.

References

1. M. Fleischmann, P. J. Hendra and A. J. McQuillan, (1974). Raman spectra of pyridine adsorbed at a silver electrode, *Chem. Phys. Lett* **26**,163–166.
2. M. Moskovits, Surface-enhanced spectroscopy(1985). *Rev. Mod. Phys.* **57**,783–826.
3. P. L. Stiles, J. A. Dieringer, N. C. Shah, R. P. Van Duyne. (2008). Surface-enhanced Raman spectroscopy. Annu. *Rev. Anal. Chem.* **1**: 601–626.
4. S. Nie, S.R. Emory, Probing single molecules and single nanoparticles by surface-enhanced Raman scattering. (1997). *Science*, **275**, 11021–11025.
5. K. Kneipp, Y. Wang, H. Kneipp, L. Perelman, I. Itzkan, R. Dasari, M. Feld (1997). Single molecule detection using surface-enhanced Raman scattering (SERS), *Physical Review Letters* **78**(9), 1667–1670.
6. K. Sokolov. G. Chumanov, T. Cotton (1998).Enhancement of molecular fluorescence near the surface of colloidal metal films *Anal. Chem.* **70,** 3898–3905.
7. N. Strekal, A. Maskevich, S. Maskevich, J.-C. Jardillier, I. Nabiev (2000). Selective enhancement of Raman or fluorescence spectra of biomolecules using specifically annealed thick gold films, *Biospectroscopy/ Biopolymers*, **57,** 325–328.
8. P. J. G. Goulet, Ricardo F. Aroca Surface-enhancement of fluorescence near noble metal nanostructures, in *Topics in Fluorescenec Spectroscopy, 8:* Radiative decay engineering (2005). C. Geddes and J. Lakowicz eds., Springer Science+Business Media, Inc Inc., new York, 223–247.
9. C.D. Geddes, K. Aslan, I. Gryczynski, J. Malicka, J. Lakowicz (2005). Radiative decay engineering. In: C.D. Geddes, J.R. Lakovicz (eds) Topics in fluorescence spectroscopy vol 8: Radiative decay engineering. Springer Science+Business Media, Inc, New York, 405–448.
10. S. Chen, R.S. Ingram, M.J. Hostetler, J.J. Pietron, R.W. Murray, T.G. Shaaf, J.T. Khoury, M.M. Alvares, R.L. Whetten, Science 280 (1998) 2098.
11. M. E. Stewart et al. (2006). Quantitative multispectral biosensing and 1D imaging using quasi-3D plasmonic crystals, *PNAS* **103**, 17143–17148.
12. D.A. Shultz (2003). Plasmon resonant particles for biological detection. *Curr. Opin. Biotechnol.* **14**, 13–22
13. N. Strekal, O. Kulakovich, V. Askirka, I. Sveklo, S. Maskevich (2008). Features of the Secondary Emission Enhancement Near Plasmonic Gold Film, *Plasmonics* **4**, 1–7.
14. D. A. Weitz, S. Garoff, J. I. Gersten and A. Nitzan (1983). The enhancement of Raman scattering, resonance Raman scattering, and fluorescence from molecules adsorbed on a rough silver surface, *J. Chem. Phys.* **78**, 5324–5338.
15. J. DeSaja-Gonzalez, R. Aroca, Y. Nagao and J. A. DeSaja (1997). Surface-enhanced fluorescence and SERRS spectra of N-octadecyl-3,4:9,10-perylenetetracarboxylic monoanhydride

on silver island films, *Spectrochimica Acta, Part A: Molecular and Biomolecular Spectroscopy* **53A**, 173–181.
16. R. F. Aroca, C. J. L. Constantino and J. Duff (2000). Surface-enhanced Raman scattering and imaging of Langmuir-Blodgett monolayers of bis(phenethylimido)perylene on silver island films, *App. Spec.* **54**, 1120–1125.
17. R. R. Chance, A. Prock and R. Silbey (1978). Molecular fluorescence and energy transfer near interfaces. *Adv. Chem. Phys.* **37**, 1–65.
18. D. A. Weitz, S. Garoff, J. I. Gersten and A. Nitzan (1983). The enhancement of Raman scattering, resonance Raman scattering, and fluorescence from molecules adsorbed on a rough silver surface, *J. Chem. Phys.* **78**, 5324–5338.
19. A. Feofanov, A. Ianoul, E. Krukov, S. Maskevich, G. Vasiliuk, L. Kivach, I. Nabiev (1997). Nondisturbing and Stable SERS-Active Substrates with Increased Contribution of Long-Range Component of Raman Enhancement Created by High-Temperature Annealing of Thick Metal Films *Anal Chem*, **69**, 3731–3740.
20. Van Duyne, R. P.; Hulteen, J. C.; Treichel, D. A. *J.* (1993). Atomic force microscopy and surface-enhanced Raman spectroscopy. I. Ag island films and Ag films over polymer nanosphere surfaces supported on glass *Chem. Phys.*, **99**, 2101–2114.
21. Semin, D. J.; Rowlen, K. L. (1994). Influence of vapor deposition parameters on SERS active Ag films morphology and optical properties *Anal. Chem.*, **66**, 4324–4331.
22. Kerker, M. (1984). *Acc. Chem. Res.*, 17, 271–277.
23. McCarthy, S. L. (1976) *J. Vac. Sci. Technol.*, 13, 135–138.
24. Aussenegg, F. R.; Leitner, A.; Lippitsch, M. E.; Reinisch, H.; Riegler, M. (1987).*Surf. Sci.*, **189/190**, 935–945.
25. Paprukailo N., Strekal N., Maskevich S (2009). Plasmonic silver films application to enhancing of staining dyes secondary emission - *Proceedings of the international conference nanomeeting*, Review and short notes, Minsk Belarus, p. 168–171.
26. Surface-Enhanced Raman scattering (1982). R.K. Chang and T.E. Furtak eds., Plenum Press, New York and London.
27. N. Strekal, V. Oskirko, V. Stepuro, A. Maskevich, S. Maskevich , I. Nabiev (1999). Chemically modified annealed thin gold films allow selective registration of SERS or fluorescence-enhanced spectra, Spectr. of Biol. Mol.: New Directions: Proc. 8th Eur. Conf. on the Spectroscopy Biological Mol., eds. J.Greve et al.– Dordrecht: Kluwer Academic Publishers, 569–570.
28. Feofanov, S. Sharonov, I. Kudelina, F. Fleury, I. Nabiev (1997). Localization and Molecular Interactions of mitoxantrone within Living K562 Cells as Probed by Confocal Spectral Imaging Analysis, *Biophys. J.* **73**, 3317–3327.
29. Kennedy B.J.; Spaeth S.; Dickey M.; Carron K.T. (1999). *J. Phys.Chem.* **103**, 3640.
30. Kennedy B.J., Spaeth S., Dickey M., Carron K.T. (1999). Determination of the Distance dependence and experimental effects for modified SERS substrates based on self-assembled monolayers formed using alkanethiols, *J. Phys. Chem. B* **103** 3640–3646.
31. R. Aroca, C. Jennings, C.J Kovacs, e.a. (1988). Fluorescent enhancement from Langmuir-Blodgett monolayers of silver island films, *Langmuir* **4**, 518–521.
32. O. Kulakovich, N. Strekal, A. Yaroshevich, S. Maskevich, S. Gaponenko, I. Nabiev, U. Woggon, M. Artemyev (2002). Enhanced luminescence of CdSe quantum dots on gold colloids, *Nano Letters* **2**, 1449–1452.
33. N. Strekal, V. Askirka, I. Sveklo, I. Nabiev, S. Maskevich (2003). Field enhancement near the annealed gold detected by optical spectroscopy with the probe biomolecules, *Physics, chemistry and application of nanostructure*, Singapore: World Scientific, 171–174.
34. J.P. Kottman, J.F. Martin (2001). Retardation-induced plasmon resonances in coupled nanoparticles, *Opt.Lett.* **26**, 1096–1098.
35. P. Gadenne, X. Quelin, S. Ducourtieux, S. Gresillon, L. Aigouy, J. C. Rivoal, V. Shalaev, A. Sarychev (2000). Direct observation of locally enhanced electromagnetic field, *Physica B* **279**, 52–58.

36. S.V. Gaponenko (2002). Possible effects of redistributed photon density of states on Raman scattering enhancement in mesoscopic structures, *Procceding SPIE* **4705**, 83–87.
37. S.V. Gaponenko (2002). Effects of photon density of states on Raman scattering in mesoscopic structures, *Physical Review B* **65**, 140303–1 -140303–4.
38. A. Ianoul, N. Strekal, S. Maskevich (2006). Imaging nanometer scale optical heterogeneities in phosphilipid monolayers deposited on metal island films, *Journal of nanoscience and nanotechnology* **6**, 61–65.
39. O. Kulakovich, N. Strekal, M. Artemyev, A. Stupak, S. Maskevich, S. Gaponenko (2006). Improved method for fluorophore deposition atop a polyelectrolyte spacer for quantitative study of distance-dependent plasmon-assisted luminescence, *Nanotechnology* **17**, 5201–5206.
40. O. S. Kulakovich, N. D. Strekal, M. V. Artemyev, A. P. Stupak, S. A. Maskevich, S. V. Gaponenko (2006). Improved fluorescent assay sensitivity using silver island films: fluorescein isothocyanate-labeled albumin as an example, *Journal of Applied Spectroscopy* **73**, 892–896.
41. J. Lakowicz (2006) Plasmonics in biology and plasmon-controlled fluorescence Plasmonics **1**, 5–33.
42. W. L. Barnes, A. Dereux, T. W. Ebbesen (2003). Surface plasmon subwavelength optics, *Nature* **424**(6950), 824–830.
43. F. Caruso, Nanoengeneering of particle surfaces (2001). *Adv. Mater.***13**, 11–22.
44. Gittings, D.I.; Caruso, F. (2001). *J. Phys. Chem. B,* **105**, 6846.

Author Index

C.D. Geddes (ed.), *Reviews in Plasmonics 2010*, Reviews in Plasmonics,
DOI 10.1007/978-1-4614-0884-0, © Springer Science+Business Media, LLC 2012

C

I

J

S

Z

Subject Index